Pet Bird Diseases and Care

Indranil Samanta · Samiran Bandyopadhyay

Pet Bird Diseases and Care

Springer

Indranil Samanta
Department of Veterinary Microbiology
West Bengal University of Animal
and Fishery Sciences
Kolkata, West Bengal
India

Samiran Bandyopadhyay
ICAR-Indian Veterinary Research Institute
Kolkata, West Bengal
India

ISBN 978-981-10-3673-6 ISBN 978-981-10-3674-3 (eBook)
DOI 10.1007/978-981-10-3674-3

Library of Congress Control Number: 2017930285

Printed on acid-free paper

This Springer imprint is published by Springer Nature
The registered company is Springer Nature Singapore Pte Ltd.
The registered company address is: 152 Beach Road, #21-01/04 Gateway East, Singapore 189721, Singapore

To the researchers and practitioners working for betterment of bird health

Foreword I

Birds are getting more and more diffused as pets worldwide. Consequently, avian species are becoming increasingly popular as patients for veterinarians. Bird medicine is already a relevant branch of pet medicine. The emotional and economic value of avian pets is often important. Pet birds, parrots in particular, are often raised in farms, which have to count on adequate veterinarian support. The veterinarian community needs to get prepared to cope with all such needs. In most faculties, in fact, the courses of avian pathology are still more oriented toward the diagnosis, cure and prevention of poultry diseases than of avian pets. More recently, however, the growing interest in clinical personnel toward such species has led to the birth of specialized courses in companion avian medicine, and in specific texts in pet birds medicine. I am honoured to present this new text addressed to veterinarians and students, which introduces very important scientific updates in avian medicine. The text is conveniently structured in sessions which treat all most relevant topics of avian medicine, and comprises a series of photographs to better illustrate all subjects. The approach is to present all useful practical information, relying on a thorough scientific treatment of all topics. The first session concerns rearing and nutrition of pet birds. The following sessions are dedicated to the description of the most important infectious diseases, metabolic disorders and pathologies due to toxicity. These will provide the scientific and technical instruments which will carry the readers to the sessions treating with diagnostic tools and therapy. In the final session, the potential zoonosis is discussed, including exhaustive guidelines for practitioners to prevent dangerous possible infections in human. This text will represent a reference book for veterinarians and interested in companion avian medicine.

Prof. Elena Circella
Avian Diseases Unit
Department of Veterinary Medicine
University of Bari
Bari, Italy

Foreword II

In recent years the contribution made by scientific research allowed the avian medicine to make significant progress and become increasingly numerous veterinarians that are dedicated to the care practices of pet birds. 'Pet bird diseases and care' is an invaluable reference resource for clinicians and a useful study guide for veterinary students indeed gives a detailed description of all aspects of the treated illness with special focus on history of diseases, etiology, host affected, pathogenesis, clinical signs, diagnosis and treatment.

The book provides clear information pertinent to medical treatment of avian species and gives important updates of the current state of knowledge and practical approaches to diagnosis and therapy. A chapter is devoted to the disorders that affect different regions and systems of the body both external and internal and above all systemic diseases. Further aspects of avian medicine discussed in the book include intoxications, malignancy, tumours and nutritional problems.

In addition, this book gives to the small animal practitioner and students a complete source for the basics of medicine avian providing detailed information on the proper management for all species of pet birds and on prevention of the most common infectious diseases, parasitic and fungal infections and zoonosis.

Prof. Maria Foti
Microbiology and Infectious Diseases Unit
Department of Veterinary Sciences
University of Messina
Messina, Italy

Preface

It is not only fine feathers that make fine birds

—Aesop

The first document of parrot as a pet was found in *Rig Veda*, an ancient Indian literature written more than 3000 years ago. Evidence of bird keeping is observed in ancient civilizations of Greece, Rome, China and Egypt. Bird keeping is an age-old practice although the science behind the diseases of the birds and their management or care is little neglected. Advancement of knowledge helps in better understanding of the subject. Amalgamation of the advancements along with the conventional knowhow in the area of pet bird diseases within the same cover was one of our best intentions. We have integrated history, etiology and classification of the pathogens along with their host preference, pathogenesis, clinical symptoms, lesion, diagnosis and treatment protocol. The book also encompasses breeds of the birds, nutritional requirements, non-infectious diseases with treatment options, toxicity and remedies, diagnostic techniques, drugs used in pet birds and their interactions, side effects and contraindications. We hope it will help the avian practitioners, students, teachers and scientists working in this fascinating area with updated information.

Moreover knowledge regarding the zoonotic infections transmissible from the pet birds is a necessity in current era after emergence and re-emergence of avian influenza, cryptococcal meningitis, chlamydiosis, *Mycobacterium genavense*, *Giardia duodenalis*, *Cryptosporidium parvum* and other zoonotic diseases. People, handling, rearing and trading the birds should have sufficient knowledge regarding dissemination potentiality of the bird borne pathogens. We have incorporated the zoonotic potentiality of each pathogen causing pet bird diseases along with prevention guidelines that can be followed by the owners.

We acknowledge the scientists and avian practitioners throughout the world for evaluation of book chapters and contribution of valuable photographs specially

Prof. Elena Circella and Dr. Petra Maria Burgmann. We offer sincere thanks to Prof. Maria Foti for her foreword. We also acknowledge friends and senior colleagues of our university/institute for their constructive suggestions. Any further suggestion for the improvement of the book will be heartily accepted.

Kolkata, India
November 2016

Indranil Samanta
Samiran Bandyopadhyay

Contents

About the Authors

Indranil Samanta obtained his Bachelor of Veterinary Sciences and Animal Husbandry degree (B.V.Sc. and A.H.) from West Bengal University of Animal and Fishery Sciences, Kolkata, India. He secured his Masters in Veterinary Sciences (M.V.Sc.) in Veterinary Bacteriology and Mycology from Indian Veterinary Research Institute, Bareilly, UP, India and Ph.D. in Veterinary Microbiology from West Bengal University of Animal and Fishery Sciences, Kolkata, India. He currently works as an Assistant Professor of Veterinary Microbiology in West Bengal University of Animal and Fishery Sciences, Kolkata, India. Previously he also worked as Assistant Professor of Veterinary Microbiology in the S.K. University of Agricultural Sciences and Technology-Kashmir, India and as a Veterinary Officer in Government of West Bengal, India. He is actively engaged in the teaching of undergraduate, postgraduate, graduate and Ph.D. scholars of Veterinary Microbiology and research related to animal health and zoonotically important microbes. He has received ten grants from national funding agencies and has supervised five postgraduate scholars. His current areas of research include on the emergence of drug-resistant, biofilm producing zoonotic and foodborne pathogens in food animals including poultry and other birds. He has published more than 100 research articles in reputed international and national journals along with review articles in international journals. He has published two textbooks entitled 'Veterinary Bacteriology' (ISBN13: 9789381450550, ISBN 10:9381450552) and 'Veterinary Mycology' (ISBN: 978-81-322-2279-8) from New India Publishing Agency and Springer, respectively. He acts as Associate Editor (Bacteriology) of BMC Veterinary Research. He is an editorial board member and reviewer of several other international and national journals. He has delivered a substantial number of talks in conferences, Government television and radio channels regarding his research and animal, poultry health related issues. He is the recipient of National Academy of Agricultural Sciences (NAAS) Associateship, American Society for Microbiology Travel Grant, Best Young Faculty Award by EET-CRS, etc.

Samiran Bandyopadhyay obtained his Bachelor of Veterinary Sciences and Animal Husbandry degree (B.V.Sc. and A.H.) from West Bengal University of Animal and Fishery Sciences, Kolkata, India. He secured his Masters in Veterinary

Sciences (M.V.Sc.) in Veterinary Medicine from Indian Council of Agricultural Research (ICAR)-Indian Veterinary Research Institute, Bareilly, UP, India and Ph.D. in Veterinary Medicine from West Bengal University of Animal and Fishery Sciences, Kolkata, India. He currently works as a Scientist in the ICAR-Indian Veterinary Research Institute, Eastern Regional Station, Kolkata, India. Previously he also worked as Scientist in ICAR-National Research Centre on Yak, Dirang, Arunachal Pradesh, India. His current areas of research include the emergence of drug-resistant zoonotic and foodborne pathogens in food animals. He has received several research grants from national funding agencies and has supervised three postgraduate scholars till date. He has published 90 research articles in reputed international and national journals. He is an editorial board member and reviewer of international and national journals. He is the recipient of the prestigious Indian National Science Academy (INSA) Medal for Young Scientists, National Academy of Agricultural Sciences (NAAS) Young Scientist Award and Fakhruddin Ali Ahmed Award from Indian Council of Agricultural Research (ICAR), New Delhi, Dr. G.P. Sen Endowment Award from Association for Public Health Veterinarians, Indian Society for Veterinary Medicine Merit award for his outstanding research work in the field of veterinary and animal science. He is also an associate of National Academy of Agricultural Sciences (NAAS), New Delhi.

Abbreviations

@	At sign
ACTH	Adrenocorticotropic hormone
ADH	Antidiuretic hormone
AIDS	Acquired immuno deficiency syndrome
AST	Antibiotic sensitivity test
ATP	Adenosine tri-phosphate
b.i.d./BID	'bis in die' (Latin) means twice a day
b.w.	Bodyweight
BGA	Brilliant green agar
BHI	Brain heart infusion
CDC	Centers for Disease Control and Prevention
CFT	Complement fixation test
CNS	Central nervous system
Cu	Copper
dl	Decilitre
°C	Degree Celsius
EDTA	Ethylenediaminetetraacetic acid
g	Relative centrifuge force
g	Gram
GI	Gastro-intestinal
h/hr	Hour
Ig G	Immunoglobulin G
Ig M	Immunoglobulin M
IMViC	Indole/Methyl red/Voges Proskauer/Citrate
IM	Intramuscular
IV	Intravenous
Kg	Kilogram
KH_2PO_4:	Potassium di-hydrogen phosphate
l	Litre
M	Molar
Mmol	Millimole
mg	Milligram
ml	Millilitre

mm	Millimetre
MHz	Megahertz
μg/mcg	Microgram
μmol	Micromol
Na_2HPO_4	Di-sodium hydrogen phosphate
NSAID	Non-steroid anti-inflammatory drug
OD	‘Omni die’ (Once daily)
P.O.	‘Per os’ (oral)
PCR	Polymerase chain reaction
PCV	Packed cell volume
PFGE	pulsed field gel electrophoresis
PGE	Prostaglandin-E
pH	pouvoir hydrogen (power of hydrogen)
ppm	Parts-per-million
PM	Post-mortem
PMN	Polymorphonuclear neutrophil
q	‘Quaque’ (Every)
RAPD	Randomly amplified polymorphic DNA
rpm	Revolutions per minute
s.i.d.	‘Semel in die’ (Once daily)
SCV	*Salmonella* containing vacuoles
SIF	*Salmonella* induced fibrils
s.w.g.	Standard wire gauge
T3SS	Type three secretion system
TSI	Triple sugar iron
U	Unit
UV	Ultraviolet
v/v	Volume/volume
XLD	Xylose-lysine-deoxycolate agar
μg	Microgram
Zn	Zinc

Pet Birds 1

Indranil Samanta

1.1 Pet Birds

Birds (class-Aves) are lightweight vertebrate creatures with feathers, beak, high metabolic rate, four-chambered heart, air sac and hard-shelled egg laying capacity. Birds can be classified into different families (Psittaciformes, Passeriformes, Columbiformes, Piciformes, Anseriformes, Galliformes etc.) containing numerous species (>10,000).

The meaning of 'pet' is narrated in popular dictionaries as 'any animal or bird that you have at home for pleasure, not for work or food'. Preference of bird species to keep as pet varies with geographical location, weather, culture and wildlife protection act of the country. Sometimes import of birds is required to meet the local demand. International trade of psittacine birds (except peach-faced lovebirds, cockatiels, Indian ring-necked parakeets, budgerigars) is regulated by 'convention on international trade in endangered species of wild fauna and flora' (CITES), an international treaty. List of bird species covered under CITES is available in its appendices (www.cites.org).

1.2 History of Bird Keeping

The first document of parrot as a pet was found in *Rig Veda*, an ancient Indian literature written more than 3000 years ago. Evidence of bird keeping is observed in ancient civilizations of Greece, Rome, China and Egypt. Raven (a crow) was mentioned as a symbol of good luck in Greek mythology and was associated with

I. Samanta and S. Bandyopadhyay, *Pet Bird Diseases and Care*,
DOI 10.1007/978-981-10-3674-3_1

God Apollo (God of prophecy). Crows are also considered as ancestors in Hindu mythology and food (*pinda*) is offered to crows during ritual of paying homage to dead persons.

Alexander the Great (327 B.C.) first imported ring-necked parakeets into European countries from India and accordingly the birds became known as Alexandrine parakeet. Possession of parakeets soon became a symbol of royalty among the wealthy people. In Roman civilization, keeping of talking parrots (*Psittacula*) was a craze among affluent people. In writings of Pliny the Elder (77 A.D.), different teaching procedures of parrots to talk was described. Egyptian hieroglyphics illustrated about keeping of pet birds such as doves and parrots.

In medieval Europe, bird keeping was considered as a royal affair and was preferred by kings and wealthy persons. Marco Polo during his world tour (1271–1295) noticed parrot keeping in different countries including southern India. During Spanish conquest of Canary island (1402–1496) by catholic monarch (Queen Isabella I and King Ferdinand II) the songbirds (canary) were discovered and they were sold to rich people of Spain. Portugese sailors brought canaries to other European countries and made bird keeping popular. Christopher Columbus during his return from voyage brought two Cuban Amazon parrots for Queen Isabella I. In 15th century, canaries were used by miners for detection of poisonous gases in the shaft of mines.

In 1418, Pope Martin V appointed two persons as parrot keepers. Henry VIII (1509–1547) of United Kingdom owned an African grey parrot and his heir Charles II (1660–1685) established an aviary of exotic birds. During the 1800s, middle-class English families started to keep budgerigars in large and decorated cages.

In United States, bird keeping started with the migration of European people in 17th century. Keeping native birds such as American goldfinch, northern cardinal and northern mockingbird became popular among the people. Import of European birds such as European starling, common chaffinch, siskin, bullfinch, canaries and song thrush began in 19th century. Canary became most popular pet bird and was commonly kept in parlours. Parrots were also imported profoundly from different countries which sometimes crossed more than 3,00,000 birds per year. Import of parrots was banned in United States in 1942 due to emergence of 'parrot fever' or human chlamydiosis. Thanks to increasing demand of treatment and care of these pet birds, specialized training of veterinarians started and 'Association of avian veterinarians' was established in 1980.

1.3 Common Breeds

In most of the countries psittacine and passerine birds are preferred as pet birds. Psittacine birds include parakeet, African grey parrot, Amazon parrot, macaw, conure, cockatoo, cockatiel, lory, lovebird and budgerigar (Table 1.1; Figs. 1.1, 1.2, and 1.3).

Table 1.1 List of common pet birds

Common name	Scientific name
Budgerigar	*Melopsittacus undulatus*
Love bird	*Agapornis* spp.
Nyasa lovebird	*Agapornis lilianae*
Black-cheeked lovebird	*Agapornis nigrigenis*
Peach-faced lovebird	*Agapornis roseicollis*
Rainbow lorikeet	*Trichogolossus haematodus*
Yellow bib lorikeet	*Lorius chlorocercus*
Goldie's lorikeet	*Psitteuteles goldiei*
Blue-streak lorikeet	*Eos reticulata*
African grey parrot	*Psittacus erithacus*
Senegal parrot	*Poicephalus senegalus*
Scarlet macaw	*Ara macao*
Blue-and-yellow macaw	*Ara ararauna*
Chestnut-fronted macaw	*Ara severa*
Red-shouldered macaw	*Ara nobilis*
Red-and-green macaw	*Ara chloropterus*
Military macaw	*Ara militaris*
Conure	*Aratinga solstitialis*
Blue-fronted Amazon	*Amazona aestiva*
White-fronted Amazon	*Amazona albifrons*
Orange-winged Amazon	*Amazona amazonica*
Yellow naped Amazon	*Amazona auropalliata*
Red lored Amazon	*Amazona autumnalis*
Vinaceous-breasted Amazon	*Amazona vinacea*
Turquoise-fronted Amazon	*Amazona aestiva*
South African Cape parrot	*Poicephalus robustus*
Australian king parrot	*Alisterus scapularis*
Red-winged parrot	*Aprosmictus erythropterus*
Senegal parrot	*Poicephalus senegalus*
Green-thighed parrot	*Pionites leucogaster*
Eclectus parrot	*Eclectus roratus*
African red-bellied parrot	*Poicephalus rufiventris*
Jardine parrot	*Poicephalus gulielmi massaicus*
Ruppell's parrot	*Poicephalus rueppellii*
Black headed parrot	*Pionites melanocephalus*
Bronze winged parrot	*Pionus chalcopterus*
Golden parakeet	*Guarouba guarouba*
Echo parakeet	*Psittacula echo*

(continued)

Table 1.1 (continued)

Common name	Scientific name
Ring-necked parakeet	*Psittacula krameri*
Alexandrine parakeet	*Psittacula eupatria*
Red-crowned parakeet	*Cyanoramphus novaezelandiae*
Rose-ringed parakeet	*Psittacula krameri*
Antipodes parakeet	*Cyanoramphus unicolor*
Vasa parrot	*Coracopsis vasa*
Crimson-fronted parakeet	*Psittacara finschi*
White cockatoo	*Cacatua alba*
Major Mitchell's cockatoo	*Cacatua leabeateri*
Sulphur-crested cockatoo	*Cacatua galerita*
Red-tailed black cockatoo	*Calyptorhynchus banksii*
Glossy black cockatoo	*Calyptorhynchus lathami*
Solomon's corella	*Cacatua ducorpsii*
Triton cockatoo	*Cacatua galerita triton*
Tanimbar corella	*Cacatua goffiniana*
Philippine cockatoo	*Cacatua haematuropygia*
Moluccan cockatoo	*Cacatua moluccensis*
Blue-eyed cockatoo	*Cacatua ophthalmica*
Yellow-crested cockatoo	*Cacatua sulphurea*
Eastern long-billed corella	*Cacatua tenuirostris*
Crimson rosella	*Platycercus elegans*
Eastern rosella	*Platycercus eximius*
Cockatiel	*Nymphicus hollandicus*
Canary	*Serinus canaria domestica*
Gold finch	*Spinus tristis*
Eurasian bullfinch	*Pyrrhula pyrrhula*
Zebra finch	*Taeniopygia guttata*
Long-tailed finch	*Poephila acuticauda*
Gouldian finch	*Erythrura gouldiae*
Bengalese finch (society finch)	*Lonchura striata domestica*
Myna	*Acridotheres* spp.
Starling	*Sturnidae* spp.
Common pigeon	*Columba livia*
Great horned owl	*Bubo virginianus*
Magpie	*Pica pica*

Fig. 1.1 Indian parrot (*Courtesy* Sudhir Kumar, India)

Passerine pet birds are canary, finch, mynah and starling (Figs. 1.4, 1.5, 1.6, 1.7, 1.8, 1.9, 1.10, and 1.11). In India, African grey parrot, finch (zebra), canary, lovebird, and pigeon (blue rock) are considered as officially permitted pet birds.

Fig. 1.2 Indian ring-necked parrot (*Courtesy* Sudhir Kumar, India)

Fig. 1.3 Indian love birds (*Courtesy* Sanjoy Sheet, ARD Department, Government of West Bengal, India)

Fig. 1.4 Asian Pied Starling (*Courtesy* Kaajal Dasgupta, Oriental Bird Club, U.K.)

Fig. 1.5 Common green magpie (*Courtesy* Kaajal Dasgupta, Oriental Bird Club, U.K.)

Fig. 1.6 Common sparrow (*Courtesy* Kaajal Dasgupta, Oriental Bird Club, U.K.)

Fig. 1.7 European gold finch (*Courtesy* Kaajal Dasgupta, Oriental Bird Club, U.K.)

Fig. 1.8 Indian paradise flycatcher (*Courtesy* Kaajal Dasgupta, Oriental Bird Club, U.K.)

Fig. 1.9 Red billed blue magpie (*Courtesy* Kaajal Dasgupta, Oriental Bird Club, U.K.)

Fig. 1.10 Ultra marine flycatcher (*Courtesy* Kaajal Dasgupta, Oriental Bird Club, U.K.)

Fig. 1.11 Yellow breasted green finch (*Courtesy* Kaajal Dasgupta, Oriental Bird Club, U.K.)

1.4 Nutritional Requirements

Nutritional requirements vary between different bird species. Caged birds in general have lower lean: fat ratio due to low physical activity and thus reduced metabolic rate, energy expenditure, and energy demands than wild birds. Among two basic types of caged birds (passerines and psittacines), passerines have 65% higher metabolic rate and energy requirements than psittacine birds.

Passerines mostly prefer seeds and they can consume 30% of their body weight daily through the feed. Psittacines consume only 10% of their body weight daily through the feed. Seeds are deficient in vitamins (A, D, E, and K), amino acids (lysine, methionine) and minerals (Ca/P). Soluble grits such as cuttlefish bone, limestone (calcium carbonate), marble and gypsum (calcium sulfate) are added in diet as a source of calcium. Insoluble grits (sand, granite) are cheaper but may cause several disorders. For additional vitamins, amino acids and other minerals, fruits, vegetables (e.g. chickpea, carrot, cabbage, chilly) and plants (e.g. lettuce) are added in diet. For development and maintenance of plumage colour, β-carotene, lutein can be added in diet specially before the breeding season. Malnutrition may cause feather disorder in birds (Fig. 1.12).

Fig. 1.12 Feather disorder due to malnutrition in a macaw (*Courtesy* Prof. Elena Circella, Avian Diseases Unit, Department of Veterinary Medicine, University of Bari, Italy)

Fig. 1.13 Seed based diet offered to pet birds (*Courtesy* Fazil Ibrahim, India)

Instead of seed based diet, pellets or mash offer more balanced nutrition to the pet birds (Fig. 1.13). Passerine birds with low body weight (except zebra finch) drink 250–300 ml/kg body weight water in a day.

Infectious Diseases

2

Indranil Samanta and Samiran Bandyopadhyay

2.1 Bacterial Diseases

2.1.1 Tuberculosis

2.1.1.1 History

Tuberculosis (Mycobacteriosis) is an ancient or 'heritage' disease which was observed even before the Neolithic age and in Egyptian mummies. In 1882, Robert Koch first isolated *Mycobacterium*, stained with alkaline methylene blue and vesuvin and established its etiologic relationship with tuberculosis. Koch described that the same human or bovine type *Mycobacterium* may cause avian tuberculosis. Strauss and Gamaleia (1891) and Maffucci (1892) cited in Cobbett (1917) illustrated that the etiological *Mycobacterium* of avian tuberculosis was different from human or bovine type. Hinshaw (1933) and Ackermann et al. (1974) reported occurrence of tuberculosis in Amazon parrot (*Amazona farinosa*). Coyle et al. (1992) proposed a new Mycobacterial species (*Mycobacterium genavense*) isolated from human patients with AIDS. Simultaneously in 1993, *Mycobacterium genavense* was reported from pet birds in Europe (Hoop et al. 1993).

2.1.1.2 Etiology

Mycobacterium avium subsp. *avium* and *M. genavense* are the most common cause of tuberculosis in pet birds. *M. genavense* is common in psittacine birds, whereas, *M. avium* subsp. *avium* mostly infects the birds kept in aviaries. *M. tuberculosis, M. bovis, M. intracellularae, M. gordonae, M. simiae, M. intermedium, M. peregrinum, M. terrae, M. avium* subsp. *paratuberculosis, M. trivial, M. fortuitum, M. diernhoferi, M. chelonae, M. smegmatis, M. flavescens, M. scrofulaceum, M. celatum, M. nonchromogenicum* and *M. marinum* are also associated etiology. *M. avium* subsp. *hominissuis* was detected in a female blue-fronted Amazon parrot (*Amazona aestiva*).

I. Samanta and S. Bandyopadhyay, *Pet Bird Diseases and Care*,
DOI 10.1007/978-981-10-3674-3_2

Mycobacterium is gram positive, straight or slightly curved rod with occasional coccobacillary, club and branched form. In the tissues, it measures 1–4 μm in length and 0.2–0.3 μm in width. It occurs singly, in pair or in bundle. It is difficult to demonstrate their gram positive nature due to high lipid content of the cell wall. The stains are relatively impermeable to the bacterial cell. They can be easily stained by Ziehl-Neelsen (ZN) or acid fast staining technique.

The genus *Mycobacterium* (Actinomycetes family) contains more than 100 species. Some are pathogenic for man and animals which grow slowly in artificial media in laboratory (slow growers) than the fast growers (*M. smegmatis*). Previously, slow growing *Mycobacterium* was sub divided into three types based on their host specificity i.e. human type, bovine type and avian type. Recent classification reveals that slow growing *Mycobacterium* is composed of several species. The most common species are *Mycobacterium tuberculosis* complex and *M. avium-intracellulare* complex (MAC). *Mycobacterium tuberculosis* complex is comprised of *M. tuberculosis, M. bovis, M. caprae, M. microrti, M. africanum* and *M. canettii*. MAC is composed of two major species—*M. avium* and *M. intracellulare. M. avium* is subdivided into four subspecies (ssp.) i.e. ssp. *avium* (Maa), ssp. *paratuberculosis* (Map), ssp. *silvaticum* and recently added ssp. *hominissuis*. MAC is considered as 'atypical mycobacteria' and members of this group are highly resistant against environmental changes such as high and low temperatures, dryness, extreme pH and common disinfectants. There are total 28 serotypes of MAC and the serotypes 1–6, 8–11 and 21 belonged to *M. avium* ssp. *avium*.

Mycobacterium genavense forms a deep branch of Mycobacterial phylogentic tree with other members such as *M. interjectum* and *M. simiae*. This group is characterized by slow growth albeit contains the signature molecule of fast growers (short helix 18). *M. genavense* can be distinguished from other slow growers by their fastidious growth and preference for liquid medium. *M. genavense* was first reported from human AIDS patients with disseminated infections (Hirschel et al. 1990).

2.1.1.3 Host Susceptibility

Among psittacine birds, grey-cheeked parakeets (*Brotogeris pyrrophterus*), amazons (*Amazona* spp.), budgerigars (*Melopsittacus undulatus*) and pinous parrots (*Pionus* spp.) are the most commonly affected species with tuberculosis. Other psittacines such as green-winged macaws (*Ara chloroptera*), cockatoos (*Cacatua spp.*), conures (*Aratinga auricapillus*, *Cyanoliseus patagonus*) and red-crowned parakeet (*Cyanoramphus novaezelandieae*) are also detected to be infected with *M. genavense* and *M. tuberculosis. M. avium* subsp. *avium* is reported to cause infection in cockatiels (*Nymphicus hollandicus*). The non-cultivable *Mycobacterium* is detected in blue and yellow macaw (*Ara ararauna*) and grey-cheeked parakeet (*Brotogeris pyrrhopterus*).

The common parrots (African grey parrot, Senegal parrot) are not considered as natural host of *M. tuberculosis* although infection in African grey parrot is reported which was transmitted from human. *M. bovis* can produce natural infection in parrots and the budgerigars are experimentally infected with *M. bovis* producing clinical syndrome within 70 days.

Among non-psittacine group of birds, canary (*Serinus canaria*), gouldian finch (*Chloebia gouldiae*) and zebra finch (*Poephila guttata castanotis*) are commonly infected with *M. genavense*. Synergistic infection of *M. genavense* and avian polyoma virus was detected in European goldfinch (*Carduelis carduelis*). Occasionally canaries are also infected with *M. tuberculosis*.

Avian tuberculosis is a disease of adult birds although occasionally detected in young (<1 year old) canaries.

2.1.1.4 Transmission

MAC is transmitted in birds chiefly by ingestion, inhalation and rarely through arthropods. *M. avium* ssp. *avium* is excreated through faeces of infected birds and survive in soil (up to 4 years), sawdust (8 months at 37 °C) and water for a long period. Bird to bird transmission in aviaries may occur through infected faeces or rarely by cannibalism. Occasionally skin abrasion acts as a route of mycobacterial infection in pet birds.

Ingestion is considered as a potential route of transmission of *M. genavense* infection in pet birds. The environment specially drinking water is an identified source of *M. genavense* infection. Lung involvement in pet birds suggests inhalation as an additional route of transmission. Birds may act as reservoir of *M. genavense*. Bird to bird transmission of *M. genavense* is rare although the possibility could not be excluded entirely. Immunosuppression plays a role in transmission of *M. genavense* in human but whether the same condition facilitates the transmission in pet birds is obscure.

Transmission of *M. tuberculosis* in pet birds (green-winged macaws, blue-fronted amazon) from human is observed due to close contact with owners suffering from tuberculosis and feeding the birds with *pre-chewed food.*

2.1.1.5 Pathogenesis

M. avium subsp. *avium*

MAC enters the host chiefly through ingestion route of transmission and become present in the intestine. The waxy cell wall of the bacteria protects it from gastric acids and enzymes. Several pathogen associated molecular patterns (PAMPs) are expressed by virulent *Mycobacterium* which can recruit 'microbicidal' macrophages through toll like receptor (TLR) mediated signaling. During *M. tuberculosis* infection in human, these PAMPs in the bacterial surface are masked with a lipid, namely phthiocerol dimycoceroserate (PDIM). The PAMPs are not recognized by the host immune system and the bacteria can avoid the reactive nitrogen species (RNS) generated by 'microbicidal' macrophages.

MAC (*M. avium* subsp. *avium*) does not contain PDIM in their surface but use a different strategy (still unexplored) to resist RNS. MAC is benefited with these RNS as commensal present in the gut are sensitive to it. Commensal mediated competitive inhibition is thus excluded and probably MAC enters M-cells like host adopted *Salmonella* to invade the underlying blood monocytes. The M-cells are specialized cells of the follicle associated epithelium and the region is relatively free from commensal mediated competitive inhibition. The invasion of monocytes is followed by bacterimia and subsequent haematogenous spread to liver, spleen and other organs.

MAC enters the macrophages (histiocytes) of periarteriolar lymphoid sheath (PALS) zone in spleen within 10 days post infection in birds. *Mycobacterium* has several virulence factors which promote their survival within macrophages using different strategies such as acid resistance, avoidance of acidification etc. MAC (*M. avium* subsp. *avium*) specifically restricts vacuole maturation and prevents the fusion of phagosome and lysosome for their survival within macrophages. Haematogenous spread of the organism leads to infection of bone marrow, lungs, air sacs, gonads and rarely, kidney and pancreas. The organs become enlarged due to accumulation of macrophages within organ parenchyma.

Granuloma formation is an attempt of host tissue to localize the infection, although *Mycobacterium* exploits it for their multiplication and further dissemination. The growth of *Mycobacterium* occurs in the macrophages present in a granuloma. The infected macrophages undergo apoptosis and leave the encased bacteria which are engulfed by newly recruited macrophages. This process of apoptosis and re-phagocytosis within a granuloma is regulated by a mycobacterial secreation system (ESX-1/ESAT6) detected in *M. tuberculosis* but absent in *M. avium* subsp. *avium*. Typical tuberculous granulomas in different organs are not frequent in *M. avium* subsp. *avium* infection, although observed in lungs and periocular region of parrots. Granulomas in different organs (liver, kidney, intestine, muscle and subcutaneous tissues) are observed in red-crowned parakeet (*Cyanoramphus novaezelandiae*) and green-winged macaw (*Ara chloroptera*) infected with *M. tuberculosis*.

Mycobacterium genevense

M. genevense infection in pet birds mostly occurs through the oral route like MAC. There is every possibility that they follow the same pattern of pathogenesis although still unexplored. *M. genevense* causes non-tuberculoid form of mycobacteriosis in pet birds albeit occurrence of granulomas are observed in glottis of amazon parrot (*Amazona ochracephala*), aorta of cockatiel (*Nymphicus hollandicus*), small intestine of canary-winged parakeet (*Brotogeris versicolurus*) and brain of spectacled amazon parrot (*Amazona albifrons*).

2.1.1.6 Clinical Symptoms

Incubation period of mycobacteriosis in pet birds is 6 months to 4 years. Clinical syndrome in psittacine birds varies widely. In acute form, sudden death without any symptom is common. In chronic form, constant loss of weight along with diarrhoea (Fig. 2.1), frequent micturition with excessively large quantity and low specific gravity of urine, depression, laboured breathing, distension of abdomen and poor feathering primarily suggests about mycobacteriosis. The condition fails to respond to common antibiotics.

Fig. 2.1 Macaw suffering from mycobacteriosis (*Courtesy* Mousam Das, Animal Resources Development Department, Government of West Bengal, India)

Cutaneous masses are sometimes observed in skin and conjunctiva. Inflammation of feather follicles (folliculitis) is occasionally observed which includes perifollicular swelling, erythema, pruritus and pain, restlessness, shivering and feather damaging behaviour.

2.1.1.7 Lesion

The liver and spleen are enlarged, mottled and whitish. Miliary abscess in liver are observed in budgerigars experimentally inoculated with *M. avium* subsp. *avium*. The intestine becomes tubular, thickened and tan coloured. Typical tuberculous granulomas in different organs are not frequent in *M. avium* subsp. *avium* infection, albeit observed in lungs and periocular region of parrots, pericardium of gang gang cockatoo (*Callocephalon fimbriatum*) and cervical esophagus of blue-fronted parrots (*Amazona aestiva*) occluding the lumen. Involvement of lung is rare in pet birds affected with mycobacteriosis. Some post mortem findings in canary (*Serinus canarius*), eurasian goldfinch (*Carduelis carduelis*) and the red siskin (*Spinus cucullatus*) reported the occurrence of lung lesions. In a rare case of *M. genevense* infection in an amazon parrot (*Amazona albifrons*), perivascular cuffs of macrophages in the grey and white matter of the brain and spinal cord, gliosis and mild vacuolation of white matter were observed.

2.1.1.8 Diagnosis

Clinical Specimens

Blood/sera, cloacal swabs, tracheal swabs, biopsies from organs can be collected as ante-mortem samples for diagnosis of mycobacteriosis in the laboratory. Post-mortem samples include vital organs such as liver, spleen, intestine, heart, lung and bone marrow. All the specimens should be immediately sent to the laboratory following the standard regulations for sending biohazardous substances. If there is delay in sending, refrigeration of the samples should be done to prevent the growth of contaminants. Addition of 0.5% boric acid may preserve the samples for 1 week.

Diagnostic Techniques

(a) Clinical signs and history of direct contact with owner and other birds suffering from tuberculosis give a tentative diagnosis
(b) *Haematological parameters*: Following haematological changes can be correlated with tuberculosis in pet birds although these changes are non-specific and are observed in other inflammatory and chronic infections also.

- Moderate to marked increases in white blood cell numbers (heterophilia, monocytosis, lymphocytosis)
- Decreased packed cell volume (PCV) (except during early stage of infection)
- Increased enzyme concentration (alanine aminotransferase, aspartate aminotransferase, lactate dehydrogenase)

(c) *Demonstration of acid-fast organisms*: Demonstration of the organisms can be done in the smears prepared from ante-mortem samples by acid-fast (or ZN) staining. *Mycobacterium* appears as red coloured slender rods singly or in bundle (Fig. 2.2). Fluorescent acid fast stain may be used for better detection. In post mortem specimens, cytoplasm of infected cells is laden with acid-fast organisms.
(d) *Gross and histopathology*: Enlarged liver, thickened intestinal loop, increased opacity in endosteal bone in the humerus, tibia, ulna, femur in advanced cases is suggestive for tuberculosis. Presence of visible granuloma is not a constant feature although may be detected in lungs. In decomposed carcasses bone marrow is the best specimen. Histopathogical findings such as presence of acid-fast bacilli in inflammatory cells can also tentatively diagnose mycobacteriosis.
(e) *Isolation of organism*: This is considered as gold standard method for confirmatory diagnosis of *Mycobacterium*. To isolate the organism, the tissue sample must be processed in proper way. Tissue samples are homogenized in pestle and mortar after keeping in the solution of hypochlorite (1:1000) for

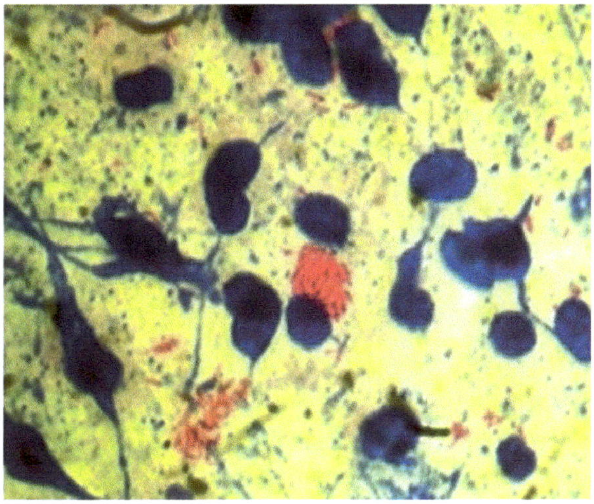

Fig. 2.2 Presence of *Mycobacterium* in Z-N stained impression smear of lung tissues (*Courtesy* Premanshu Dandapat, Eastern Regional Station, Indian Veterinary Research Institute, Kolkata, India)

4–16 h. It is decontaminated by the addition of acid (5% oxalic acid), alkali (2–4% sodium hydroxide) or detergent (0.375–0.75% hexadecylpyridinium-chloride, HPC). The acid or alkali mixture is neutralized, centrifuged and the sediment is used for culture. *M. avium* subsp. *avium* can be isolated in Dorset egg medium, Lowenstein-Jensen (LJ), Herrold's and Middlebrook's (7H10,7H11) medium containing pyruvate. Glycerol and 'mycobactin' is also added as growth enhancer. Mycobactin extracted from the environmental Mycobacteria, acts as siderophore helping in acquisition of iron. Mycobactin is produced by all cultivable Mycobacteria except *M. avium* subsp. *avium* and *M. avium* subsp. *paratuberculosis.* Incubation period is 8 week at 40 °C. Even it can grow at 42–43 °C. It produces smooth or rough type of colonies. In liquid culture a radiometric method using ^{14}C labelled substrate can be used for rapid detection (BACTEC system).

M. genavense isolation is difficult although can be done on special media with prolonged incubation for 2–9 months. Conventionally, after decontamination of the samples with 2(N) NaOH, the samples can be inoculated in Herrold's egg yolk medium with and without mycobactin J and Sula's liquid medium and incubated at 37 °C. The growth is periodically checked in every two weeks. The successful isolation of *M. genavense* from blue headed parrot (*Pionus menstruus*) was observed in Herrold's egg yolk medium without mycobactin J after 270 days of incubation. In specialized laboratories, newly developed liquid culture systems [manual mycobacteria growth indicator tube (M-MGIT), BACTEC system) are used for confirmatory isolation of *M. genavense.*

(f) *Immunological/serological tests*: Use of tuberculin test with purified protein derivative of *M. avium* subsp. *avium* correlates poorly with clinical disease in psittacine birds. ELISA was experimentally developed with *M. fortuitum*, *M. vaccae*, and *M. avium* antigens for detection of antibodies against *M. avium* subsp. *avium* in birds. However ELISA produces variable results in different species of birds. Immunological or serological tests for detection of mycobacteriosis in pet birds are not routinely followed.

(g) *Molecular biology*: Polymerase chain reaction (PCR) can detect *Mycobacterium* from fresh samples, faeces and paraffin-blocked tissues. The species-specific PCR targets *IS901* and *hsp65* genes for detection of *M. avium* subsp. *avium* and *M. genavense*, respectively. Differentiation of both the species can be done by sequencing of the 16SrRNA gene. A nested polymerase chain reaction (PCR) from the consensus sequences of the *hsp65* gene, followed by analysis with restriction enzymes can also differentiate *M. avium* and *M. genavense*. Real-time TaqMan PCR assay is developed to detect *hsp65* gene of *M. genavense* and MAC subsp. Other recent technologies such as GenoType assay and DNA microarrays can be used for diagnosis of avian tuberculosis. For detection of genetic diversity among the strains of *M. avium* subsp. *avium*, mycobacterial interspersed repetitive units—variable-number tandem-repeat markers (MIRU-VNTR) typing can be successfully used.

2.1.1.9 Zoonosis

M. avium subsp. *avium* is considered as a potential zoonotic risk in the immunocompromised persons albeit majority of human infection is caused by another member of *M. avium* group (*M. avium* subsp. *hominissuis*). *M. genavense* is associated with gastrointestinal or pulmonary mycobacteriosis in immunosuppressed patients associated with AIDS. Other species of mycobacteria associated with pet bird causes opportunistic and sporadic infections in human.

The cases of reverse zoonosis (anthroponosis) are also reported where *M. tuberculosis* was transmitted from the infected owners to their pet birds (green-winged macaws, blue-fronted amazon).

2.1.1.10 Treatment and Control Strategy

Successful treatment of pet birds is reported with various combinations of anti-tuberculous drugs at highest tolerable dose for a prolonged period (9 months or more). Single anti-tuberculous drug is not preferred due to possibility of resistance development. Emergence of multidrug-resistant tuberculosis (MDR-TB; resistant to isoniazid and rifampicin) and extremely drug-resistant tuberculosis (XDR-TB; in addition to being multidrug-resistant the bacteria are resistant to fluoroquinolone and 1 of 3 antibiotics such as capreomycin, kanamycin and amikacin) is a global problem now a days. In most of the cases, dose is fixed on the basis of human paediatric studies because pharmacokinetic properties of anti-tuberculous drugs in pet birds are still unknown. The anti-tuberculous drugs and antibiotics used against *M. avium* infections in pet birds are isoniazid, rifampin, rifabutin, ethambutol, clofazimine, ciprofloxacin, enrofloxacin, streptomycin and amikacin.

Successful therapy of *M. genavense* infections with combinations of moxifloxacin, clarythromycin, ethambutol and amikacin in humans has been reported but there is no specific drug recommended for *M. genavense* infection. Treatment with clarithromycin, rifampin, and ethambutolm against *M. marinum* infection in a blue-fronted Amazon parrot was not successful.

Experimentally, different combinations of anti-tuberculous drugs and antibiotics such as isoniazid (30 mg/kg) + ethambutol (30 mg/kg) + rifampin (45 mg/kg) for 12–18 months, clofazimine (6 mg/kg) + ethambutol (30 mg/kg) + rifampin (45 mg/kg) for 9–18 months, ciprofloxacin (80 mg/kg) + ethambutol (30 mg/kg) + rifampin (45 mg/kg) for 9–12 months was used successfully against confirmed cases of tuberculosis in different pet birds (grey-cheeked parakeet, double yellow-headed Amazon, lilac-crowned Amazon). Although combination of azithromycin (43 mg/kg), rifampin (45 mg/kg), and ethambutol (30 mg/kg) administered orally once daily for 180 days in ring-neck doves (*Streptopelia risoria*) naturally infected with *M. avium* subsp. *avium* failed to eradicate the infection. Further, treatment with combination of clarithromycin (61 mg/kg bw), moxifloxacin (25 mg/kg bw) and ethambutol (60 mg/kg bw) administered in budgerigars experimentally infected with *M. avium* subsp. *avium* by crop gavage every 12 h for 18 weeks significantly improved the situation but failed to recover completely. Combination of minocycline (10 mg/kg p.o. b.i.d.) and clarithromycin (10 mg/kg p.o. s.i.d.) significantly reduced oral plaques in blue penguins (*Eudyptula minor*) naturally infected with *M. intracellulare*.

Due to zoonotic potential specially for children and elderly persons and immunocompromised patients, prolonged treatment and poor success rate, difficulty of drug administration to avian patients maintaining proper doses, natural and acquired antimicrobial resistance, poor owner compliance and moreover, lack of a proper treatment schedule, the debate exists regarding advice of treating pet birds against tuberculosis. Nevertheless, euthanasia is the preferred measure in the prevention of tuberculosis in pet birds in relation to human health. For prevention of further transmission, removal of all organic matter and debris from cages, washing the cages and surroundings properly with disinfectants and maintenance of biosecurity measures are required. Chlorohexidine and quaternary ammonium compounds can act as mycobacteriostatic disinfectants.

2.1.2 Salmonellosis

2.1.2.1 History

D.E. Salmon (1885) first isolated *Salmonella* from the infected pigs. It was considered as a cause of 'hog cholera' until the discovery of the etiological virus. The nomenclature of the bacteria ('Salmonella') was done in memory of Salmon. In 1889, Klein (United Kingdom) first isolated *Salmonella* Gallinarum from chickens with 'fowl typhoid'. Loeffler first described *Salmonella* Typhimurium from a natural outbreak of typhoid like infection in mice. Among the pet birds, Salmonellosis was first described in ducks (Vandervort 1954; Keymer 1958), and parakeets (Kaye et al. 1961; Madewell and McChesney 1975).

2.1.2.2 Etiology

The genus *Salmonella* is classified under the family Enterobacteriaceae that belongs to the order Enterobacteriales. There are two major species under the genus *Salmonella* i.e. *S. enterica* (>2440 serovars) and *S. bongori* (20 serovars). A third proposed species is *S. subterranea*, yet to be recognized as a true species. *Salmonella enterica* is considered as type species of the genus at present. *S. enterica* has 6 subspecies (ssp.): *salamae, arizonae, diarizonae, houtenae, indica, enterica.* Most of the pathogenic *Salmonella* are designated as 'serovar' under the *S. enterica* ssp. *enterica.* Virulent serovars are: Typhi, Typhimurium, Dublin, Choleraesuis, Pullorum, Gallinarum, Abortusovis.

Salmonella Typhimurium (var. *copenhagen*) is the most frequently isolated serovar from different outbreaks in psittacine birds. In passerine birds (perching/song birds), *S.* Typhimurium phage types DT40, DT41, DT56, DT160 are adapted. It acts as either a primary pathogen or it causes sub-clinical infection in young and immunocompromised birds. If the density of flock is high and the quantity of available feed is low, most of the birds become debilitated and are susceptible to *S.* Typhimurium infection.

Other sub-species and serovars of *Salmonella* isolated from pet birds include *S. houtenae, S. arizonae, S.* Rissen, *S.* Enteritidis, *S.* Pullorum, *S.* Gallinarum, *S.* Newport, *S.* Panama, *S.* Rublislaw, *S.* Aberdeen, *S.* Thompson and *S.* Wasenaar. Among them, *S.* Gallinarum can infect canaries, ring dove, pheasants, peacocks and peafowl. A novel *Salmonella* serovar (*S.* Pajala) was isolated from Peregrine falcon (*Falco peregrinus*) nestlings.

2.1.2.3 Host Susceptibility

Clinical outbreaks of salmonellosis are frequently detected in passerine and psittacine birds. Among passerine birds, finches (Fringillidae) and sparrows (Passeridae) seem to be particularly susceptible to *Salmonella* spp. infection. Salmonellosis is reported from canary (*Serinus canaria*), eurasian siskin (*Carduelis spinus*), zebra finch (*Taeniopygia guttata*), bengalese finch (*Lonchura striata domestica*) and picoplat (*Sporophila intermedia*). Fatal outbreaks with high mortality were reported in psittacine birds such as lories and lorikeets (*Trichoglossus*, *Lorius*, *Eos* spp.), budgerigars (*Melopsittacus undulatus*), parakeets (*Psephotus* spp., *Psittacula* spp.), and sulphur crested cockatoo (*Cacatua galerita galerita*).

The birds with caecum [e.g. macaw (*Ara* sp.), amazon parrot (*Amazona* sp.)] can asymptomatically carry *Salmonella* spp. like poultry. Both free-ranging and captive blue-fronted amazon parrots (*Amazona aestiva*) are detected to carry *Salmonella* spp. Although, lilac-crowned amazon parrots (*Amazona finschi Schlater*) was found died due to *S.* Enteritidis infection. Moreover, *S.* Typhimurium is detected as a primary pathogen causing death of blue and gold macaws (*Ara ararauna*).

Occasionally, raptors or hunter birds (e.g. falcon, red kite), game birds [e.g. red-legged partridge (*Alectoris rufa*)], free-ranging sparrow (*Passer domesticus*), gull (Laridae), wild birds such as temminck's seedeater (*Sporophila falcirostris*), chestnut-capped blackbird (*Chrysomus ruficapillus*), brown-headed cowbirds

(*Molothrus ater*), white-throated sparrows (*Zonotrichia albicollis*) can also harbour *Salmonella* spp.

2.1.2.4 Transmission

Infected pet birds, rodents, reptiles, wild birds, contaminated water, feed and eggs act as source of *Salmonella* spp. When the pet birds are gathered in an exhibition the healthy birds come in direct contact to the infected birds. The rodents, reptiles and wild birds having access to the open-air aviary can contaminate the place. The bacteria can survive for extensive periods on wood and dust and can live for 28 months in avian faeces. The contaminated places become a constant source of infection. In a pet shop, iguana (*Iguana iguana*) was identified as a source of *Salmonella* spp. infection in a cockatoo.

Ingestion of contaminated feed and drinking water is the major route of transmission in pet birds. Sometimes, the infection is also transmitted by owners or attendants through their contaminated hands, feet and clothes. Trans-ovarian transmission is common in poultry, although, is not frequently observed in pet birds.

2.1.2.5 Pathogenesis

Salmonella Typhimurium

Following oral route of transmission, *Salmonella* is deposited in the intestine, where they invade enterocytes. The bacteria can invade the epithelial cells throughout the intestine, although, caeca and ileocaecal junction are the preferred site. In the intestine, low pH, peristalsis, intestinal mucus, lysozyme in secreations and moreover normal microbial flora try to destabilize the bacterial colonization. Normal microflora in adults prevents the *Salmonella* colonization by occupying their receptors, known as 'competitive exclusion'. Young are more susceptible as their intestinal microfloral range is not fully developed.

Bacterial fimbriae, lipopolysaccharide (LPS), pathogenicity island (SP-1-T3SS associated proteins) help in adherence with the enterocytes. The interaction between the T3SS proteins (SipA, SipC) with the actin cytoskeleton of the enterocytes causes cytoskeletal rearrangements to generate an uneven surface (membrane ruffle). The organisms are trapped within the ruffled membrane and are internalized by the enterocytes. Within the enterocytes, the bacteria reside in a membrane bound vacuole i.e. *Salmonella* containing vacuoles (SCV). As the SCV matures, it migrates from the luminal border of the enterocyte to the basal membrane. In the basal membrane, the SCV enter the macrophages associated with peyer's patches in the sub mucosal space. The SCV never fuses with lysosome within macrophages and thus avoid phago-lysosome fusion which is necessary to kill the bacteria. Additional factors such as SP-2-T3SS (SipC protein), SP-3 associated proteins also help in intracellular survival. The formation of *Salmonella* induced fibrils (SIF) help in bacterial replication in an unknown way. It is evident that major portion of the infection is cleared by the macrophages, only certain part can survive leading to chronic infection or carrier state with persistent faecal shedding. Sometimes, invasion of *Salmonella* takes place beyond the intestine which causes bacterimia, survival and replication of the bacteria in reticulo-endothelial cells of liver and spleen. In passerine birds (canary, finch and starling), esophagus and crop are the preferred site for bacterial colonization after bacterimia.

2.1.2.6 Clinical Symptoms

Acute and chronic form of salmonellosis is detected in pet birds. In acute form, huge mortality without any prior cardinal signs was observed in a flock of canaries. A *S.* Typhimurium infected macaw (*Ara ararauna*) showed depression, anorexia, delay in the emptying of crops (ingluvies), laboured breathing and diarrhoea for 3–4 days before death. Greenish-yellow diarrhoea was observed in adult budgerigers (*Melopsittacus undulatus*) infected with *S.* Gallinarum.

The chronic form shows numerous general symptoms such as anorexia, diarrhoea, dyspnoea, lethargy, cachexia, ruffled feathers, subcutaneous granuloma, crop stasis, conjunctivitis, arthritis and panophthalmitis. Adult budgerigars with chronic salmonellosis showed loss of condition, unwillingness to fly, inability to perch, and gathering in the bottom of the cage (Fig. 2.3). Drop in egg production and increased embryonic mortality rate was observed. In pigeons, in addition to the general syndrome, polyuria, arthritis of the joint, nervous symptoms and dermatitis followed by death was detected.

Stress associated with environment, season, change of diet and housing, transport, breeding, concurrent infection, and introduction of new birds without quarantine are the major predisposing factors for salmonellosis in pet birds.

Fig. 2.3 Sick budgerigar (*Courtesy* LafeberVet)

2.1.2.7 Lesion

In acute form of salmonellosis in pet birds, no specific gross lesion is observed. In chronic infection, congestion of lung and intestine, spleenomegaly, hepatomegaly, liver with necrotic foci is observed (Fig. 2.4). The passerines birds infected with *S.* Typhimurium shows the granulomatous lesions like pseudotuberculosis. Esophagitis and ingluvitis with necrotic plaques in esophagus and crop, respectively, are consistently detected in wild or free-range passerine birds. In European goldfinch (*Carduelis carduelis*), instead of esophagus lesions are sometimes developed in the subcutaneous tissues. In canaries, congestion of vital organs and necrotic foci on the liver with a nodular and miliary appearance is detected. Intestinal content become dark in colour due to haemorrhagic diasthesis. In lories and lorikeets, petechial hemorrhages on the serosal surface of the proventriculus, ventriculus, and cardiac muscle along with atrial dilation are observed. Sometimes, bacterial emboli occur in liver, spleen, lung, kidney, and proventriculus. In *S.* Typhimurium infected macaws, pectoral muscle atrophy, fibrinous exudate on intestinal mucosa, white-greyish nodules on intestinal serosa, myocardium, lungs and ingluvies mucosa are commonly observed. In young budgerigars, fibrin deposit is observed as a white, thick layer on the pericardium. Petechial haemorrhages are detected on the surfaces of pericardium, myocardium, gizzard, duodenum and ileo-caecum. In garden birds infected with *S.* Typhimurium, esophageal ulcers, granulomata in soft tissues, hepatomegaly and spleenomegaly are observed.

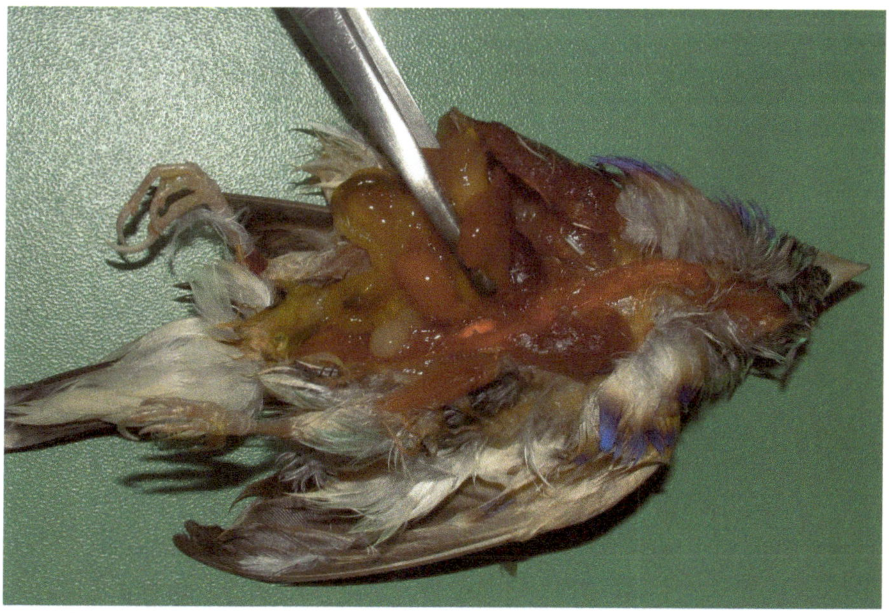

Fig. 2.4 Splenomegaly in Gouldian finch due to salmonellosis (*Courtesy* Prof. Elena Circella, Avian Diseases Unit, Department of Veterinary Medicine, University of Bari, Italy)

Microscopic inspection of histopathological sections shows necrosis of parenchymatous organs specially the liver with granulocyte infiltration and fibrin deposition. In young and adult budgerigars, the blood vessel walls become hyalinized in appearance with various-sized microthrombi. The chronic cases are characterized by formation of a granuloma. A typical granuloma consists a necrotic centre which is surrounded by granulocytes and macrophages containing *Salmonella* spp. Multinucleated giant cells are found in chronic infection. In passerine birds, epithelial surface of the esophagus is ulcerated and it forms a thick layer of necrotic cellular debris composed of degenerated and intact leukocytes with gram-negative bacteria. Infiltration of heterophils, lymphocytes and plasma cells occur into the underlying sub-mucosa.

2.1.2.8 Diagnosis

Clinical Specimens

Clinical samples include faeces or cloacal swabs, blood/serum of live birds and affected tissues, such as liver, spleen, heart, intestine/caeca, lung, esophagus/crop, brain and kidney in 10% buffered formalin. Before collection of cloacal swabs, pericloacal asepsis with iodized alcohol is performed. Blood samples are collected from jugular, wing or ulnar vein. The environmental samples, such as pooled faeces, litter and dust from the cages, feed and drinking water should be examined to know about an outbreak, if any. Specimens should be collected before antibiotic treatment of the birds. After death, the collection should be done immediately from fresh carcasses. For 'pre-enrichment', swabs should be collected in buffered peptone water. Pre-enrichment in buffered peptone water helps in survival of *Salmonella* from freezing, heating and desiccation. The cold chain (4–5 °C) should be maintained during transportation of the samples to the laboratory.

Diagnostic Techniques

(a) Clinical signs and lesions after necropsy, history of direct contact with infected birds give a tentative diagnosis
(b) *Direct Examination*: An impression smear prepared from clinical samples such as cloacal swab/faeces/tissues, is stained by Gram's Method. *Salmonella* appears as gram negative small rods with no distinct characteristics. The tissue samples of heart, lung, liver, spleen, kidney, and intestine are fixed in 10% buffered formalin, embedded in paraffin, sectioned at 3 mm, and stained with hematoxylin-eosin and periodic acid-Schiff for direct examination of the bacteria.
(c) *Isolation of bacteria from clinical samples*: Clinical samples require pre-enrichment and enrichment for growth. For pre-enrichment, the samples collected in buffered peptone water are kept at 40 °C for 24 h. The pre-enriched clinical samples are transferred into enrichment medium such as selenite or tetrathionate broth and are incubated at 40 °C for another 24 h.

From the supernatant, the samples are plated in brilliant green agar (BGA) or xylose-lysine-deoxycolate agar (XLD) and are re-incubated for another 24 h at 40 °C. Convex, pale red, translucent colonies in BGA and red coloured colonies with black centres in XLD agar are presumably diagnosed as *Salmonella* spp. *S.* Pullorum produces small, paler colonies than other salmonellae in BGA. The suspected colonies can be confirmed by different biochemical tests such as catalase, oxidase, IMViC, TSI, carbohydrate fermentation profile and lysine decarboxylase test.

(d) *Serological tests*:

  (i) Rapid whole blood/serum agglutination test: It can be used for rapid detection of *Salmonella* spp. with crystal violet stained or unstained *Salmonella* polyvalent 'O'-antigen. Equal amount of suspected whole blood or serum is gently mixed with the antigen on a white tile. Tile is agitated gently for 2 min and is observed for reading. In a positive case clumping of antigen is visible within 2 min. The antigen is commercially available or it can be obtained from Veterinary institutes in different countries.
  (ii) Tube agglutination test
  (iii) Immunodiffusion
  (iv) Immunofluorescence
  (v) ELISA

(e) *Molecular Biology*: For rapid and reliable detection of *Salmonella*, conventional PCR based diagnostic techniques targeting *invA* or other genes are used. Genus and serovar (*S.* Enteritidis and *S.* Typhimurium) specific real-time PCR system have also developed. For phylogenetic analysis of the *Salmonella* isolates, pulsed field gel electrophoresis (PFGE) and randomly amplified polymorphic DNA (RAPD) can be used.

2.1.2.9 Zoonosis

Zoonotic transmission of *Salmonella* spp. to human from parakeet kept as a pet was documented. A salmonellosis outbreak associated with dissection of an owl was reported among the elementary school children. The possibility of children infection (below 5 years age) is more in those families who rear a pet bird (odd ratio: 2.7) or a lizard (odd ratio: 3). In open air aviaries and children's zoos, the transmission of *Salmonella* spp. was reported between wild birds, pet birds and human. Special care for designing such aviaries should be adopted.

2.1.2.10 Treatment and Control Strategy

Antibiotics against *Salmonella* spp. in infected pet birds can be administered after doing the sensitivity of the bacterial isolates. Antimicrobial resistance of *Salmonella* spp. is a global concern at present. Successful treatment of infected canaries with 10% (w/v) enrofloxacin solution provided as 200 mg/l in drinking water for 5–7 days was observed. Treatment with kanamycin, gentamicin, trimethoprim/sulfamethoxazole suspension along with anti-diarrhoeals such as daolin and pectin combination is recommended.

General control and prevention strategies such as isolation of diseased birds from the rest of the flock, cleaning and disinfection of cages, water and feed utensils with 10% (v/v) solution of sodium hypochlorite or commercial disinfectants are recommended. If the feedstuffs are suspected it should be replaced with new batch immediately.

2.1.3 Chlamydiosis

2.1.3.1 History

Halberstaedter and von Prowazek (1907), a Czech zoologist, first described *Chlamydia*, although it was actively infecting people for centuries before its discovery. First description of *Chlamydia* associated trachoma in human was found in the Ebers papyrus dated around 1550 B.C. In modern times, Ritter (1879) first described *Chlamydia psittaci* infection in human acquired from parrots. During 1890–1930, numerous outbreaks of human psittacosis occurred in Europe, North and South America, associated with parrots and other pet birds. In 1929–30, pandemic psittacosis outbreaks in human were reported due to import of infected psittacine birds from South America to Europe and North America.

In 1930, Levinthal, Coles, and Lillie, independently described the properties of the pathogen, and accordingly *Chlamydia* was known as Levinthal-Coles-Lillie (LCL) agent. Moulder (1962) first revealed the structural and chemical composition of *C. psittaci*. Hatch (1975) demonstrated the requirement of adenosine-triphosphate (ATP) supplementation for growth of *Chlamydia*.

In literatures, first description of chlamydiosis in parakeets was reported from Germany (Strauch and Rott 1958). Further study after few years also revealed the presence of *Chlamydia* in parrots and other parakeets in Germany (Schmittdiel 1966).

Human psittacosis outbreaks specially in persons associated with poultry, turkey and duck industry were reported from United States and European countries during 1980–90.

2.1.3.2 Etiology

All the Chlamydiae are placed under the order Chlamydiales, and family Chlamydiaceae. Based on cluster analysis of 16S and 23S rRNA genes, the family Chlamydiaceae was divided into two genera i.e. *Chlamydia* and *Chlamydophila*. Recent genome comparison study of the two genera proposed to unite the *Chlamydia* in a single genus. The latest edition of the Bergey's manual of systematic bacteriology also described the single genus of *Chlamydia*. Pathogenic species under the genus *Chlamydia* are *C. psittaci, C. trachomatis, C. suis, C. muridarum, C. abortus, C. caviae, C. felis, C. pecorum* and *C. pneumoniae.* Among the pathogenic species, *C. psittaci* is mostly associated with avian chlamydiosis (psittacosis) in pet birds, ornithosis in poultry, and zoonotic infection in human. The 16S rRNA gene based phylogenetic study indicates the presence of a distinct cluster

Table 2.1 Serovars and genotypes of *C. psittaci* and their hosts

Serovar	Host	Genotypes	Host
A	*Psittaciformes* (cockatoos, parrots, parakeets, lories)	A	*Psittaciformes* (cockatoos, parrots, parakeets, lories)
B	*Columbiformes* (pigeons and doves)	B	*Columbiformes* (pigeons and doves)
C	*Anseriformes* (ducks and geese)	C	*Anseriformes* (ducks and geese)
D	Turkeys	D	Turkeys
E	Pigeons, turkeys, ratites	E	Pigeons, turkeys, ratites
F	Psittacines	F	Psittacines and turkeys
		E/B	Ducks, turkeys, pigeon
		WC	cattle
		M56	rodents

of *C. psittaci* strains which are associated with chlamydiosis in Psittaciformes (cockatoos, parrots, parakeets, lories etc.) and Columbiformes (pigeons) birds. Although, other Chlamydial species such as *C. abortus, C. trachomatis* and *C. pecorum* are occasionally detected in brown skua, parrots, parakeets, and pigeons.

Three new avian species of *Chlamydia*, namely *C. ibidis* sp. nov., *C. avium* sp. nov., and *C. gallinacea* sp. nov. are proposed. Among them, *C. ibidis* and *C. avium* are isolated from feral sacred ibis (*Threskiornis aethiopicus*), psittacines, and pigeons.

Earlier, avian isolates of *C. psittaci* was divided into six serovars (serotypes) which can infect different species of birds (A–F, Table 2.1). Based on major outer membrane protein (*ompA*) sequence, *C. psittaci* is currently divided into 15 genotypes. Among them, nine genotypes (A–F, E/B, M56, WC) are associated with different species of birds and mammals (Table 2.1).

2.1.3.3 Host Susceptibility

Avian strains of *C. psittaci* are detected in more than 460 species of birds under 30 orders. The pet birds belonged to the order Psittaciformes (cockatoos, parrots, parakeets, lories etc.) and Columbiformes (pigeons) are most susceptible to *C. psittaci* infection. In parrots, the worldwide prevalence of *C. psittaci* varies from 16–81%. The infection in pet birds is reported from Europe, Brazil, Africa, USA, Iran and India. Free ranging Galapagos doves (*Zenaida galapagoensis*) and rock doves (*Columba livia*) in Spain; monk parakeets (*Myiopsitta monachus*), Amazon parrots, red-tailed Amazon (*Amazona brasiliensis*) in Brazil; ring necked parakeet (*Psittacula krameri*), Alexandrine parakeet (*Psittacula eupatria*), African grey parrot (*Psittacus erithacus*), Timneh grey parrot (*Psittacus erithacus timneh*) in Iran were reported to be infected with *C. psittaci*. In India, *C. psittaci* was isolated from pigeons (*Columba livh*), parrots (*Psittacula krameri*) and crows (*Corvus splendens*). Infection of Passeriformes birds is not common, although, canaries were detected to be infected with *C. psittaci* in Croatia.

C. psittaci was also detected in healthy asymptomatic birds such as in *Ara macao*, and *Amazona ochrocephala* in Costarica, and free-living Hyacinth macaw (*Anodorhynchus hyacinthinus*) and blue-fronted parrot (*Amazona aestiva*) in Brazil. The syndrome was not expressed either due to infection with low virulent strain or resistance of some bird species.

The seroprervalence studies revealed the presence of *C. psittaci* antibodies in macaws (*Ara macao*, *Ara ambigua*), hyacinth macaws (*Anodorhynchus hyacinthinus*), budgerigars (*Melopsittacus undulatus*), lovebirds (*Agapornis* sp.), cockatiels (*Nymphicus hollandicus*), Alexandrine parakeets (*Psittacula eupatria*), Eurasian siskins (*Carduelis spinus*), oriental skylarks (*Alauda arvensis*), and black-tailed grosbeaks (*Coccothraustes migratorius*) in different countries. The presence of *C. psittaci* antibodies indicates the exposure of the birds to the organism. In a study in China, highest seroprevalence was observed in cockatiel which was followed by Alexandrine parakeets, lovebirds, and budgerigars. It seems that lovebirds and budgerigars among the psittacine birds are relatively resistant against *C. psittaci* infection, although, the reason is unexplored.

Among the wild predator birds, white-tailed sea eagle (*Haliaeetus albicilla*) and the peregrine falcon (*Falco peregrinus*) are detected to be infected with *C. psittaci.*

2.1.3.4 Transmission

Inhalation of contaminated dust, airborne particles from the feathers and ingestion are major ways of *C. psittaci* transmission in the birds. Direct contact during close proximity with the infected birds also helps in transmission. Throughout the breeding season, specially during incubation of eggs, male psittacine birds prefer to feed the females by regurgitation. In this process the feeds are often mixed with secreations of the crop, pharynx and nasal cavity. Transmission of *C. psittaci* is observed from parent birds to their nestlings during feeding.

Asymptomatic carrier birds infected with *C. psittaci*, excreate the organisms through faeces, nasal and lacrimal discharge, oropharyngeal mucus, crop milk and other secreations. Shedding is increased during coexisting infections and stress conditions such as shipment, breeding, crowding, chilling and nutritional deficiencies. When the excreated faecal material dries, the organisms are aerosolized. Elementary bodies (infections form) of *C. psittaci* survive in the dried faeces for several months, in the contaminated feed for up to two months, on glass for 15 days, and in straw for 20 days.

Mechanical transmission of *C. psittaci* by biting arthropods such as flies, mites and lice are observed. Vertical transmission is infrequently observed in parakeets, seagulls, snow geese and poultry.

Zoonotic transmission of *C. psittaci* in human occurs mostly through inhalation of contaminated dust, feathers and aerosolized excreations. Direct contact with infected pet birds or their cages, utensils, beddings contaminated with discharges can transmit the bacteria. Sometimes, biting of the infected birds also helps in transmission. Person to person spread is rarely reported although possible through inhalation.

2.1.3.5 Pathogenesis

Chlamydia follows a unique life cycle with tri-phasic developmental stages. The infectious form (elementary body, EB) is extremely small (250–350 nm in diameter), pear to spherical shaped particle with electron dense irregular nucleoid. It has rigid cell wall with disulphide cross linkage among the cysteine rich amino acids of outer membrane proteins (OMP). This form can survive for a prolonged period in the environment. After transmission, infection starts with the attachment of elementary bodies to the host cell membrane. In birds, apical surface of columnar epithelial cells in intestine acts as preferred site for attachment of *C. psittaci*. Primary attachment of EB takes place by electrostatic interactions, most likely with glycosaminoglycan (GAG) moieties on the host cell surface. This reversible binding is followed by receptor mediated irreversible attachment. Protein disulfide isomerase (PDI), present in the host cell membranes and causing disulphide reduction, helps in attachment of EBs.

Chlamydial major outer membrane protein (MOMP) mostly acts as adhesin to bind with the host cellular receptor. Following receptor mediated attachment, the EB enters the cell via endocytosis [microfilament dependent/independent process (clathrin mediated)]. Some Chlamydial strains enter the host cells through cholesterol-rich lipid raft domains. Among different pathways, *C. psittaci* strains prefer to use clathrin-coated vesicles for cellular entry. *C. psittaci* elementary bodies contain rosette like long projections (Matsumoto's projection) on their surface which acts as type three secreation system (T3SS). *C. psittaci*-T3SS helps in introduction of Chlamydial proteins into the host cell cytoplasm. These T3SS-injected proteins interact with host cellular proteins and cause modulation of host cell function.

After cellular entry, vesicles containing EBs escape the lysosomes in the host cell cytoplasm and reach near the nucleus within 8–12 h after entry. In *C. psittaci* infection, IncB proteins (T3SS-effector protein) interact with host cell proteins (dynein motor proteins) for intracellular transport of vesicles with EBs into the nuclear zone. The EB is converted into reticulate bodies (RB) in this nuclear zone. The EB loses its electron dense DNA core and its cell wall loses its rigidity due to break of disulphide bridges. The reticulate bodies are non-infectious form, larger in size (500–2000 nm diameter) and metabolically active. RBs multiply by binary fission and start genus specific protein synthesis. The structural reorientation started and RBs are transformed into intermediate bodies (IB, 300–1000 nm in diameter). A central electron-dense core with radially arranged nucleoid fibres surrounding the core is observed in the IBs. The IBs are converted into progeny EBs within a vesicle after 30 h of the entry of initial EBs. Chlamydial micro colony with 100–500 EBs within the vesicle is called 'inclusion body' and it is generated after 48–50 h. The inclusion bodies move to the golgi apparatus region with the help of host dynein proteins. EB is released to attack new cells by rupturing the vesicles and the cycle is repeated. The signal for release of EBs is yet unknown but it is associated with host cellular apoptosis. Suppression of host cellular apoptosis can induce persistent Chlamydial infection.

Intracellular survival of EBs depends on escape from the lysosomal breakdown process. The EBs can induce delayed maturation of lysosomes as an escape mechanism. The intracellular inclusion bodies are covered with a mesh of host cytoskeletal filaments which prevent the exit of the content and consequent activation of the host immune system. Close attachment of *C. psittaci* inclusions with the mitochondria helps in acquisition of ATP because they cannot synthesize it. Moreover, intracellular survival of the inclusions depends on acquisition of lipids such as sphingomyelin, phosphatidylinositol and phosphatidylcholine. Golgi apparatus of the host cells act as major source of lipids for the inclusions and often the golgi apparatus are fragmented to provide the lipid.

Some non-replicating reticulate bodies persist within the host cytoplasm and produce latent infection. The growth cycle of *Chlamydia* within the body of the host is disrupted due to nutritional deprivation, treatment with antibiotic and activated immune system. In disrupted growth cycle, reticulate bodies are converted into enlarged pleiotrophic 'aberrant' RBs. The aberrant RB contains chromosome but the genes associated with growth (genes encoding membrane proteins, transcription regulators, cell division factors, EB-RB differentiation factors) are not expressed. Further, the genes encoding chlamydia protein associated with death domains (CADD) are down regulated which causes suppression of host cell apoptosis and persistence of infection. Interaction of host cellular protein (G3BP1) and chlamydial IncA (T3SS-effector protein) also suppress host cellular apoptosis. When the inducers of the disrupted growth cycle (antibiotic, immune system products) are removed, the aberrant RB is again converted into normal RB and they can complete the growth cycle.

2.1.3.6 Clinical Symptoms

In birds, chlamydiosis has an incubation period of 3–10 days. Clinical symptoms are not specific. General syndrome such as loss of condition, anorexia, fever, diarrhoea, respiratory problems, nasal and ocular discharges are observed. Expression of syndrome and associated mortality (up to 80%) depends on virulence of *C. psittaci* strains, age, species, nutritional and immune status of the pet birds. Occasionally, sub-clinical *C. psittaci* infection without visible syndrome is observed in birds. During stress conditions, the sub-clinical infection is activated with increased shedding of *C. psittaci*.

2.1.3.7 Lesion

The pet birds with avian chlamydiosis do not show any pathognomonic gross lesion. Conjunctivitis, lateral nasal adenitis, sinusitis, fibrinous airsacculitis, lung congestion, fibrinous pneumonia, pericarditis with presence of fibrinous cover, peritonitis, hepatitis with multifocal necrosis and splenitis are observed (Figs. 2.5 and 2.6). In pigeons, conjunctivitis, swollen eyelids, rhinitis, presence of fibrinous exudates over peritoneum, air sac and pericardium, enlarged, soft and dark coloured liver and spleen are observed. The budgerigars, infected with *C. psittaci* and Reovirus, showed distinct cachexia, hepatomegaly and spleenomegaly. The livers become enlarged, mottled and tan-brown in colour. Other lesions include uric acid

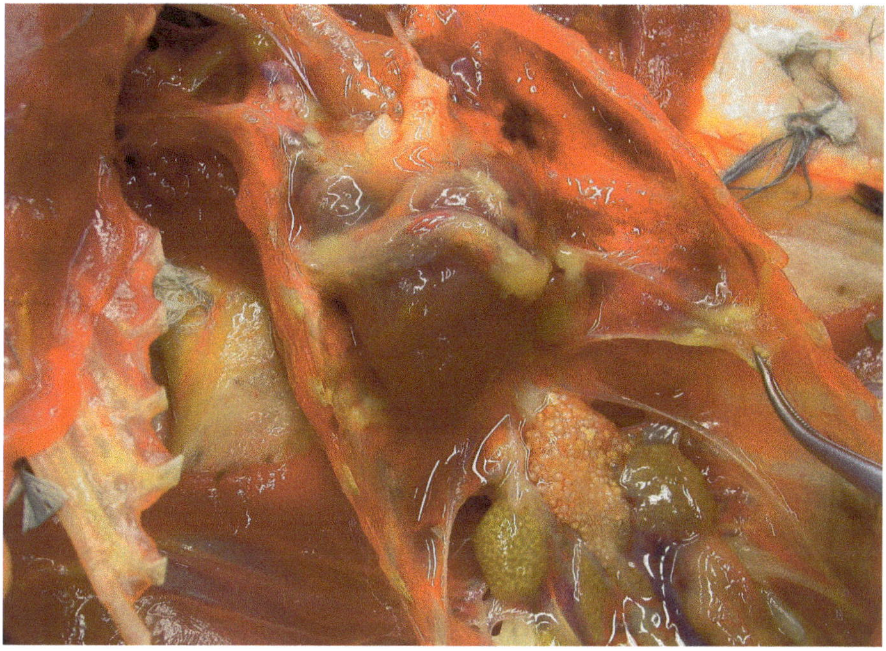

Fig. 2.5 Pericarditis with presence of fibrinous cover in a bird with avian chlamydiosis (*Courtesy* Prof. Richard Hoop, University of Zurich, Switzerland)

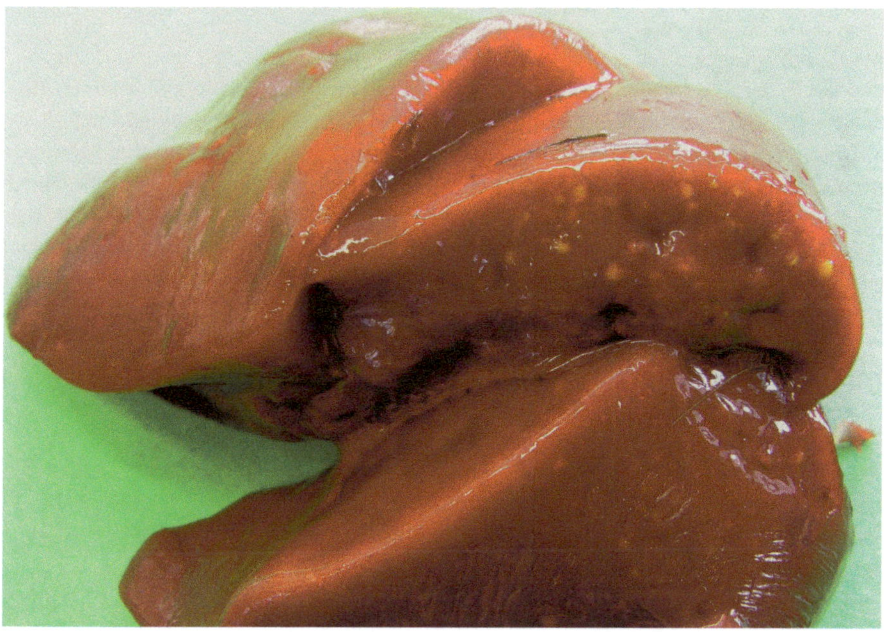

Fig. 2.6 Hepatitis with multifocal necrosis in a bird with avian chlamydiosis (*Courtesy* Prof. Richard Hoop, University of Zurich, Switzerland)

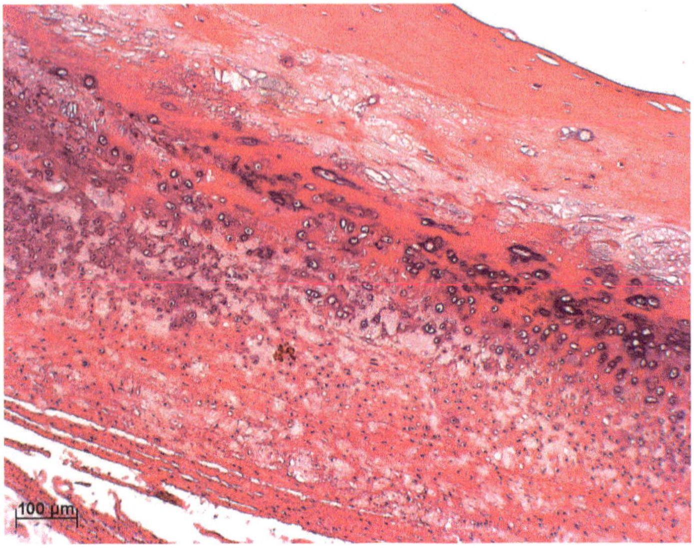

Fig. 2.7 Histopathological section of the blood vessel of a bird with atherosclerosis (*Courtesy* Prof. Richard Hoop, University of Zurich, Switzerland)

deposit in kidney, conjunctivitis and air-sacculitis. Carrier birds with asymptomatic infection do not show any gross lesion.

Atherosclerosis is considered as a well defined ailment specially in aged pet birds. African gray parrots, macaws and Amazon parrots are most susceptible to this condition. Sudden death without prior symptom is the cardinal sign of atherosclerosis. Like human, the risk factors for atherosclerosis include high cholesterol and triglyceride concentrations, sex, age, species, obesity and inactivity, and moreover, *C. psittaci* infection. Arteriosclerotic plaques are observed between the intima and internal elastic lamina of the blood vessels in many species of birds (Fig. 2.7). The plaques are composed of fibrous tissues and are observed as pale yellowish areas at the thickened portion of intima. In severe cases, the plaques become circumferential lesion which cause narrowing of the lumen and reduced blood flow.

2.1.3.8 Diagnosis

Clinical Specimens

From live birds, pharyngeal/choanal slit swabs, conjunctival swabs and nasal swabs can be collected aseptically as ante-mortem samples. Faeces or cloacal swabs are less preferred because shedding of *Chlamydia* is not consistent. Post mortem samples collected from the dead birds include lungs, spleen, liver and air sacs.

Chlamydiae are relatively labile organisms and special precautions are required for their detection. Samples should be maintained in cold chamber and processed immediately after collection. The tissue samples can be preserved at −80 °C for prolonged period. DNA extracted from the tissue samples can be stored in stabilization buffer. For successful isolation of *Chlamydia*, the clinical samples should be collected in special *Chlamydia* transport medium such as 2SP (0.2 M sucrose phosphate medium containing 10 μg/ml of gentamicin, 25 U of nystatin and 25 μg/ml of vancomycin) and SPG (75 g of sucrose, 0.52 g of KH_2PO_4, 1.22 g of Na_2HPO_4, 0.72 g of glutamic acid and water in 1litre, pH 7.4–7.6) supplemented with bovine serum albumin, streptomycin, vancomycin and nystatin. Broad spectrum antibiotics like tetracycline, chloramphenicol, macrolides, sulphonamides, penicillin should not be added as they have anti-chlamydial effect.

Diagnostic Techniques

(a) *Direct examination*: Smears prepared from collected faecal samples, conjunctiva or impression smears of tissue samples can be stained with Macchiavello, Castaneda, Giemsa, Giménez, modified Gimenez (PVK stain), Stamp, modified Z-N, and methylene blue for demonstration of Chlamydial inclusion bodies. Giemsa stain is more useful in the smears prepared from conjunctival scrapings. The inclusion bodies appear purple/blue with Giemsa, Castaneda and methylene blue stain and red with Macchiavello, Giménez, Stamp, and modified Z-N stains.

(b) *Isolation of Chlamydia from clinical samples*: Isolation of *Chlamydia* can be done in the yolk sacs of embryonated hen eggs, laboratory animals and cell culture. Fertile chicken eggs (6–8 days old) are inoculated through the yolk sac route. The embryo dying three or more days after incubation is examined for chlamydial inclusions. Mice are ideal laboratory animal for isolation of *Chlamydia*. The mice usually die within ten days of intranasal, intracerebral or intraperitoneal inoculation and the EBs can be isolated from viscera and peritoneal exudates. Cell lines treated with a metabolic inhibitor (cycloheximide at 2 μg/ml) can be used for isolation of Chlamydiae. McCoy, HeLa, monkey kidney cells, L-929, Buffalo Green Monkey (BGM) cells, mouse fibroblast cells, fish and lizard cells are used. The inoculated cells should be incubated at 35–37 °C for 48–72 h and the intracytoplasmic inclusion bodies are detected by staining (Giemsa) or fluorescein conjugated monoclonal antibody. Isolation by cell culture is still considered as gold standard method for detection of *Chlamydia*. However, it requires biosafety level-3 (BSL3) laboratory with expertise.

    The clinical samples should be decontaminated by antibiotics like gentamicin (50 μg/ml), vancomycin (75 μg/ml) and nystatin (500 unit/ml) before inoculation into eggs, animals or cell lines. Transport medium (2SP) can be used as buffer.

(c) *Detection of C. psittaci antigen*: ELISA based antigen detection kits are available for detection of *C. trachomatis* infection in human. The same kit can be used for detection of *C. psittaci* because the two species share common antigen. However, minimum 600 elementary bodies are needed in the samples for detection.

(d) *Serological tests*: Serological tests can be used as supplementary diagnostic tests along with detection of antigen or isolation. Presence of antibodies in the host cannot confirm active infection. Sometimes, false negative results are produced if the samples are collected before development of antibody or during treatment with antibiotics. The serological methods such as micro immunofluorescence (MIF) test, ELISA, CFT, elementary body agglutination (EBA) tests are used for detection of anti-*Chlamydia* antibodies. MIF is more sensitive and can detect all types of immunoglobulins in the sera. ELISA based tests using whole organism, LPS, lipoglycoprotein of *Chlamydia* as antigen are sensitive but less specific for detection of *C. psittaci*. Whereas, ELISA with recombinant major outer membrane protein (MOMP) of *C. psittaci* as antigen, can more specifically detect *C. psittaci*. CFT can detect anti-*Chlamydia* Ig G only, not Ig M. Further, CFT is tedious, time consuming, less sensitive test and the antigens (complement fixing) are not commercially available. The EBA test can detect anti-*Chlamydia* Ig M only, and as a consequence, infection in early stage can only be diagnosed.

(e) *Molecular biology*: PCR is a specific, sensitive, and rapid technique to detect *C. psittaci*. Successful application of PCR depends on quality of extracted DNA from the clinical samples. Guanidine-detergent lysing solution should be used for lysis of eukaryotic host cells and *Chlamydia* for extraction of DNA. The 16S rRNA gene is conserved in the genus *Chlamydia* and is a suitable target gene for detection of *Chlamydia* up to species level. Major outer membrane protein (*ompA*) is used as a target gene in nested PCR, although, variations exist in the MOMP gene sequence among *C. psittaci* strains. SYBR green-based real time PCR targeting *ompA*, 23S rRNA gene, inclusion membrane protein A gene (*incA*), molecular cysteine-rich protein gene (*envB*) of *C. psittaci* and microarray-based detection assays are also developed for detection of *Chlamydia*.

2.1.3.9 Zoonosis

Human psittacosis cases are reported in Europe, USA, South America, Japan and Australia. Other than the persons who rear the birds in their home, occupational risk groups such as veterinarians, pet shop workers, avian quarantine workers, poultry processing plant workers, bird breeders, and farm workers are most susceptible. Even psittacosis outbreak was detected among custom officers in some countries due to their exposure to imported parakeets in the airport. Incubation period in human is 5–14 days. Clinical syndrome in human includes fever, chills, headache, pneumonia, renal disorders, and miscarriages in pregnant women. All the vital organs are affected with the progression of infection and endocarditis, hepatitis, myocarditis, arthritis and encephalitis are reported. Ocular infection with follicular kerato-conjunctivitis is also observed.

2.1.3.10 Treatment and Control Strategy

Doxycycline, tetracycline and enrofloxacin were successfully used in budgerigars and psittacine birds to cure avian chlamydiosis. Doxycycline is the drug of choice for the birds and the treatment should be continued for 45 days. It may induce toxicity in some bird species and produce signs of depression, inactivity, anorexia, greenish or yellowish urine. Use of the drug in those birds should be stopped immediately and supportive symptomatic treatment should be started. Recommended dose of doxycycline in feed is 300 mg/kg feed for 45 days. In drinking water, 400 mg of doxycycline hyclate/litre of water will maintain therapeutic concentration in psittacine birds. Administration of the drug through the feed or drinking water is suitable for aviaries. For pet bird owners, oral administration of the capsule in individual bird is appropriate. Recommended oral dose of the drug is 40–50 mg/kg body weight in every 24 h for cockatiels, Senegal parrots, blue-fronted, orange-winged Amazon parrots; 25 mg/kg body weight in every 24 h for African gray parrots, blue and gold macaws, green-winged macaws; and 25–50 mg/kg body weight in every 24 h for other psittacine birds. Injectable doxycycline is administered at doses of 75–100 mg/kg body weight, intramuscularly (pectoral muscle), in every 5–7 days for the first 30 days and subsequently in every 5 days for the rest of the treatment period. Long acting oxytetracycline can be injected sub-cutaneously at the dose of 75 mg/kg body weight in every 3 days in cockatoos, blue-fronted and orange-winged Amazon parrots, and blue and gold macaws. The oxytetracycline injection causes irritation at the site. If tetracycline is orally administered or used in feed, dietary calcium sources (mineral block, oyster shell, supplemented pellets) should be reduced.

To control the psittacosis infection in aviaries, general precautionary measures, such as quarantine of newly introduced birds for 30 days and periodical testing for *C. psittaci* infection, separation of birds after return from bird shows or fairs, rodent control, control of exposure to wild birds, regular disinfection of the cages and utensils, proper ventilation to reduce aerosol load within the unit should be followed. Use of prophylactic antibiotic is not recommended as it may produce resistant bacteria. Recommended disinfectants for *C. psittaci* infection are 1:1,000 dilution of quaternary ammonium compounds, 70% isopropyl alcohol, 1% lysol, and chlorophenols. Use of vacuum cleaner in the aviaries is not preferred as it will aerosolize infectious particles.

No vaccine is commercially available for the pet birds against *C. psittaci* infection. Experimental DNA vaccination in budgerigars with plasmid DNA expressing MOMP of *C. psittaci* was found effective.

2.1.4 Campylobacteriosis

2.1.4.1 History

In 1886, Escherich observed *Campylobacter* like organisms in stool samples of children with diarrhoea. In 1913, McFaydean and Stockman first isolated and identified *Campylobacter* spp. in foetal tissues of aborted sheep. Confirmatory tests were also carried out by Smith in 1918 when similar organisms were isolated from aborted bovine foetuses. In this period, the bacteria were known as *Vibrio foetus*. In 1963, due to certain differentiating characteristics, the bacteria were separated from Vibrionaceae family and the new genus *Campylobacter* ('curved rod') under Campylobacteriaceae family was proposed.

2.1.4.2 Etiology

Campylobacter spp. is gram negative, comma shaped rods specially in infected tissues and young cultures. When two bacterial cells are found together in a microscopic field, occasionally it looks like 'S' or 'wing of gull' ('flying seagull'). They are motile by single unipolar/bipolar unsheathed flagella. Motility is darting or corkscrew type, best observed by dark field microscopy.

Campylobacteriaceae family contains four genera namely *Campylobacter, Arcobacter, Sulfurospirillum* and *Thiovulum*. Currently there are 18 species and 6 sub species of the genus *Campylobacter*. Important species and sub species of *Campylobacter* are—*Campylobacter jejuni* ssp. *jejuni*, *C. jejuni* ssp. *doyeli*, *C. coli*, *C. lari*, *C. fetus* ssp. *venerealis*, *C. fetus* ssp. *fetus*. Thermotolerant *Campylobacter* (*C. jejuni* ssp. *jejuni*, *C. coli*, *C. lari*, some strains of *C. upsaliensis*) is isolated from pet birds with or without clinical syndrome. The thermotolerant species requires higher temperature for their growth (42 °C) which is provided by the pet birds due to high body temperature. Other than thermotolerant *Campylobacter* species, *C. fetus* and *C. intestinalis* have also isolated from parrots.

2.1.4.3 Host Susceptibility

The pet birds such as tropical finches (juvenile Estrildidae), canaries, pigeons, parakeets [except red-crowned parakeet (*Cyanoramphus novaezelandiae*), dusky-headed parakeet (*Aratinga weddellii*), orange winged parrot (*Amazona amazonica*), red bellied macaws (*Ara manilata*)], emu, ostriches, waterfowls (mallard duck, shoveler duck, green-winged teal duck) are detected to harbour *Campylobacter* spp. In a study in Peruvian Amazon, parrots (*Ara*, *Brotogeris* and *Pionites*) were detected to be infected with *Campylobacter* spp.

Wild and free-living birds, for instance, sparrows, crows, waders, black-headed gull (*Larus ridibundus*), sparrow hawk (*Accipter nisus*), jackdaw (*Corvus monedula*), hooded crow (*Corvus cornix*), dunnock (*Prunella modularis*), yellowhammer (*Emberiza citrinella*), white wagtail (*Motacilla alba*), dunlin (*Calidris alpina*), curlew sandpiper (*Calidris ferruginea*), bald ibis (*Geronticus eremita*), little stint (*Calidris minuta*), broad-billed sandpiper (*Limicola falcinellus*), ruff (*Philomachus pugnax*), wood sandpiper (*Tringa glareola*), long-eared owl (*Asio otus*), starling (*Sturnus vulgaris*), reed warbler (*Acrocephalus scirpaceus*), winter wren (*Troglodytes troglodytes*), redwing (*Turdus iliacus*), blackbird (*Turdus merula*), song thrush (*Turdus philomelos*), fieldfare (*Turdus pilaris*), blackbirds (*Turdus merula*), thrush (*Turdus viscivorus*) can act as reservoir of *C. jejuni* in nature. Certain clonal lineages of *C. jejuni* and species of wild birds are positively associated. Among the raptors (birds of prey), only hawks were detected to carry *C. jejuni* in their gut.

Possession of *Campylobacter* spp. in birds depends on feeding habits. Gulls and crows have higher possession rate than the pigeons due to their preference for sewages. The shoveler ducks (*Spatula clypeata*) have higher carriage rate than green-winged teal duck (*Anas acuta*) because they prefer bottom sediments of aquatic environments containing molluscs as a feed.

2.1.4.4 Transmission

Direct and indirect contact with infected birds and vectors (house flies, beetles, cockroaches, mealworms) are major ways of *C. jejuni* transmission in pet birds. *C. jejuni* is sensitive to oxygen and cannot grow below 31–32 °C temperature. So, they cannot survive in feed and drinking water for a prolonged period. Presence of *C. jejuni* in drinking water acts as an indicator for faecal contamination from wild birds or livestock. Sometimes, *Campylobacter* spp. can make a symbiosis with aquatic protozoa and survive in the environmental water.

In human, major source of *C. jejuni* is contaminated poultry and its products, pork (with intact skin), beef, mutton and raw milk. Consumption of undercooked meat, milk or their products and handling poultry are the key ways of transmission. Direct or indirect contact with infected pet birds may play a role in zoonotic transmission of *C. jejuni*, although, not recorded in scientific literatures.

2.1.4.5 Pathogenesis

In poultry, after transmission by faecal-oral route, *C. jejuni* colonizes at the mucous layer of caecal and cloacal crypts. The colonization is mediated by adhesin proteins like CadF (*Campylobacter* adhesin to fibronectin), PEB (Periplasmic/membrane-associated protein), CapA (*Campylobacter* adhesion protein *A*), JlpA (jejuni lipoprotein *A*), CiaB (*Campylobacter* invasin antigen B), flagella, and lipopolsaccharide (LPS). Occasional invasion of the intestinal epithelium takes place. No gross or microscopic lesions and clinical signs are produced in poultry during *C. jejuni* colonization or invasion. Similar type of *C. jejuni* colonization takes place in psittacines and canaries and they mostly act as asymptomatic carriers. Severe clinical signs and lesions are produced in tropical finches, especially in juvenile Estrildidae. The precise mechanism of *C. jejuni* infection in pet birds is still unexplored.

2.1.4.6 Clinical Symptoms

No clinical signs and lesions are detected in canaries, psittacines, free-living (migrating passerines) and wild birds, and they act as asymptomatic reservoir of *C. jejuni*. In finches [juvenile Estrildidae, Gouldian finch (*Chloebia gouldiae*)], symptoms include sitting posture with its head under the wings, yellow droppings due to undigested starch (amylum), lethargy, and retarded moulting. High rate of mortality is observed among fledglings. In young ostriches, green coloured urination is the predominant sign.

Recent study indicates the possible synergistic role of *C. jejuni* in proventricular dilatation disease (PDD) in parrots caused by avian bornavirus.

2.1.4.7 Lesion

In tropical finches infected with *C. jejuni*, distinct cachexia, congestion in gastrointestinal tract, and presence of yellow coloured amylum or undigested seeds in gastrointestinal tract are the lesions. In sub-acute cases, hepatitis with focal necrosis and mucoid haemorrhagic enteritis is observed.

2.1.4.8 Diagnosis

Clinical Specimens

Fresh droppings (without urine) and cloacal swabs can be collected as clinical specimens. Post mortem samples include intestine or intestinal contents and liver.

Diagnostic Techniques

(a) *Direct Examination*: A smear can be prepared from clinical samples and stained by dilute carbol fuchsin (DCF). *Campylobacter* spp. appears as pink coloured small curved rod arranged in a pair. Occasionally, the bacteria produce characteristic 'S' or 'wing of gull' appearance. The bacteria can also be demonstrated by wet mounts of collected droppings by phase contrast or dark field microscopy. Darting motility of the organisms is suggestive for *Campylobacter* spp.

(b) *Isolation of bacteria from clinical samples*: The selective media for *Campylobacter* isolation is broadly categorized into two types: charcoal based and blood based. Charcoal and blood components remove toxic derivatives of oxygen from the media. Examples of selective media are: modified charcoal, cefoperazone, deoxycholate agar (mCCDA), Karmali agar or CSM (charcoal-selective medium), Preston agar, Skirrow agar, Butzler agar and Campy-cefex agar. Commonly used non-selective media for isolation of *Campylobacter* spp. are blood agar with or without 0.1% sodium thioglycolate and antimicrobials (cephalosporins, trimethoprim, polymyxin, vancomycin, bacitracin, actidione, colistin, nystatin). Optimum growth condition for thermotolerant *Campylobacter* spp. (*C. jejuni*, *C. coli*) are 42 °C temperature for 24–48 h, pH 6.8, and 3–5% CO_2 with 3–15% O_2. *C. jejuni* produces non-haemolytic, finely granular, irregular margin, flat, greyish colonies. On charcoal based media, the colonies may produce 'metallic sheen'. Thermotolerant *Campylobacter* spp. can be confirmed up to species level by staining, colony characteristics, and biochemical properties. Hippurate test can primarily differentiate *C. jejuni* and *C. coli*, but, it should be further confirmed by other tests.

(c) *Detection of Campylobacter antigen*: ELISA based kits are available for detection of *Campylobacter* antigen from human stool samples. They can be used for detection of *Campylobacter* antigen from droppings collected from the suspected pet birds, although yet not evaluated.

(d) *Molecular biology*: PCR can be used in combination with cultural technique for rapid detection of *C. jejuni*. A multiplex PCR is developed for detection of thermotolerant *Campylobacter* spp. such as *C. jejuni* (23SRNA gene); *C. coli*, *C. lari*, and *C. upsaliensis* (*glyA*).

2.1.4.9 Zoonosis

Infection with *C. jejuni* in human causes watery or bloody diarrhoea, elevated body temperature, abdominal pain, nausea, and vomition. Septicaemia develops in a few diarrhoeic cases (0.15%) which may cause enlargement of the liver and spleen, endocarditis, arthritis, and meningitis. Rarely, a complicated auto-immune response is developed as a sequel, known as Guillain-Barré syndrome. It is a demyelating disorder which results muscle weakness and neuromuscular paralysis.

2.1.4.10 Treatment and Control Strategy

In mild infection, treatment with antibiotics is not recommended due to possibility of antibiotic resistance development. In severe cases, several antibiotics such as clindamycin, gentamicin, tetracyclines, erythromycin, cephalothin, and fluoroquinolones (nalidixic acid) can be used under the supervision of a qualified veterinarian. Choice of antibiotic depends on sensitivity of the *C. jejuni* isolates, availability in suitable form, and species of the birds.

In aviaries or personal collection, implementation of biosecurity practices such as regular cleaning and use of fly repellents in the cages is effective to prevent the introduction of *Campylobacter* spp. No vaccine against *Campylobacter* spp. is currently available for birds.

2.1.5 Lyme Disease

2.1.5.1 History

Lyme disease is a tick-borne, multi-system disorder of human and animals and it is characterized by swelling of joints, pain, lameness, fever, lethargy, anorexia, nephropathy with renal failure, myocarditis, cardiac arrest and CNS involvement. The etiological agent is maintained in tick and several birds and animals. Clinical description of Lyme disease was first documented by Arvid Afzelius, a Swedish dermatologist. The disease was first identified in 1975 (or 1976), among the people suffering with suspected juvenile rheumatoid arthritis in the area of Lyme, Connecticut, United States. Hence it is known as 'Lyme disease' or 'Lyme Borreliosis'. The causative agent, *Borrelia burgdorferi* sensu lato (s.l.) was identified in 1982.

2.1.5.2 Etiology

Borrelia spirochete is a gram negative, spiral organism with linear chromosome. The life cycle of *Borrelia* requires arthropod vectors and mammalian hosts. It belongs to the family Spirochetaceae under the order Spirochaetales. *Borrelia* spirochetes comprise three distinct species groups i.e. Lyme borreliosis group

Table 2.2 Distribution of *B. burgdorferi* s.l

Genospecies	Distribution	Reservoir host
B. burgdorferi sensu stricto	USA, Europe (Germany, Poland, Denmark, UK), Asia (Japan)	Rodent, birds
B. afzelii	Europe (Denmark, Germany, Poland, UK), Scandinavia, Asia (Japan)	Rodents
B. garinii	Western Europe, Asia (Japan)	Birds (blackbirds, song thrushes), rodents
B. valaisiana	Ireland, UK, Denmark, Germany, Poland, Netherlands, Scandinavia, Switzerland, Italy	Birds (pheasants)
B. lusitaniae	Europe	Lizard

(*Borrelia burgdorferi* sensu lato; hard tick transmission), relapsing fever group (*B. duttonii*, *B. hermsii*; soft ticks transmission) and a third group phylogenetically similar with relapsing fever group but transmitted by hard ticks (*B. theileri*, *B. lonestari*, *B. miyamotoi*).

Lyme disease in human and animals (dogs, horses) is mostly caused by *Borrelia burgdorferi* sensu lato (s.l.). *Borrelia miyamotoi* is recently detected to produce Lyme disease like syndrome in human.

Borrelia burgdorferi s.l. can be divided into 15 genomic groups or genospecies (*B. burgdorferi* sensu stricto, *B. garinii*, *B. afzelii*, *B. japonica*, '*B. andersonii*', *B. tanukii*, *B. turdi*, *B. valaisiana*, *B. lusitaniae*, '*B. bissettii*', *B. sinica*, '*B. californiensis*', *B. spielmanii*, *B. americana*, *B. carolinensis*). The pathogenic genospecies [*B. burgdorferi* sensu stricto (outer surface protein A-OspA type 1), *B. afzelii* (OspA type 2), *B. garinii* (OspA types 3–8), *B. valaisiana*, *B. lusitaniae*] and their distribution in different continents are described in Table 2.2.

2.1.5.3 Host Susceptibility

Borrelia burgdorferi is maintained in nature through a cycle. In the cycle, hard ticks (*Ixodes* spp, *Haemaphysalis* spp.) and small mammals (birds, rodents) act as vector and reservoir host, respectively (Table 2.2). The serum complement of the reservoir hosts determines host preference of *B. burgdorferi*. The bird associated genospecies are resistant to the bird complement but susceptible to the rodent complement.

Ixodes scapularis and *I. pacificus* in USA and Canada, *I. ricinus* in Europe and *I. persulcatus* in Asia (Japan) act as major vectors of *Borrelia burgdorferi* s.l. Occasionally, other species of ticks such as *I. uriae*, *I. affinis*, *I. dammini*, *I. frontalis*, *I. angustus* Neumann, *I. spinipalpis* Hadwen and Nuttall, *I. auritulus* Neumann, *I. pacificus* Cooley and Kohls are also associated. All of these ticks cannot parasitize human to transmit the spirochete, but, they can act as maintenance host (e.g. *I. affinis*, *I. dentatus*). *Ixodes persulcatus* in Japan and *I. scapularis* in USA was detected to act as vector of *Borrelia miyamotoi*.

Different stages of *Ixodes* ticks (larva, nymph and adult) can attach with three different hosts to take the blood meal and after engorgement they drop off the host in the environment. Immatured ticks (larva or nymph) prefer to stay in moist areas such as vegetative mat of the forest floor or meadow. Ground-feeding birds (passerines, game birds, sea birds), rodents, lizards act as preferred hosts of the imamatured ticks (Table 2.3). Although, in comparison to rodents, tick infestation in migratory passerine birds is 20–30 times less, but the birds can transmit the infection for long distances. Sometimes, reservoir birds generate mutant and more virulent strains of *Borrelia*. Passerine birds in mixed coniferous (evergreen) forest were more infected with *B. burgdorferi* s.l. than the birds in alder swamp forest. Experimentally, Mallard ducks (*Anas platyrhynchos platyrhynchos*) are susceptible to *B. burgdorferi* infection and the ducks shed the organism in the droppings. They may transmit the infection without the help of tick vectors. The study indicated that psittacine birds such as yellow naped amazon parrots (*Amazona auropalliata*) are generally not infected with *B. burgdorferi* s.l. More studies are needed to explore their resistance status against *Borrelia* infection.

The ticks normally attach with eyelid, head, neck and ventral feather of passerine birds during blood meal (Fig. 2.8). The immature ticks take a blood meal for 2–4 days from their preferred hosts. In adult stage, the ticks attach with the tip of the grasses to get adhere with a large mammalian host. The adult ticks take a blood meal for 5–6 days. The ticks itself have less mobility but they can be carried by their hosts specially the migratory passerine birds across the countries. The seabird tick (*I. uriae*) is observed to disseminate *Borrelia burgdorferi* s.l. from one hemisphere to another (trans-hemispheric transmission). In Canada, passerine birds move northward during spring for breeding and nesting and they disseminate ticks with the pathogens.

B. burgdorferi is transmitted to immatured ticks from infected birds, rodents and lizards along with the blood meal. The spirochete after transmission multiplies in the gut of the ticks. When the immature ticks molt into adult stage, the numbers of *Borrelia* spirochete is decreased. During attachment of adult tick with large mammalian hosts, the multiplication of spirochete restarts and the number is increased. The expression of *B. burgdorferi* outer surface protein (osp) is also changed from ospA to ospC. The ospC helps in transmission of *Borrelia* from the mid gut to the salivary glands of ticks. Thus, *B. burgdorferi* is transmitted transstadially from larva to nymph and from nymph to adult. Rarely, within the tick population, *B. burgdorferi* is transmitted transovarially. When the adult ticks bite a new host, the spirochetes are transmitted from the salivary glands. Possibility of *B. burgdorferi* transmission by the adult ticks is more than the nymph and larvae, because the adult ticks have two blood meals in different hosts.

Sometimes, a single species of tick is infected with more than one numbers of *Borrelia* genospecies (e.g. *B. garinii* and *B. valaisiana*) due to superinfection of the already infected ticks during their consecutive blood meals. Occasionally, two different species of ticks (*I. scapularis* and *I. affinis*) may attach with the same *Borrelia burgdorferi* infected host. Co-transmission occurs between the infected and naive nymphs or larvae.

Table 2.3 Birds as a reservoir host of *B. burgdorferi* s.l. in different countries

Tick species infected with *Borrelia*	Birds as a reservoir host	Country	Reference
–	Tree pipit (Anthus trivialis)	Poland	Gryczynska et al. (2004)
	Dunnock (*Prunella modularis*)		
	Chaffinch (*Fringilla coelebs*)		
	Song thrush (*Turdus philomelos*)		
	Nuthatch (*Sitta europea*)		
	Hawfinch (*Coccothraustes coccothraustes*)		
	Robin (*Erithacus rebecula*)		
	Eurasian Blackbird (*Turdus merula*)		
	Wren (*Troglodytes troglodytes*)		
Ixodes ricinus	Eurasian blackbirds (*Turdus merula*)	Italy	Mannelli et al. (2005)
Ixodes pacificus	Rio Grande wild turkeys (*Meleagris gallopavo intermedia*)	California, USA	Lane et al. (2006)
Ixodes ricinus	Eurasian blackbird (*Turdus merula*)	Czech Republic	Dubska et al. (2009)
	Song thrush (*Turdus philomelos*)		
	Great tit (*Parus major*)		
Ixodes scapularis	White-throated sparrow (*Zonotrichia albicollis*)	Canada	Scott et al. (2010)
	Common yellowthroat (*Geothlypis trichas*)		
	American robin (*Turdus migratorius*)		
	Song sparrow (*Melospiza melodia*)		
	Swainson's thrush (*Catharus ustulatus*)		
	Fox sparrow (*Passerella iliaca*)		
Ixodes ricinus	Eurasian blackbird (*Turdus merula*)	Norway	Hasle (2011)
	Song thrush (*Turdus philomelos*)		
	Redwing (*Turdus iliacus*)		
Ixodes spp.	Eurasian blackbird (*Turdus merula*)	France	Socolovschi et al. (2012)
Ixodes ricinus	Eurasian blackbird (*Turdus merula*)	Spain	Palomar et al. (2012)
I. frontalis	European robin (*Erithacus rubecula*)		
	Song thrush (*Turdus philomelos*)		
	Eurasian wren (*Troglodytes troglodytes*)		
	Eurasian jay (*Garrulus glandarius*)		
Ixodes ricinus	Eurasian blackbird (*Turdus merula*)	Portugal	Norte et al. (2013)
	Song thrush (*Turdus philomelos*)		
	Great tit (*Parus major*)		
	Common chaffinch (*Fringilla coelebs*)		
Ixodes affinis	Carolina Wrens (*Thyrothorus ludovicianus*)	Virginia, USA	Heller et al. (2015)
	Brown Thrashers (*Toxostoma rufum*)		
	American Robin (*Turdus migratorius*)		
	Eastern Towhee (*Pipilo erythrophthalmus*)		
	Northern Cardinal (*Cardinalis cardinalis*)		
	White-throated Sparrow (*Zonotrichia albicollis*)		
	Swainson's Thrush (*Catharus ustulatus*)		

(continued)

Table 2.3 (continued)

Tick species infected with *Borrelia*	Birds as a reservoir host	Country	Reference
Ixodes frontalis	European robin (*Erithacus rubecula*)	Czech Republic	Literak et al. (2015)
	Common chaffinch (*Fringilla coelebs*)		
	Eurasian blackcap (*Sylvia atricapilla*)		
	Eurasian blackbird (*Turdus merula*)		
Ixodes spp.	Nightingale (*Luscinia megarhynchos*)	Germany	
	Dunnock (*Prunella modularis*)		
	Chiffchaff (*Phylloscopus collybita*)		
	Reed warbler (*Acrocephalus scirpaceus*)		
Ixodes uriae	Kittiwake (*Rissa tridactyla*)	France	Duneau et al. (2008)
	Puffin (*Fratercula* spp.)		
	Guillemot (*Uria* spp., *Cepphus* spp.)		
	Fulmar (*Fulmarus* spp.)		

Fig. 2.8 Greenfinch (*Carduelis chloris*) infested with ticks (*Courtesy* Ola Nordsteien, Jomfruland bird observatory, Norway)

2.1.5.4 Transmission

Migratory birds help in spreading *Borrelia* infected ticks in distant places, from which the infection can be further transmitted to human or other mammals. The migratory birds can act as carrier of previously infected tick or transovarially infected larvae. Occasionally, transmission of *Borrelia* occurs from the infected ticks to uninfected ticks during their co-feeding from the same birds. The migratory birds not only import the infected ticks in a locality, but also, there is a possibility that local ticks get the infection during attachment with the birds. After a long journey, the birds prefer to take rest in some places for a few days. Recently, role of cottontail rabbit in this transmission cycle is also explored.

2.1.5.5 Clinical Symptoms

During carriage of *B. burgdorferi* most of the birds do not show any clinical symptom or lesion. Experimental inoculation of *B. burgdorferi* in Canary finches (*Serinus canaria*) produced only a brief episode of diarrhoea. Natural infestation of *B burgdorferi* infected tick (*Ixodes auritulus*) results gasping, lameness and death in fledgling American robin (*Turdus migratorius*).

2.1.5.6 Diagnosis

Clinical Specimens

For collection of suspected ticks from the migratory birds, the birds are caught by mist nets and are observed carefully for the presence of ticks in head, neck, and beak by magnifying glasses. The ticks are collected by a blunt forcep and are placed in 70% ethanol. They should be labelled properly indicating species of bird and tick, and date of collection. For identification of bird and ticks up to the species level, expertise is needed.

From the birds, suspected for *B. burgdorferi* infection, heparinized blood and tissues from liver, spleen, kidneys can be collected after post mortem.

Diagnostic Techniques

(a) *Direct Examination*: Dark field Microscopy or Giemsa stain can directly demonstrate the *Borrelia* spirochete in the blood film, liver/spleen smears. FAT can be used for direct examination of the smears.
(b) *Isolation of bacteria from clinical samples*: Isolation of *Borrelia* is difficult due to its slow and fastidious growth and microaerophilic requirements. Modified BSK (Barbour-Stoenner-Kelly) medium is used for isolation of *B. burgdorferi* s.l. It is an enriched serum broth containing the antibiotics like kanamycin and 5-flurouracil. The media after inoculation is incubated at 33–34 °C for 3 weeks. The collected heparinized blood sample (0.02 ml) or triturated tissue sample (0.1 ml) can be added in BSK medium (7–8 ml).
(c) *Serological tests*: ELISA based kits for detection of total immunoglobulin, Ig G, Ig M against *B. burgdorferi* s.l is available for human. However, studies in animals (dogs), indicated that results of serological tests are inconclusive. The antibodies may be produced due to earlier exposure specially in those areas where infected tick bite is a common phenomenon. Such kind of serological studies are not conducted in birds probably due to this uncertainty.

(d) *Molecular biology*: The whole blood samples collected from the birds can be used for DNA extraction. Whereas, from the ticks, DNA is extracted by spin column technique. PCR for consensus *flaB* gene of *Borrelia* and the spacer region between the 5S and 23S rRNA genes can be carried out to confirm *Borrelia burgdorferi* s.l.

2.1.5.7 Zoonosis

Zoonotic transmission of *Borrelia burgdorferi* s.l. occurs from the bites of infected ticks. The persons during outdoor recreation, professionals such as wildlife and forest caretakers are at high risk. Migratory birds passively maintain the infection in nature. No direct transmission of *Borrelia burgdorferi* s.l. from the birds to human is evidenced. The infection in human initiates with a red coloured allergic pimple (erythema migrans), and it is followed by fever, headache, fatigue, muscle and joint pain. In severe cases, meningitis, unilateral facial nerve palsy and renal failure occur.

2.1.5.8 Control Strategy

Products that kill or repel ticks (e.g. permethrin) can be used in the habitat to reduce the tick density. However, these acaricides may cause environmental hazard and they are only recommended during epidemic situation.

2.1.6 Others

2.1.6.1 Yersiniosis

Yersinia spp. was first isolated by Alexandre Yersin in 1894 in HongKong. He was sent by Louis Pasteur to investigate about plague outbreak there. In Japan, S. Kitasato also independently isolated the bacteria a few days later. Previously, it was known as *Pasteurella pestis* in honour of Pasteur. Later, in memory of Yersin, the bacterium was renamed as *Yersinia pestis*. In 1976, *Yersinia pseudotuberculosis* was isolated from a sick palm dove (*Streptopelia senegalensis*) in Israel. *Yersinia enterocolitica* was first detected in budgerigars in 1980.

Yersinia is gram negative, short rods or coccobacilli shaped bacteria. They show bipolar staining characteristics ('safety pin appearance') when stained with Leishman's or Wright or Giemsa stain. The genus *Yersinia* is classified under the family Enterobacteriaceae that belongs to the order Enterobacteriales. *Yersinia* consists of 11 numbers of species. Among them, *Y. pseudotuberculosis* and *Y. enterocolitica* are commonly associated with psittacine and passerine bird infection.

Mynahs are most susceptible to *Yersinia* spp. infection. Yersiniosis is also reported from canaries (*Serinus canaria*), zebra finch (*Poephila guttata*), kaka (*Nestor meriondalis*), rainbow lorikeet (*Trichoglossus mollucanus*), budgerigar (*Melopsittacus undulatus*), New Zealand wood pigeons (*Hemiphaga novaeseelandiae*), blue-fronted Amazon (*Amazona aestiva*), yellow-headed Amazon (*Amazona oratrix*), Eurasian collared dove (*Streptopelia decaocto*) and cockatoo

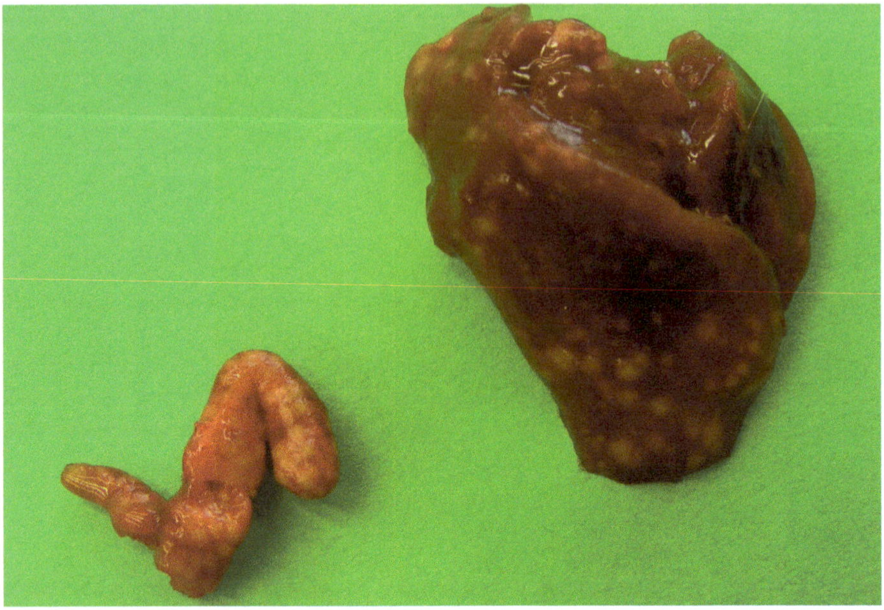

Fig. 2.9 Yellow coloured foci in vital organ of a weaver bird suffering with Yersiniosis (*Courtesy* Prof. Richard Hoop, University of Zurich, Switzerland)

(*Cacatua alba*). Rodents and wild birds are major reservoir of infection and the feed and water are often contaminated with rodent urine or faeces. Ingestion of contaminated feed and water is the key route of transmission.

High mortality and non-specific clinical signs such as ruffling of feathers, depression, diarrhoea, and biliverdin in the urine are observed in the birds. The infection is acute and mostly enteric in passerine birds. Chronic infection takes place in psittacines and pigeons, and it produces hepatitis, splenitis, pneumonia, nephritis and enteritis. The liver becomes dark, swollen and congested. Yellow coloured foci (bacterial granulomata) are found in the liver, spleen, lungs, kidneys, intestines and heart (Fig. 2.9). Microscopically, these foci are composed of necrosed hepatocytes and splenic pulp with fibrin and *Yersinia* colonies. Accumulation of iron in the liver (hepatic haemosiderosis) may act as a predisposing factor for systemic *Yersinia* spp. infection.

A smear can be prepared from the collected tissues of liver, spleen, kidney, intestine and stained by Leishman's, Wright, and Giemsa stain. *Yersinia* shows typical 'bipolar characteristics' (safety-pin appearance). *Yersinia* can be isolated in blood agar, nutrient agar, MacConkey's agar, brilliant green agar (*Y. enterocolitica*). The selective medium is CIN agar which contains antibiotics such as cefsulodin (15 mg/l), irgasin (4 mg/l) and novobiocin (2.5 mg/l). 'Cold enrichment' method can be followed for primary isolation of *Y. pseudotuberculosis* and *Y. enterocolitica* from clinical samples. The samples are kept in sterile phosphate buffered saline (PBS) or nutrient broth at 4 °C for 3 weeks. Subculture in MacConkey's or CIN agar is done at weekly interval.

Amoxicillin in drinking water or soft food is the choice of treatment. In unresponsive cases, treatment should be carried out on the basis of antibiotic sensitivity test.

2.1.6.2 Mycoplasmosis

Albert Bernhard Frank (1889), a German Biologist, first coined the term *Mycoplasma* which is originated from the Greek word *mykes* (fungus) and *plasma* (formed). Earlier *Mycoplasma* was known as pleuropneumonia-like organisms (PPLO). Adler (1957) first reported isolation of PPLO from the air sac of a parakeet bird.

Mycoplasma is the smallest pathogenic bacteria (0.3–0.8 μm) and is pleomorphic in shape due to absence of the rigid cell wall. In stained smears, they appear as ring, globules, filaments or elementary bodies. The cell membrane is constituted of trilaminar structure enriched with phosphoprotein, lipoprotein, glycolipid, phospholipid and sterol moieties. *Mycoplasma* belongs to the class Mollicutes, order Mycoplasmatales, and family Mycoplasmataceae. The family comprises of two genera i.e. *Mycoplasma* and *Ureaplasma*. Among different species under the genus *Mycoplasma, M. gallisepticum, M. iowae*, and *M. sturni* are associated with pet bird infection.

An epidemic of Mycoplasmal conjunctivitis was noticed in house finches (*Carpodacus mexicanus*) in USA in 1994. Other birds such as budgerigars, cockatiel, canary, yellow-naped Amazon parrot, pigeons, pea-fowls (*Pavo cristatus*), fledgling cliff swallows, European starling (*Sturnus vulgaris*), chukar partridges (*Alectoris chukar*), ring-necked pheasants, purple finches (*Carpodacus purpureus*), evening grosbeaks (*Coccothraustes vespertinus*), pine grosbeaks (*Pinicola enucleator*) are also reported to be infected. Concomitant Mycoplasmal infection with other bacteria and protozoa (*Cryptosporidium* spp.) was detected in Amazon parrots and fledgling cliff swallows. Experimentally, American goldfinch (*Carduelis tristis*) carried *M. gallisepticum* for prolonged period without showing any clinical sign. House sparrows (*Passer domesticus*) are transiently infected experimentally with *M. gallisepticum* for a short period. In United States, tufted titmice (*Baeolophus bicolor*) bird was reported as non-clinical carriers of *M. gallisepticum*.

Feeders or any focal point where the birds gather, act as a source of *M. gallisepticum* infection, because the infected birds excreate their droplets in the feeder. Statistical correlation (multivariate analysis) was established between presence of tube style feeders, non-breeding period and low environmental temperature and *M. gallisepticum* infection in house finches. Vertical way is a rare possibility of Mycoplasmal transmission in birds.

Clinically the infected birds show variable symptoms ranging from serous nasal discharge, sinusitis, swollen eyes with discharge, conjunctivitis and blindness (Fig. 2.10). In fledgling cliff swallows and European starling (*Sturnus vulgaris*) infected with *M. sturni*, bilateral conjunctivitis, episcleritis, epiphora, hyperaemia of palpebrae and nictitans are observed. Gross lesions in birds include congestion of

Fig. 2.10 Mycoplasmal conjunctivitis in a pea-fowl (*Courtesy* M. Scott Echols, Medical Center for Birds, California)

mucosa and accumulation of exudates in nasal sinus, trachea, bronchi, and air sacs. Air sac congestion was also detected in budgerigars experimentally challenged with *M. gallisepticum*. Histological investigation in birds reveals the presence of mucous gland hyperplasia and thickened mucous membrane of the respiratory tract with mononuclear cell infiltration. In European starling birds, ulceration of mucous membrane and absence of epithelial hyperplasia and lymphoplasmacytic infiltration was observed. In fledgling cliff swallows, lymphoplasmacytic conjunctivitis, rhinitis and infraorbital sinusitis with follicular lymphoid hyperplasia were detected.

Conjuntival swabs and head, lung, and spleen in 10% buffered formalin after post-mortem, can be collected as clinical samples from the suspected birds. The smears prepared from clinical specimens are stained with Giemsa. *M. gallisepticum* appears as coccoid organisms having 0.25–0.5 μm in size. Contrast phase microscopy, dark phase illumination techniques can be used for their direct visualization. *M. gallisepticum* can be isolated in specialized medium supplemented with 10–15% heat-inactivated avian, swine or horse serum. Change in broth colour indicates positive growth after incubation at 37 °C for 3–5 days. Serum plate agglutination test is the rapid serological test for detection of *M. gallisepticum* antibodies in birds, although, sometimes it produces false positive result due to cross reaction. PCR targeting 16SrRNA gene and loop-mediated isothermal amplification (LAMP) assay based on a gene within the pyruvate dehydrogenase complex (*pdhA*) are developed for rapid detection of *M. gallisepticum*.

In house finches, application of oral tylosin (1 mg/ml drinking water for 21 days) and ciprofloxacin eye drop (for 7 days) successfully treated conjunctivitis associated with *M. gallisepticum*. Doxycycline (40–50 mg/kg body weight, orally) is also recommended for Mycoplasmal infection in cockatiels and amazons.

2.1.6.3 *Pasteurella multocida*, *Gallibacterium* spp., *Volucribacter* spp.

Bollinger (1878) first reported the isolation of *Pasteurella* like organisms from cattle and wild animals. Louis Pasteur (1880) conducted more comprehensive studies on fowl cholera and its etiological agent. Trevisan (1887) coined the name *Pasteurella* for the bipolar organisms described earlier by Pasteur and others. Lignières (1900) proposed the specific name for each species of *Pasteurella* according to their host preference, such as *Pasteurella aviseptica* for fowls, *P. suiseptica* for pigs, *P. boviseptica* for bovines, *P. oviseptica* for *ovines* and *P. leptiseptica* for rabbits. Rosenbusch and Merchant (1939) proposed a single species *Pasteurella multocida* and it is in use till date. Miringa (1975) described pasteurellosis in African grey parrots.

P. multocida is a gram-negative, non motile, non spore-forming short rod or coccobacillus bacterium. In fresh cultures and animal tissues, it produces typical bipolar staining characteristics, particularly with Leishman or methylene blue stain. *Pasteurella* belongs to the family *Pasteurellaceae*. Other avian pathogens such as *Gallibacterium* (*Pasteurella anatis*) and *Volucribacter* are also members of the same family. *P. multocida* can be classified into 6 capsular types (A–F) and 16 somatic types (1–16).

Psittacines [parrots, red-fronted conure (*Aratinga wagleri*)], passerine birds, owls, raptors and waterfowls (ducks) suffer from pasteurellosis. *P. multocida* is also isolated from eye swabs of healthy psittacine birds. Unclassified members of the *Pasteurellaceae* family were isolated from lesions in domestic goose (*Anser anser forma*), Fischer's lovebird (*Agapornis fischer*), parrots (*Amazona* spp.), macaws (*Ara macao*), rock dove (*Columba livia*), budgerigars (*Melanopsittacus undulates*), and African gray parrot (*Psittacus erithacus*).

Contaminated environment (e.g. water) is the major source of *P. multocida* infection. Mechanical transmission by blood sucking arthropods and cat-bite is possible. In cat-bite cases, dermatitis and myositis develops rapidly and it is followed by septicaemia and death. In psittacine birds *P. multocida* serotype 3 and 4 are associated with septicaemia and cutaneous lesions, respectively. In African gray parrots, *P. multocida* produced obstruction of nares and dyspnoea, due to formation of intranasal caseous and fibrinous plugs along with other bacteria. In *P. multocida* infected budgerigars, crop inflammation and apathy was observed. *Gallibacterium melopsittaci* are associated with septicaemia and salpingitis in budgerigars and parakeets. *Volucribacter psittacicida* causes respiratory tract infections, septicaemia, crop inflammation, and diarrhoea in psittacine birds.

A smear can be prepared from collected blood sample or the nasal swabs and it is stained by Leishman or methylene blue or Gram's stain. *Pasteurella* appears as gram negative non-sporing coccobacilli with typical bipolar staining characteristics. *P. multocida* can be isolated in dextrose-starch agar, casein-sucrose-yeast (CSY) medium with 5% blood (bovine or sheep). *P. multocida* specific PCR (PM-PCR) helps in rapid and confirmatory detection from clinical samples.

Treatment of avian pasteurellosis with ampicillin (150–200 mg/kg body weight for pigeons, Amazon parrots) and tiamulin fumarate (25–50 mg/kg body weight, oral) are recommended.

2.1.6.4 *Escherichia Coli*

Escherichia coli was first isolated by Theodor Escherich in 1885 from the faeces of human infants. It was named in honour of the German pediatrician and its major natural habitat i.e. colon. In 1978, *E. coli* was detected in faecal samples collected from psittacine birds. Raphael and Iverson (1980) described *E. coli* associated coligranuloma in Amazon parrot along with psittacosis.

E. coli is gram negative, short rods, varying form coccoid shape to long filamentous forms (Fig. 2.11). They occur singly, in pair or in short chain. They are non-spore forming and mostly motile by peritrichous flagella. The genus *Escherichia* is classified under the family Enterobacteriaceae that belongs to the order Enterobacteriales. There are total 6 species under the genus *Escherichia.* Among them, *Escherichia coli* are the important pathogen.

The gastro-intestinal tract of all vertebrates including birds is the most common natural habitat of *E. coli*. The studies revealed that healthy parrots (31%), cockatoos (*Cacatua* spp., 60%) and shore birds carried *E. coli* in their intestine. In healthy passerine birds, *E. coli* are not considered as a major intestinal flora. Psittacines imported or illegally traded from other countries and shore birds act as source of *E. coli.* Majority of these psittacine *E. coli* isolates possessed antimicrobial

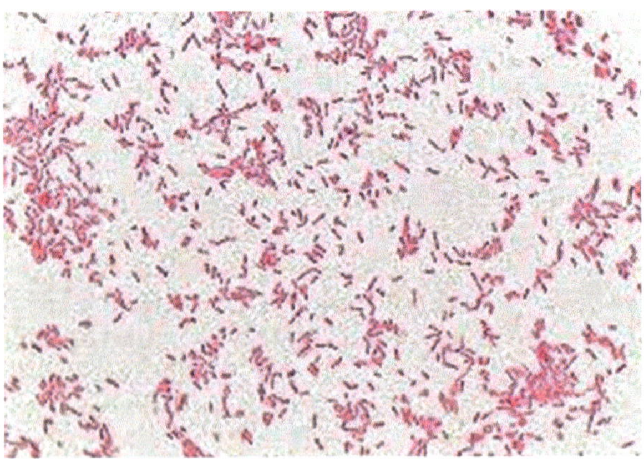

Fig. 2.11 Gram stained *Escherichia coli* isolated from bird (100×)

resistance due to the exposure of the birds to the prophylactic antibiotics after their capture.

In pet birds, *E. coli* is transmitted by contaminated feed, drinking water, aerosols, and fomites. The stress conditions like transport, dietary change and extreme climate also help to establish the infection. In adult canaries and finches, *E. coli* are most common secondary pathogens associated with epizootic mortality. Non-specific clinical signs and lesions such as lethargy, rhinitis and conjunctivitis are detected. *E. coli* as a primary pathogen is reported from a hyacinth macaw (*Anodorhynchus hyacinthinus*), died due to septicaemia and enteritis with hemorrhages in different organs, and a kakapo (*Strigops habroptilus*) with exudative cloacitis. Recently, attaching-effacing *E. coli* is detected as a primary pathogen in a captive flock of budgerigars (*Melopsittacus undulatus*). Common lesions in budgerigars include hepatitis, enteritis, and attaching and effacing lesions along the intestinal tract.

In nestlings of canaries and finches, *E. coli* is considered as most important cause of diarrhoea, dehydration, cachexia and mortality. Appearance of young birds and their mothers became dirty, wet and yellowish ('sweating disease').

Isolation of *E. coli* from the clinical samples is the major diagnostic technique. Blood agar, MacConkey's agar are choice of the medium for isolation. After overnight incubation in MacConkey's agar, characteristic pink coloured colonies are transferred into eosine methylene blue (EMB) agar for detection of 'metallic sheen' (Fig. 2.12). The isolates are further confirmed by different biochemical tests. Pathogenicity of the *E. coli* isolates from clinical samples should be confirmed as they are present as normal bacterial flora within the body. Virulence of the isolates can be ascertained by ligated loop assay, cell culture cytotoxicity assay, typing and detection of toxin by serological or DNA based methods.

E. coli infections can be treated with ampicillin sodium, amoxicillin/clavulanate, cephalexin, oxytetracycline and spectinomycin. In unresponsive cases, antibiotic

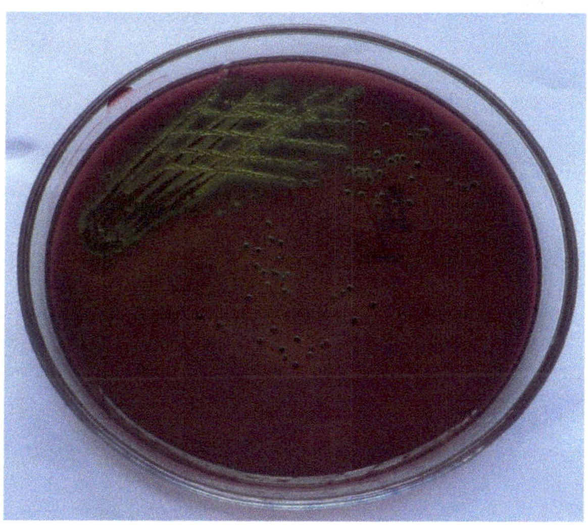

Fig. 2.12 Metallic sheen in EMB agar by *Escherichia coli* isolates

Table 2.4 Other important bacterial infection of pet birds

Bacteria	Susceptible hosts	Clinical signs/gross lesions	Treatment
Staphylococcus spp.	Resident of skin in healthy birds such as African Grey Parrots, Budgerigars and Cockatiels. Detected as primary pathogen in hyacinth macaw, lovebird and passerine birds	In hyacinth macaws, ulcerative dermatitis and septicaemia associated endocarditis are reported. In lovebirds, septicaemia followed by blindness and CNS disorder was detected. In passerines, abscesses, dermatitis, bumble foot, conjunctivitis, sinusitis, arthritis, pneumonia, and death occurs	Amoxicillin/clavulanate, Piperacillin sodium/tazobactam Sodium, Tiamulin fumarate
Enterococcus spp.	Commensal flora of gastro intestinal tract	Wound infection, septicaemia and death in psittacine birds. In canaries, tracheitis, pneumonia, air sac infections, dyspnoea and loss of normal voices occur	Treatment should be carried out on the basis of sensitivity test
Pseudomonas spp.	Parrots and passerine birds. Contaminated feed and water are major source of infection	Upper respiratory tract infection, cellulitis and conjunctivitis in parrots. Pneumonia and aerosacculitis in passerine birds. Necrotizing hepatitis is also detected in pt birds	Enrofloxacin, amikacin sulfate. In unresponsive cases, treatment should be carried out on the basis of sensitivity test
Clostridium spp.	Parrots, blue and yellow macaw, lories, rainbow lorikeet, pigeon, ostrich	Necrotic enteritis, ulcerative enteritis with dilatation of the small intestine and presence of yellow foci, necrotizing hepatitis, myocarditis, and ventriculitis	Treatment should be carried out on the basis of sensitivity test
Bordetella avium	Psittacine, turkey, finches	'Lockjaw syndrome', upper respiratory tract infection	Tiamulin fumarate
Klebsiella spp.	Psittacine	Pneumonia	Ampicillin sodium

should be selected after sensitivity test of the etiological *E. coli* isolates. In nestlings, antibiotics are administered in drinking water and egg food from one day before hatching up to 6 days after hatching. Extra drinking water should be provided to prevent dehydration.

Other bacterial infections of pet birds are described in Table 2.4.

2.2 Viral Diseases

2.2.1 Newcastle Disease

2.2.1.1 History

Newcastle disease in virulent form was first observed in poultry in Java (Indonesia) and NewCastle-on-Tyne (United Kingdom) in 1926. The disease was prevalent in Korea and Scotland prior to 1926 but was not documented in details. Doyle first coined the term 'Newcastle disease' to describe the infection in poultry according to the name of the place. The infection spread rapidly among the poultry in the subsequent years in Asian countries. In India, it was first described in poultry in 1927 in Ranikhet, Uttarakhand. Consequently, the infection is still known as 'Ranikhet disaese' in India and neighbouring countries.

Malbrant (1942) reported an outbreak in Australian parrots and red-headed lovebirds (*Agaporius pullaria pullaria* L.) in Africa suspected to be suffered and died off Newcastle disease. Zuydam (1952) first isolated Newcastle disease virus (NDV) from parakeet (*Psittacula krameri borealis* Nearn) and ospray birds (*Pandion haliaetus*) in The Netherlands. The first major outbreak of Newcastle disease among grey parrots (*Psittacus erithacus* L.) occurred in Kenya in 1955 (Scott et al. 1956). Subsequently, in 1960, NDV was isolated from grey parrots (*Psittacus erithacus* L.) in Kenya (Scott and Winmill 1960). In 1970s, Pigeon paramyxovirus-1 (PPMV-1), a variant of NDV, was discovered in the Middle East countries.

2.2.1.2 Etiology

NDV belongs to the *Avulavirus* (Avian Paramyxovirus-1) genus within the *Paramyxoviridae* family in the order *Mononegavirales*. The virus is an enveloped, single-stranded, un-segmented, negative-sense RNA virus. The virion has a genome of 15 kb in size and the genome comprises of the genes which encode nucleocapsid protein (NP), phosphoprotein (P), matrix protein (M), fusion protein (F), hemagglutinin-neuraminidase (HN) and polymerase enzyme (L).

On the basis of virulence, NDV can be classified into velogenic (highly virulent, ICPI > 1.5), mesogenic (intermediate, ICPI > 0.7 but ≤ 1.5) and lentogenic strains (non-pathogenic, ICPI ≤ 0.7). Intracerebral pathogenicity index (ICPI) is OIE recommended in vivo test for determination of NDV virulence. The strains differ in virulence also differ in amino acid sequence at the cleavage site of precursor fusion protein (F0). Velogenic strains have more than two basic amino acids (arginine or lysine at positions 113–116) and phenylalanine at the position 117. On the basis of fusion protein (F) and polymerase enzyme (L) nucleotide sequence, NDV is classified into two major classes (class I and II) with a single genotype under class I and 18 genotypes under class II. Distribution of NDV genotypes and sub-genotypes in different species of birds is described in Table 2.5. The virulence of class I NDV isolates is low (except one isolate from Ireland) and they are mostly maintained by

Table 2.5 Distribution, hosts, virulence of different NDV class, genotypes and sub-genotypes

NDV class	NDV genotype/sub-genotype	Virulence	Host	Country
Class I	Genotype 1, sub-genotype 1a	Lentogen	Domestic ducks (Anatidae spp.), chickens (Gallus gallus domesticus), geese	China
	Genotype 1, sub-genotype 1b	Lentogen	Chickens, ducks, black swan, peafowl, egret, heron	China
	Genotype 1, sub-genotype 1c	Lentogen	Ducks, swans, geese, shorebirds	United States, Europe, China
Class II	Genotype I, sub-genotype Ia	Lentogen	Chickens	Australia, China, Colombia, Malaysia, and South Korea
	Genotype I, sub-genotype Ib	Lentogen	Wild and domestic waterfowls	China, Japan, Luxembourg, Madagascar, Nigeria, Russia, South Korea, Ukraine, and the United States
	Genotype I, sub-genotype Ic	Lentogen	Shorebirds, waterfowl, gulls	Japan, Mexico, Russia, Sweden, and the United States
	Genotype II	Velogen	Poultry, domestic waterfowl	United States, Africa, Asia, Europe, South America
	Genotype III	Velogen	Poultry, domestic waterfowl	Australia, Japan, UK, Taiwan, Zimbabwe, and Singapore, Pakistan, China
	Genotype IV	Velogen	Poultry, pigeons	Europe, Sudan (Africa), HongKong (Asia), Russia
	Genotype V	Velogen	Psittacine birds, poultry, ducks	Central and South America, Europe, Africa
	Genotype V, sub-genotype Va	Velogen	Double-crested cormorants, pelicans, gulls	United States
	Genotype V, sub-genotype Vb	Velogen	Poultry, caged-birds	United States, Brazil, Central America, Africa
	Genotype V, sub-genotype Vc	Velogen	Poultry, caged-birds, tree-ducks, quails	Mexico and Central America
	Genotype V, sub-genotype Vd	Velogen or mesogen	Poultry	Kenya, Uganda
	Genotype VI, sub genotype VIa	Velogen or mesogen	Columbidae birds	Asia, Europe, Middle East, United States
	Genotype VI, sub genotype VIb	Velogen or mesogen	Pigeons, kestrels, falcons, cockatoos, budgerigars, pheasants, swans, robin	Argentina, China, Italy, United States, Europe, South Africa

(continued)

Table 2.5 (continued)

NDV class	NDV genotype/sub-genotype	Virulence	Host	Country
	Genotype VI, sub genotype VIc	Velogen or mesogen	Chickens, Columbidae birds	East Asia
	Genotype VI, sub genotype VId	Velogen or mesogen	Poultry	Europe
	Genotype VI, sub genotype VIe	Velogen or mesogen	Pigeon, poultry	China
	Genotype VI, sub genotype VIf	Velogen or mesogen	Pigeon	United States
	Genotype VI, sub genotype VIg	Velogen or mesogen	Pigeon, dove	Nigeria, Russia, Ukraine
	Genotype VI, sub genotype VIh	Velogen or mesogen	Pigeon	Argentina
	Genotype VI, sub genotype VIi	Velogen or mesogen	Collared doves, pigeons	Italy, Nigeria
	Genotype VII, sub genotype VIIa	Velogen	Poultry	Western Europe
	Genotype VII, sub genotype VIIb	Velogen	Poultry, wild birds	China, Vietnam, Israel, Europe, Turkey, South Africa, Mozambique, Kazakhstan, the Far East, the Middle East, India
	Genotype VII, sub genotype VIIc	Velogen	Poultry	Czech Republic, Switzerland
	Genotype VII, sub genotype VIId	Velogen	Poultry, wild birds (crested ibis)	China, South Korea, Colombia, Israel, South Africa, Ukraine, Venezuela, Europe, Kazakhstan
	Genotype VII, sub genotype VIIe	Velogen	Chickens, domestic waterfowl	China, Japan, Taiwan, Vietnam
	Genotype VII, sub genotype VIIf	Velogen	Poultry, pigeon	China
	Genotype VII, sub genotype VIIg	Recombinant strains		
	Genotype VII, sub genotype VIIh	Velogen	Poultry, wild egret	Bali, Indonesia, Malaysia, China
	Genotype VII, sub genotype VIIi	Velogen	Chicken, koklass pheasants (*Pucrasia macrolopha*), peafowl	Indonesia, Israel, Pakistan, Eastern Europe
	Genotype VIII	Velogen	Chicken, turkey	Argentina, China, Malaysia, South Africa, Singapore, Italy

(continued)

Table 2.5 (continued)

NDV class	NDV genotype/sub-genotype	Virulence	Host	Country
	Genotype IX	Velogen	Poultry, whooper swan (*Cygnus cygnus*), spotted necked dove (*Streptopelia chinensis*), green peafowl (*Pavo muticus*), white-cheeked starling (*Sturnus cineraceus*), Eurasian blackbird (*Turdus merula*)	China
	Genotype X	Lentogen	Wild waterfowl, turkey	United States, Argentina
	Genotype XI	Velogen	Chickens, wild birds	Madagascar
	Genotype XII	Velogen	Chickens, geese	South America, China
	Genotype XIII	Velogen	Cockatoo (family Cacatuidae)	India
	Genotype XIII, subgenotype XIIIa	Velogen	Chicken, wild little tern (*Sterna albifrons*)	Europe, Asia, Middle-East, Russia
	Genotype XIII, subgenotype XIIIb	Velogen	Chicken, Japanese Quail	India, Pakistan
	Genotype XIV, subgenotype XIVa	Velogen	Chicken	Nigeria
	Genotype XIV, subgenotype XIVb	Velogen	Chickens, turkeys, guinea fowl	Nigeria
	Genotype XV	Velogen	Chickens, geese	China
	Genotype XVI	Velogen	Chicken	Mexico, Dominican Republic
	Genotype XVII	Velogen	Poultry	West Africa
	Genotype XVIII, sub genotype XVIIIa	Velogen	Chickens, guinea fowl	Ivory Coast, Mali, and Mauritania
	Genotype XVIII, sub genotype XVIIIb	Velogen	Poultry, wild village weaver (*Ploceus cucullatus*)	Ivory Coast, Mali, Nigeria, Togo

poultry and waterfowls. Whereas, class II NDV isolates are highly virulent and are associated with fatal outbreaks in poultry, pet birds and wild waterfowls.

2.2.1.3 Host Susceptibility

Velogenic or mesogenic strains of NDV were detected in psittacine birds (cockatoos, budgerigars, macaw, lory, parrot, love bird, conure, yellow-headed Amazon parrots, yellow-naped Amazon parrots), pelicans, gulls, kestrels, falcons, white crested laughing thrush, pheasants, swans, robin, peafowl, whooper swan, spotted necked dove, white-cheeked starling, Eurasian blackbird, wild little tern, wild village weaver, mynah (*Gracula religiosa*), drongo (*Dicrurus* spp.) and partridges (family Phasianidae) (Table 2.5). These wild and pet birds may act as reservoir or spillover host for poultry. Occasional outbreaks of NDV were observed in racing pigeons and double-crested cormorants (*Phalacrocorax auritus*). The lentogenic strains are common in gulls, waterfowls, shorebirds, peafowl, egret and heron (Table 2.5) without any clinical symptoms. From neuronal tissues of parrots, lentogenic NDV strains were isolated.

Migratory birds may act as 'global reservoir' of NDV which transmits the infection into distant places across the continents. Generation of new variety of NDV (multi-recombinant) having the sequences from different putative parents of distant places suggests co-replication of the virus in the nature. In a natural NDV multi-recombinant strain, isolated from cockatoo in Indonesia (cockatoo/Indonesia/14698/90), parental lineage from NDV isolate of anhinga or snakebird (*Anhinga anhinga*) of United States and a NDV vaccine strain was detected.

2.2.1.4 Transmission

In pet birds, horizontal transmission through direct contact with infected birds or ingestion of contaminated feed and water are the major ways of NDV transmission. Sometimes, infection of parrots occurs from live animal market during their direct contact with poultry. Use of contaminated or improperly attenuated live vaccines in poultry against NDV is another possible source of infection in nature and pet birds. Experimentally, yellow-headed Amazon parrot was infected with velogenic NDV strain by nebulization. Common houseflies (*Musca domestica*) act as mechanical vector for NDV transmission, although, yet to be validated in pet birds.

Imported exotic birds including the psittacines may act as reservoir and they excreate the virus in the faecal matter for prolonged period without showing clinical signs. Legally or illegally trafficked pet birds and migratory birds thus introduce the virus into the countries which produces a persistent risk for poultry. Moreover, the study has also shown the possibility of spreading NDV in a native wildlife population through NDV contaminated illegal wildlife trade.

2.2.1.5 Pathogenesis

After transmission of the virus into the susceptible host, cellular entry requires activation of fusion protein (F0) present in the viral envelope. The F0 is activated by post-translational cleavage by the host protease enzyme. The post-translational cleavage varies with the amino acid sequence present in the cleavage site and the type of host protease enzyme. In lentogenic NDV strains, cleavage site contains monobasic amino acid sequence at the C-terminus and leucine at the N-terminus (^{112}G-R/K-Q-G-R-L^{117}). The cleavage site of velogenic and mesogenic strains contains multibasic amino acid sequence at the C-terminus and a phenylalanine at the N-terminus (^{112}R/G/KR-Q/K-K/R-R-F^{117}). The cleavage site of lentogenic strains is cleaved by the protease present in respiratory and intestinal tract only. Velogenic strains are cleaved by ubiquitous protease present in all vital organs. Moreover, activation of HN (HN0) protein and other viral proteins such as V, NP, P, and L also play role in pathogenesis.

2.2.1.6 Clinical Symptoms

Psittacine birds suffering with NDV mostly show respiratory signs (rhinitis, conjunctivitis), greenish watery diarrhea (green staining around the vent), lethargy, drooping of wing, torticollis, waving movement of head and neck and limb paralysis. Experimental inoculation of six species of pet birds (budgerigar, yellow-headed Amazon parrot, halfmoon conure, hill mynah, black-headed nun, canary) with velogenic NDV strain produced ruffled plumage, conjunctivitis, ataxia, wing tremors, paralysis of the extremities, and tremors of the head. Neurological signs are more common in parakeets.

Mortality can reach as high as 100%, but typically ranges between 20–80% depending upon the virulence of the virus strain, host species, age, and immune status.

2.2.1.7 Lesion

Petechial haemorrhages are often observed in viscera of pet birds suffering with Newcastle disease. In naturally infected cockatiels and parrots, diffuse spongiosis of gray and white matter, neuronal necrosis, perivascular infiltration of mononuclear cells are detected. The lamina propria of the proventriculus shows infiltration of mononuclear cells and ulceration. Accumulation of haemosiderin is detected in the cytoplasm of mononuclear cells. Depletion of lymphoid cells is observed in spleen and bursa of Fabricious.

2.2.1.8 Diagnosis

Clinical Specimens

Tracheal and cloacal swabs from live birds and the organs such as liver, brain, spleen, kidney after post mortem can be collected as clinical specimens. Collection of cloacal swabs from small birds is a complicated process which may causes injury of the vent. Fresh faeces collection is an alternative approach. During transport, the samples should be kept in isotonic phosphate buffered saline (pH 7.0–7.4) or brain heart infusion broth with antibiotics such as penicillin (2000 units/ml), streptomycin (2 mg/ml), gentamicin (50 μg/ml), and mycostatin (1000 units/ml). The samples can be preserved at 4 °C for four days.

Diagnostic Techniques

(a) Clinical signs and lesions after necropsy, history of direct contact with infected birds give a tentative diagnosis.
(b) *Isolation of virus from clinical samples*: The clinical samples collected in isotonic phosphate buffered saline with antibiotics are centrifuged (1000 g) and the supernatant fluid is inoculated into 9–11 days old embryonated hen's eggs (specific pathogen free) by allantoic sac route or the cell lines such as chicken embryo kidney (CEK) cells, chicken embryo fibroblast (CEF) cells. The inoculated eggs are incubated at 37 °C for 4–5 days. After incubation,

eggs are kept at 4 °C. The embryo will die in positive samples and the allantoic fluids are collected to detect the haemagglutination activity of the viral isolate. The isolate is further confirmed by haemagglutination inhibition (HI) test. Virulence of the isolate should be detected by intracerebral pathogenicity index (ICPI) or amino acid sequencing of fusion protein to confirm a ND outbreak.

(c) *Serological tests*: Virus neutralization test, plaque neutralization, hemagglutination-inhibition (HI), single radial immunodiffusion, agar gel immunodiffusion (AGID), enzyme-linked immunosorbent assay (ELISA) are employed for detection of NDV antibodies. However, these tests can not differentiate the infection caused by velogenic and lentogenic viral isolates. Serological tests can give a tentative diagnosis of NDV exposure in the birds.

(d) *Molecular biology*: Real-time reverse-transcriptase polymerase chain reaction (rRT-PCR) can be used for detection of fusion protein, matrix protein and RNA-dependent RNA polymerase enzyme either from the viral isolate or from the collected tissues and faeces of suspected birds. It can also confirm the virulence of the isolates. In conventional reverse-transcriptase polymerase chain reaction (RT-PCR), cloning and sequencing of PCR products will confirm the pathogenic potentiality of the isolate.

2.2.1.9 Zoonosis

Zoonotic transmission of NDV is possible which causes acute conjunctivitis, malaise, and sinusitis in susceptible human. The flu like symptoms resolves automatically within 1–3 weeks. Direct transmission of NDV into human from pet birds is still not documented possibly due to similarity of symptoms with common flu and its auto recovery. The pet birds as reservoir of NDV become a potential hazard for poultry population where the human may act as intermediate host. Human to human transmission of NDV is not documented.

2.2.1.10 Treatment and Control Strategy

No effective treatment for pet birds against NDV infection is documented. To control the infection in pet birds, exposure to live bird market, wild birds, and migratory birds should be restricted. Infected birds should be kept separately and general hygiene practices should be followed to avoid the contamination of feed and drinking water.

Vaccination is an effective measure to control NDV infection in commercial and domestic poultry. Vaccination in pet birds is not recommended because it cannot eliminate the carrier birds. Vaccination with modified live virus may produce the infection in pet birds. However, experimental use of inactivated NDV vaccine produced a humoral response in wild house sparrows with doses above 0.05 ml per bird. The humoral response was produced within 4–6 weeks after experimental inoculation of the vaccine.

2.2.2 Avian Influenza Infection

2.2.2.1 History

In 1878, avian influenza (earlier known as fowl plague) was described for the first time in Italy by Perroncito. Centanni and Savonuzzi (1901) first observed the role of filterable agent (virus) as etiology of avian influenza. The definitive etiological correlation with Influenzavirus A was established later (1955). Highly pathogenic avian influenza (H5 subtype) was first reported from chickens in Scotland (1959) and common terns (*Sterna hirundo*) in 1961 from South Africa. During 1972–80, Influenzavirus A was reported from exotic birds including budgerigars (*Melopsittacus undulatus*), migratory ducks, and pelagic seabird (shearwater).

2.2.2.2 Etiology

Avian Influenza virus (AIV) belongs to the family Orthomyxoviridae, genus *Influenzavirus A*. The virions are sensitive to heat (56 °C for 30 min exposure), acid (pH 3.0) and lipid solvents. So they are easily destroyed under common environmental conditions. The virions are enveloped and pleomorphic, spherical to filamentous in shape. The virion surface is covered with two types of glycoprotein projections, known as haemagglutinin (HA, rod shaped trimer protein) and neuraminidase (NA, mushroom shaped tetramer protein). Other constituent viral proteins are nucleoprotein (NP), matrix proteins (M1, M2), polymerase basic (PB1, PB2), polymerase acidic (PA), and non structural proteins (NS2). The viral genome is linear, negative sense single stranded RNA and it contains eight numbers of segments. During genetic reassortment, these segments are exchanged between two viral strains to generate a mutant one (antigenic shift). The point mutation in HA and NA gene causes antigenic drift which can generate a new viral strain.

On the basis of HA and NA gene sequences, *Influenzavirus A* has 16 HA (H1–16) and 9 NA (N1–9) subtypes. Genetic reassortment between the subtypes theoretically can produce 144 types of combinations. On the basis of pathogenicity, AIV can be further differentiated into two categories i.e. highly pathogenic avian influenza (HPAI, e.g. H5 and H7 subtypes) and low pathogenic avian influenza (LPAI, e.g. H9N2). HPAI causes 90–100% mortality in birds and LPAI infection is mostly restricted within the respiratory system. Further molecular variations of HA gene can differentiate HPAI (H5N1) into several clades (first order–fourth order clade).

2.2.2.3 Host Susceptibility

Other than commercial and domestic poultry, birds belong to the order Anseriformes (waterfowls) and Charadriiformes (shorebirds, gulls, auks, terns, waders) are considered as major reservoir of avian influenza virus (Table 2.6). H3 and H6 subtypes are common in waterfowls (Anseriformes), and H4, H9, H11, and H13 subtypes are widespread in Charadriiformes. Among the Anseriformes, mallard (*Anas platyrhynchos*) and northern pintail ducks (*Anas acuta*) in United States and bar headed geese (*Anser indicus*) in India are considered as major reservoirs.

Natural H5N1 outbreak among the migratory waterfowls such as bar-headed geese (*Anser indicus*), great cormorants (*Phalacrocorax carbo*), Pallas's gulls (*Larus ichthyaetus*), brown-headed gulls (*Larus brunnicephalus*), ruddy shelducks (*Tadorna ferruginea*) is also detected in China.

Role of Passeriformes (canary, finch, starling, sparrow) as AIV reservoir is uncertain. A study in France with large numbers of wild passerine birds failed to detect AIV. Although, wild passerine birds such as Eurasian tree sparrow (*Passer montanus*) in China and golden crowned kinglet (*Regulus satrapa*), fox Sparrow (*Passerella iliaca*), western tanager (*Piranga ludoviciana*), northern waterthrush (*Seiurus noveboracensis*), Cassin's finch (*Carpodacus cassinii*) in United States, and flycatchers (family Muscicapidae) in Central Africa is detected to possess AIV infection. Experimentally, society finches (*Lonchura striata domestica*), zebra finches (*Taeniopygia guttata*), house sparrows (*Passer domesticus*) and starlings are found susceptible to AIV (H5N1, H7N9, H7N7). The sialic acid receptors for both avian influenza (α 2, 3) and human influenza (α 2, 6) viruses are present in house sparrows (*Passer domesticus*) and starlings (*Sturnus vulgaris*). In contrast, Eurasian tree sparrows contain abundance of sialic acid receptors (α 2, 6) for AIV.

Detection of AIV is rare in psittacine birds and as such the psittacines are not considered as potential reservoir of AIV. Limited reports of viral isolation such as H9N2 subtype from Indian ring-necked parakeets and H5N2 subtype (Mexican lineage) from red-lored amazon parrot is available (Table 2.6). Experimentally, parakeets (*Melopsittacus undulates*) are found susceptible to human H7N9 isolate and after inoculation, development of clinical signs and shedding of the virus into water troughs is observed. On the other hand, psittacine isolate can also replicate in chicken, duck and turkeys and transmission into healthy cage mates is observed.

2.2.2.4 Transmission

Water mediated transmission of AIV is possible in waterfowls and shorebirds due to their exposure to the contaminated water. Lower temperature maintained in water bodies helps in survival of AIV for prolonged period. Possibility of AIV transmission in waterfowl is more during their assembly in water bodies associated with post-breeding and pre-migratory molt. Migratory shorebirds and other Charadriiformes birds mostly transmit the infection when they congregate to feed and roost at places *en route* of migration. Thus the migratory birds become a possible source of infection in pet birds living in open air aviaries.

Possibility of water mediated transmission is low in terrestrial and passerine birds. Detection of AIV in passerines is possible, when the virus is maintained in local poultry population or the passerines share a common habitat with infected waterfowls. This kind of AIV transmission dynamics was observed during H5N1outbreaks (2005–10) in wild birds in China, Russia and Mongolia. Consumption of infected bird carcass by raptors (bird of prey) is another possible way of transmission.

In psittacine birds, AIV infection is transmitted by direct contact with infected birds if kept together after capture or during quarantine. International trade of exotic birds can transmit the AIV infection from one continent to another.

Table 2.6 Avian Influenza virus (AIV) subtypes documented in waterfowls, passerines, and wild birds in different countries

Bird (family)	AIV subtype	Country
Pigeons (*Columba livia*), starlings (*Sturnus vulgars*)	H9N3	Iraq
Fringillidae, Parulidae, Turdidae, Tyrannidae	H1N1	Ontario, Quebec
Passerines (Fringillidae, Timaliidae)	H6N4, H8N3, H11N3, H11N8, H12N6, H2N3, H11N6, H12N2, H1N3, H9N3, H3N1, H3N3, H6N3, H7N3, H8N6, H10N3, H11N1, H12N3, H12N4, H13N1, H2N1, H14N3, H8N1, H1N1, H14N3, H2N8, H6N8, H12N1, H12N3, H13N2, H15N4, H5N5, H7N1, H14N4, H11N5, H8N7	Slovakia
Columbidae, Corvidae, Estrillidae, Laniidae, Muscicapidae, Passeridae, Strunidae, Timaliidae, Zosteropidae	H5N1	Hong Kong
Alaudidae, Hirundinidae, Locustellidae, Motacillidae, Passeridae, Ploceidae, Pycnonotidae, Sylviidae	H5	South Africa
House sparrow (*Passer domesticus*)	H9N2	Iran
Cuculidae, Emberizidae, Fringillidae, Hirundinidae, Motacillidae, Muscicapidae, Paridae, Passeridae, Remizidae, Sylviidae, Timaliidae, Turdidae	H10N2, H9N2, H7N5, H13N1, H2N5, H6N5, H12N2, H12N5, H13N1, H9N5, H11N3, H12N3, H10N2, H11N5, H11N2, H12N5, H13N3, H3N2, H12N1, H3N5, H9N5, H7N2, H9N2, H13N2, H10N3, H12N5, H6N5, H10N6, H7N6, H1N6	Slovakia
Anseriformes, Charadriiformes, Passeriformes, Falconiformes, Ciconiiformes, Columbiformes	H5N1	Europe
Magpies (*Pica pica sericea*)	H5N1	South Korea
Passeridae (tree sparrow)	H5N1	Indonesia
Columbidae, Dicruridae, Emberizidae, Strunidae	H5N1	Thailand
Dicruridae, Pycnonotidae, Timaliidae	H5, H6, H9	Vietnam
Strunidae	H1N1, H7N7	Israel
Strunidae	H7N7	Victoria (Australia)
Passeridae (tree sparrow)	H5N1	China
Columbidae	H5N1	Japan
Ostrich	H7N1	South Africa
Ostrich	H5N9	South Africa

(continued)

Table 2.6 (continued)

Bird (family)	AIV subtype	Country
Ostrich	H5N2	Zimbabwe
Emu, Casowaries	H5N9	The Netherlands
Rhea, Emu	H3N2, H4N2, H5N2, H7N1, H4N6, H5N9, H10N4, H7N3, H10N7	United States
Sparrow (*Arremonops* spp.)	H7	Lebanon
Mediterranean Gull (*Larus melanocephalus*)	H9N2	France
Psittacine [red-lored amazon parrot (*Amazona autumnalis autumnalis*)]	H5N2	United States
Indian ring-necked parakeets	H9N2	HongKong
Passerines (song birds)	H1N1, H4N6, H5N1, H5N2, H7N9, H9N2	Asia, Europe, North America
Parrot	H5N2	Guangdong (South China)
Bar-headed geese (*Anser indicus*), great cormorants (*Phalacrocorax carbo*), Pallas's gulls (*Larus ichthyaetus*), brown-headed gulls (*Larus brunnicephalus*), ruddy shelducks (*Tadorna ferruginea*)	H5N1	China
Domestic geese	H5N1	China
Sparrow	H7N7	Australia
Pigeon	H7N9	China
Ring-necked parakeet	H9N2	Japan

Further transmission of AIV from wild birds to local poultry population is possible. In HPAI infection, most of the wild birds die except in H5N1-HPAI (Guangdong lineage) infection. The virus of Guangdong lineage can persist in wild birds and is transmitted to the poultry. Sometimes, LPAI maintained in the wild birds is transmitted into poultry as observed during HPAI outbreaks in poultry in United States. The source of the virus was confirmed as LPAI from wild birds which undergo several mutations to generate HPAI strain.

2.2.2.5 Pathogenesis

Pathogenesis of AIV in psittacine and passerine birds is still unexplored. Experimental inoculation of HPAI (H5N1) in finches, sparrows and budgerigars indicated the existence of variations in pathogenesis of AIV from the gallinaceous poultry. Neurotropism of HPAI in nongallinaceous birds is identified as the major cause of mortality. Experiments with migratory passerines speculated that the virus enters the central nervous system via the cerebrospinal fluid.

Localization of virus is also detected in other tissues such as heart, pancreas, spleen, nasal epithelium, and reproductive organs of nongallinaceous birds. Muti-organ failure or dysfunction is also identified as additional factor for mortality of the affected birds.

2.2.2.6 Clinical Symptoms

Non-specific clinical symptoms such as neurological signs (head between legs), depression, ruffled feathers, and standing at the bottom of the cage are observed in pet birds with AIV infection (Fig. 2.13). In a natural outbreak of LPAI infection (H5N2) in a red-lored amazon parrot (*Amazona autumnalis autumnalis*), lethargy, diarrhoea and dehydration are noted. Sudden onset, depression, neurological symptoms are detected in finches and budgerigars experimentally inoculated with a chicken isolate of HPAI (H5N1). In wild migratory passerine birds (blackcap, red-billed quelea) experimentally inoculated with HPAI (H5N1), sudden death, ruffled feathers, lethargy, and neurological disorders (ataxia) are observed.

Fig. 2.13 Head between legs in a parrot (*Courtesy* Sanjoy Shit, Animal Resources Development Department, Government of West Bengal, India)

2.2.2.7 Lesion

In experimentally inoculated passerine and psittacine birds (zebra finches, house finches, budgerigars) with HPAI (H5N1), carcass dehydration, spleenomegaly with mottling of parenchyma, accumulation of watery faeces in cloaca is observed. In house finches and budgerigars, vents are pasted with faeces and bile tinged urates.

In wild migratory passerine birds (blackcap, red-billed quelea) experimentally inoculated with HPAI (H5N1), lung congestion, pancreatic necrosis with multiple, white foci on pancreas are major gross lesions.

2.2.2.8 Diagnosis

Clinical Specimens

Cloacal swabs from live birds and tissue specimens from heart, pancreas, spleen, and brain after post-mortem can be collected as clinical specimens. The samples should be transported in isotonic phosphate-buffered-saline (PBS, pH 7.0–7.4) with antibiotics such as penicillin (2000 units/ml), streptomycin (2 mg/ml), gentamycin (50 μg/ml), mycostatin (1000 units/ml) and protein (5% cattle serum, 0.5% bovine albumen). The specimens can be preserved at 4 °C for 4 days and at −80 °C for extended period.

Diagnostic Techniques

The laboratory should have at least biocontainment level 3 facilities and official clearance from the concerned authority to handle the AIV suspected samples.

(a) *Isolation of virus from clinical samples*: The clinical samples collected in isotonic phosphate buffered saline with antibiotics are centrifuged (1000 g) and the supernatant fluid is inoculated into 9–11 days old embryonated hen's eggs (specific pathogen free) by allantoic sac route. The inoculated eggs are incubated at 37 °C for 4–5 days. After incubation, eggs are kept at 4 °C. The embryo will die in positive samples and the allantoic fluids are collected to detect the haemagglutination activity of the viral isolate. Use of Madin-Darby canine kidney (MDCK) cells in place of eggs is an alternative approach. The viral isolate is confirmed by haemagglutination inhibition test (HI), neuraminidase inhibition test (NI), immunodiffusion test and antigen-capture enzyme-linked immunosorbent assay (ELISA). Differentiation of HPAI and LPAI can be performed by chicken inoculation test, intravenous pathogenicity index (HPAI has the index > 1.2), and amino acid sequencing of haemagglutinin protein.

(b) *Serological tests*: Hemagglutination-inhibition (HI), agar gel immunodiffusion (AGID), virus neutralization, enzyme-linked immunosorbent assay (ELISA) are employed for detection of AIV antibodies. The sera collected from the birds other than chicken sometimes agglutinate chicken red blood cells used in HI test (idiosyncrasy). To avoid idiosyncrasy, sera of the suspected birds should be adsorbed with chicken RBC before conducting the test. Immunodiffusion test (AGID) can detect the antibodies against nucleoprotein

(NP) and matrix proteins (M1, M2) of AIV which are antigenically similar in all isolates. AGID cannot differentiate between HPAI and LPAI isolates.

(c) *Antigen detection tests*: ELISA (antigen capture) based commercial kits are available to detect the nucleoprotein (NP) of AIV. The kits are mostly intended for use in poultry, yet not validated in other species of birds.

(d) *Molecular biology*: Real-time reverse-transcriptase polymerase chain reaction (rRT-PCR), light upon extension PCR (LUX-PCR) and conventional reverse-transcriptase-PCR can be used for detection of H5 or H7 subtype of AIV. Although false negative results are obtained in cloacal swabs and faecal samples due to presence of PCR inhibitors. To avoid the high cost and expertise needed in PCR based techniques, isothermal techniques such as rapid isothermal nucleic acid detection assay (RIDA), loop-mediated isothermal amplification (LAMP) and nucleic acid sequence-based amplification (NASBA) are used for detection of AIV. Further development in the diagnostic approach used proximity ligation assay (PLA) for detection of AIV in chicken, and DNA-microarray based technique for characterization of AIV in wild and domestic birds. In future, rapid isothermal nucleic acid detection assay-lateral flow (RIDA-LF) and immunoassay-based biosensors will be a better choice due to less dependence on instrumentation and rapid process of a good numbers of samples in less time, respectively.

2.2.2.9 Zoonosis

World Health Organization (WHO) has reported more than 600 human infections with HPAI (H5N1) since 2003, of which 60% infected people died. Transmission of H5N1 infection in human occurred during close contact with birds or contaminated environments. Keeping pet birds in household also increased the seroconversion of the owners during a H7N7 outbreak. Religious ceremonies, such as 'merit release' among Buddhists, in which a passerine bird is purchased, kissed and released, may increase the transmission possibility of AIV among the human. No report of human to human transmission of AIV is reported so far.

2.2.2.10 Treatment and Control Strategy

No effective treatment for pet birds against AIV infection is documented. To control the infection in pet birds, exposure to live bird market, wild birds, and migratory birds should be restricted. Infected birds should be kept separately and general hygiene practices should be followed to avoid the contamination of feed and drinking water. The cages or aviaries should be cleaned with formaldehyde, gluteraldehyde, beta-propiolactone, binary ethylenimine, quaternary ammonium disinfectants, sodium hypochlorite, dilute acids, and hydroxylamine after an AIV outbreak.

Vaccination against AIV in birds is a controversial issue. Due to high mutating capability of the virus, instead of control, vaccination with live virus may cause more damage to the birds. Experimentally, inactivated recombinant H5N3 vaccine was used in several types of zoo birds (Anseriformes, Charadriiformes,

Ciconiiformes, Columbiformes, Coraciiformes, Falconiformes, Galliformes, Gruiformes, Passeriformes, Psittaciformes) which produced strong antibody titer against H5 subtype in all the birds except in Psittaciformes. Prime-boost strategy of vaccination (priming with H5N9 and booster with H5N3) produced strong antibody titer in Psittaciformes.

2.2.3 West Nile Virus Infection

2.2.3.1 History

West Nile virus (WNV) was first isolated from a woman suffering with fever and other complications in Uganda (1937). In Africa, Middle East and European countries, WNV was mostly known to cause sub-clinical and self-limiting infections in horses and human during 1960s. Later in 1990s, higher frequency of WNV infection was noticed among human, farm animals, pet animals and birds of prey. In 1999, a fatal outbreak of WNV was detected among birds, horses and human in New York, USA. During 2008–10, WNV was isolated from a clinically infected sun conure (*Aratinga solstitialis*) and green-winged macaw (*Ara chloropterus*) in United States.

2.2.3.2 Etiology

West Nile virus is an arthropod-borne (arbovirus), enveloped virus which belongs to family *Flaviviridae* and genus *Flavivirus*. The virion is 50 nm in diameter with icosahedral symmetry. No spikes or peplomers are present on the virion surface. Genome of the virus is positive sense single-stranded RNA. The genome (11 kbp) encodes envelope protein (E), membrane precursor protein (prM), capsid protein (C) and non-structural proteins (NS1, NS2a, NS2b, NS3, NS4a, NS4b, NS5). The structural proteins help in formation of virion and the non-structural proteins help in viral replication and evasion of host immune response.

Seven lineages of WNV have identified and lineages 1 (clade Ia and Ib) and 2 are considered as major lineages. Lineage 1is mostly distributed in Europe (clade Ia), America (clade Ia), the Middle East (clade Ia), India, Africa (clade Ia) and Australia (clade Ib, Kunjin virus). Lineage 2 is widespread in South Africa, Madagascar and Europe. Both the lineages of WNV have neurotropism property, although, viruses belong to lineage 1 (clade Ia) are more virulent than the clade Ib and lineage 2 viruses.

2.2.3.3 Host Susceptibility

WNV is identified in 326 species of birds with or without clinical symptoms. The most susceptible birds to WNV infection are crows (*Corvus* spp.), ravens (*Corvus corax*), jays (*Garrulus* spp.), magpies (*Pica* spp.), owls (*Strigiformes* spp.), and some raptors (Spanish imperial eagle, goshawk, golden eagle, sparrow hawk, gyrfalcon). The passerine birds and the mosquitoes (*Culex* spp.) are considered as major host and vector of WNV, respectively. Crows are more exposed to WNV infection due to their communal roosting (perching) behaviour. After sunset, during the communal roosting, the mosquitoes (*Culex* spp.) mostly feed on the birds.

Migratory passerines (American Robins) can transmit the infection in distant places, whereas, resident passerines (house sparrows), crows [American crows (*Corvus brachyrhynchis*)] and other birds act as local amplifying host of WNV. WNV is detected from resident birds such as Columbiformes (*Columbina talpacoti*), Coraciiformes (*Melanerpes aurifrons*), Piciformes (*Cardinalis cardinalis*, *Molothrus aeneus*), and Passeriformes (*Myiozetetes similis*, *Sporophila torqueola*, *Thamnophilus doliatus*, *Tiaris olivaceus*, *Tyrannus melancholicus*). Seroprevalence of WNV is observed among passerines such as Northern Cardinals (*Cardinalis cardinalis*) and Carolina Wren (*Thryothorus ludovicianus*). Studies indicated that presence of non-passerine birds (enormous numbers) in a population can reduce the virus amplification and human transmission risk due to diversity of hosts (dilution effect).

Among the psittacine birds, seroprevalence of WNV is detected in budgerigars (*Melopsittacus undulatus*), cockatiels (*Nymphicus hollandicus*), cockatoos (*Cacatua* spp.), macaws (*Ara* spp.), parrots (*Amazona*, *Rhynchopsitta*, *Poicephalus*, *Psittacus* spp.), pacific parrotlets (*Forpus coelestis*), canary-winged parakeet (*Brotogeris versicolurus*), rosellas (*Platycercus* spp.), lories and lorikeets (*Eos*, *Lorius*, *Pseudeos*, *Trichoglossus* spp.) and blue-crowned conure (*Thectocercus acuticaudata*). Red-legged partridges (*Alectoris rufa*) are resistant to natural WNV infection, although they can be experimentally infected with the virus.

Other than birds, human, horses, sheep, alpacas, dogs, cats, white-tailed deer, reindeer, squirrels, chipmunks, bats, and alligators are susceptible to WNV. Human and horses are dead-end-hosts of the virus as sufficient amount of virions are not maintained in the blood to infect the mosquitoes feeding on the hosts.

2.2.3.4 Transmission

Maintenance of WNV throughout the world takes place by an enzootic cycle ('rural cycle'). The susceptible birds and mosquitoes (*Culex* spp.) are two major components of the cycle. The birds maintain the virus by acting as reservoir and the mosquitoes act as vector and spread the virus into new hosts. Microfilarial infection of mosquitoes can hasten viral replication and rapid transmission of the virus (microfilarial enhancement of arboviral transmission). When the mosquitoes introduce the virus into the human habitats, the 'urban cycle' begins. In endemic zones, urban cycle begins with mortality of wild birds (summer to autumn) and the cycle ends with human and horse infection (dead-end-hosts).

In winter months, when the adult mosquitoes are mostly inactive, vertical transmission of the virus takes place to sustain in the vector population ('overwintering strategy'). Sometimes, WNV is re-introduced into the vector population through migratory birds and rarely by human transport (mosquitoes on aeroplanes).

Other than mosquito bites, WNV is rarely transmitted by oral route (ingestion of infected prey, drinking water) and direct contact in birds, cats and other vertebrates. In psittacine birds, feathers are identified as an important source of WNV. Association of testes in psittacine birds suggests the possibility of sexual transmission.

2.2.3.5 Pathogenesis

Following the WNV transmission through the mosquitoes in human and rodents, the virus primarily enters dendritic cells (langerhans cells) via receptor mediated endocytosis. The cell surface proteins (DC-SIGN, integrin) act as WNV receptors. The dendritic cells carry the virus into draining lymph nodes where viral multiplication occurs. Following genomic replication and translation, the progeny virions are matured through ER-golgi secreation pathway and are released by exocytosis into the blood circulation. Transient viraemia develops and different vital organs such as liver, spleen, kidney are infected. Neuroinvasion can take place through direct infection with or without breakdown of blood-brain barrier or virus transport along peripheral neurons. Certain host proteins such as Drak2 (death-associated protein-kinaserelated 2), ICAM-1 (intercellular adhesion molecule), MIP (macrophage migration inhibitory factor) and MMP-9 (matrix metaloproeinase 9) help in altering blood-brain barrier permeability. Sometimes, host innate immune response (TLR3) mediated up regulation of tumor necrosis factor alpha (TNFα) causes capillary leakage and increased permeability of blood-brain barrier.

In contrast, in birds, viraemia develops within 30–45 min of mosquito bite without any local virus multiplication in the lymph nodes. Positive correlation of peak viraemia and bird mortality is observed. In birds, WNV prefers to replicate primarily in spleen and mononuclear phagocytic cells and are disseminated into vital organs (liver, kidney, heart). Different lineages (1 and 2) of WNV have different tissue tropism in avian hosts. Lineage 1 virions prefer to infect liver and myocardium, whereas, lineage 2 of WNV mostly infect spleen, kidney and liver in goshawks. Depending upon the viral load in blood WNV may infect central nervous system in birds. Exact mechanism of neuroinvasion in birds is unexplored. Role of endothelial cells and immune cells in neuroinvasion is predicted. Death of the birds occurs due to WNV associated lesions and secondary infections with bacteria, fungi and parasites. Pathogenicity of WNV infections in birds is influenced by route of viral transmission, host defense, age and species of birds.

The virus can persist in different organs of birds such as spleen, kidney, eye, brain and skin. Detection of WNV in a hawk (birds of prey) during winter months, when mostly the mosquitoes are inactive and unable to transmit the infection, revealed the possibility of persistent viral infection. Experimentally, persistent WNV infection is produced in ducks, pigeons and immunocompromised mice. Effect of persistent WNV infection in health status of the birds is indistinct. Low level of WNV infection is detected in carcasses of rock pigeons (*Columba livia*) and mourning doves (*Zenaida macroura*), although, WNV is not confirmed as a cause of death. Whereas, in a kea (*Nestor notabilis* Gould), a large mountain parrot, natural persistence of WNV in central nervous system for more than 6 years is detected. This persistence is associated with death of the bird after prolonged incubation period.

2.2.3.6 Clinical Symptoms

Non-specific clinical signs such as depression, anorexia, dehydration and ruffled feathers are observed in birds. In complicated cases, neurological signs, for instance, convulsions, ataxia, abnormal head postures and movements, tremors, paresis, and uncoordinated flight are detected. The neurological signs do not always correlate with the lesions in the brain (neuronal necrosis). Partial or complete blindness develops in raptors and owls. Sequelaes of viral neuroinvasion are observed in long-lived birds (e.g. raptors). In raptors, feather pulp abnormalities and abnormal molt can persist up to 4 years as sequelae of WNV infection.

In naturally infected psittacine birds (rosellas, conures, lorikeets, cockatoos, caiques, parakeets), sudden death without any symptoms or non-specific signs like loss of weight, anorexia, lethargy, depression, and weakness are noted. Specific neurological signs consisted of rolling over, legs stretched backward, stumbling and disorientation.

2.2.3.7 Lesion

No pathognomonic lesion of WNV infection is detected in birds. In highly susceptible birds (crows) sudden death without any gross lesion is observed. In passerines, necrosis and mild inflammation in the heart, spleen, liver, kidney, mild encephalitic lesions and absence of neuronal necrosis is detected. In naturally infected psittacines (rosellas, conures, lorikeets, cockatoos, caiques, parakeets), splenomegaly, hepatomegaly, mottled pale liver with multifocal petechiae, diffuse pallor in kidneys, myocardial pallor, petechiae on the gizzard serosa are observed. In long-lived birds (raptors) due to chronic WNV infection, hemorrhages, petechiae and congestion in vital organs, splenomegaly, hepatomegaly, myocardial pallor, pale mottling in the liver, spleen, kidney, cerebral atrophy and malacia are detected. In central nervous system, gliosis, perivascular cuffing and glial nodules are major microscopic findings.

2.2.3.8 Diagnosis

Clinical Specimens

From the dead birds after post mortem, brain, heart, liver and kidney can be collected as clinical specimens for laboratory confirmation. The laboratories should have containment level 3 to handle the samples suspected for WNV infection

Diagnostic Techniques

(a) *Isolation of virus from clinical samples*: WNV can be isolated in rabbit kidney (RK-13) and Vero cells or in embryonated chicken eggs. Several passages in cell line are required to observe the cytopathic effects. The virus isolates are confirmed by indirect FAT or PCR.
(b) *Immunological tests*: The tissues collected from the suspected birds and fixed with formalin can be stained by immunohistochemical (IHC) staining for identification of WNV antigen.
(c) *Serological tests*: Hemagglutination inhibition (HI), plaque reduction neutralization (PRN), IgM capture ELISA can be used for detection of antibodies against WNV in avian serum.
(d) *Molecular biology*: Reverse transcriptase-nested PCR (E-protein as target) and real-time RT-PCR can be used for detection of WNV from avian tissues.

2.2.3.9 Zoonosis

Sporadic outbreaks of WNV occurred in human in the mediterranean region, Africa and Europe before 1994. Severe WNV outbreaks with neuroinvasion took place in human throughout the world after 1994. Global warming associated increased temperature and prolonged rainfall helps in breeding of mosquitoes and spreading of arboviral diseases such as WNV. In human, WNV transmission can occur through biting of infected mosquitoes, blood transfusion, organ transplantation, breast milk, and intrauterine route. The clinical presentation ranges from asymptomatic (80% of infections) to encephalitis/paralysis and death (1% of infections). Sometimes, flu like symptoms such as fever, headache, malaise, myalgia, fatigue, skin rash, lymphadenopathy, vomiting, and diarrhea are observed.

2.2.3.10 Treatment and Control Strategy

No effective treatment for pet birds against WNV infection is documented. To control the infection in pet birds, exposure to mosquitoes should be restricted.

Maternal antibodies can protect the house sparrow chicks up to 3 days post hatch. No specific vaccine against WNV is available to use in birds. In some countries, equine vaccine is used, although not licensed. Successful use of recombinant subunit WNV vaccine is reported from experimental geese. Experimental vaccination in thick-billed parrots (*Rhynchopsitta pachyrhyncha*) produced detectable antibody titer against WNV.

2.2.4 Usutu Virus Infection

2.2.4.1 History

Usutu virus (USUV) was first detected in a mosquito (*Culex neavei*) in 1959 in South Africa. The isolated virus is currently considered as a reference strain of USUV (SouthAfrica-1959). Subsequently, the virus was detected in different bird and mosquito species in Africa. In recent decade, USUV was identified in passerine birds in Austria (2001), in different birds and mosquitoes in Hungary (2005), Spain (2006), Switzerland (2006) and Italy (2009).

2.2.4.2 Etiology

Usutu virus belongs to the genus *Flavivirus* (Japanese encephalitis serocomplex) under the family *Flaviviridae*. Different species of *Flavivirus* such as USUV, West Nile virus (WNV), Japanese encephalitis virus (JEV), Murray Valley encephalitis virus (MVEV) and Saint Louis encephalitis virus are originated from same ancestral virus. Nucleotide and amino acid sequencing revealed that MVEV, among Japanese encephalitis serocomplex, are the closest relative of USUV.

USUV is a small (40–60 nm), spherical, enveloped virus with positive sense single stranded RNA genome (11 Kbp). The genome has a cap at 5′ end but no poly-A-tail at 3′ end. The genome can encode three structural proteins (core protein, pre-membrane and envelope protein) and eight non-structural (NS1, NS2A, NS2B, NS3, NS4A, 2K, NS4B, NS5) proteins.

2.2.4.3 Host Susceptibility

Eurasian blackbirds (*Turdus merula*) mostly suffer from USUV infection. Several other families of birds (Accipitriformes, Anseriformes, Caprimulgiformes, Charadriiformes, Ciconiiformes, Columbiformes, Coraciiformes, Galliformes, Passeriformes, Piciformes, Strigiformes) of different European countries are also susceptible to USUV infection (Table 2.7). Migratory birds such as whitethroat (*Sylvia communis*), lesser whitethroat (*Sylvia curruca*), garden warbler (*Sylvia*

Table 2.7 Distribution and susceptible hosts of Usutu virus infection

Birds	Country	Year
Eurasian blackbird (*Turdus merula*)	Italy, Germany, Hungary, Austria	2001–11
Great grey owl (*Strix nebulosa*)	Germany, Austria	2001–11
Chicken (*Gallus gallus domesticus*)	UK, Italy, Switzerland	2006–09
Humboldt penguin (*Spheniscus humboldti*)	Switzerland	2006–07
Greater flamingo (*Phoenicopterus ruber*)	Switzerland	2006–07
Laughing kookaburra (*Dacelo novaeguineae*)	Switzerland	2006–07
White stork (*Ciconia ciconia*)	Austria	2006–07
Marabou stork (*Leptoptilos crumeriiferus*)	Austria	2006–07
Egyptian vulture (*Neophron percnopterus*)	Austria	2006–07
Eurasian eagle owl (*Bubo bubo*)	Austria	2006–07
Ural owl (*Strix uralensis*)	Austria	2006–07
Snowy owl (*Bubo scandiacus*)	Austria	2006–07
boreal owls (*Aegolius funerius*)	Italy	2007
Eurasian pygmy owl (*Glaucidium passerinum*)	Italy	2007
Northern hawk-owl (*Surnia ulula*)	Italy	2007
Grey heron (*Ardea cinerea*)	Germany, Italy	2010–11
Eurasian bee-eater (*Merops apiaster*)	Germany, Italy	2010–11
House sparrow (*Passer domesticus*)	Germany, Italy	2010–11
Eurasian green woodpecker (*Picus viridis*)	Germany, Italy	2010–11
Common starling (*Sturnus vulgaris*)	Germany, Italy	2010–11
Partridge (*Alectoris rufa*)	Italy	2010–11
Long-eared owl (*Asio otus*)	Italy	2010–11
Nightjar (*Caprimulgus europaeus*)	Italy	2010–11
Eurasian jay (*Garrulus glandarius*)	Italy	2010–11
Yellow-legged gull (*Larus michahellis*)	Italy	2010–11
Eurasian magpie (*Pica pica*) Eurasian collared dove (*Streptopelica decaocto*)	Italy	2010–11
Common kingfisher (*Alcedo atthis*)	Germany	2011
Canary (*Serinus canaria domestica*)	Germany	2011
Song thrushes (*Turdus philomelos*)	Spain	2012
Domestic pigeon (*Columba livia domestica*)	Greece	2014
Great spotted woodpecker (*Dendrocopos major*)	Belgium	2014
Bullfinch (*Pyrrhula pyrrhula*)	Belgium	2014

borin), kestrel (*Falco tinnunculus*), marsh harrier (*Circus aeruginosus*), house martin (*Delichon urbica*), reed warbler (*Acrocephalus scirpaceus*), pied flycatcher (*Ficedula hypoleuca*), barn-swallow (*Hirundo rustica*) are detected to carry antibodies against USUV.

Most of these bird families are susceptible to both USUV and West Nile virus (WNV) infection. Evidence of both viral infection was detected in Eurasian blackbirds, Eurasian blackcaps, European robins, magpies etc.

2.2.4.4 Transmission

USUV is maintained in nature by a mosquito-bird cycle in which mosquitoes act as vector and the birds act as amplifying host. The mosquitoes occasionally spread the virus into other hosts (incidental) such as human, horses and rodents. USUV is detected in several mosquitoes such as *Culex pipiens*, *Culex neavei*, *Culex perexiguus*, *Aedes albopictus*, *Aedes caspius*, *Anopheles maculipennis*, *Culex perfuscus*, *Coquillettidia aurites*, *Mansonia Africana*. Among them, *C. pipiens* and *C. neavei* are considered as competent vectors for USUV. Although the migratory birds are sometimes infected with USUV the evidence of their role in transmission of infection is still missing.

Trematode infestation in blackbirds is detected as a predisposing factor for USUV infection.

2.2.4.5 Pathogenesis

After introduction of USUV into the body of the host, viraemia lasts for a short period (2 days). The tissue tropism of the virus is almost similar to WNV infection. The virus is detected in brain, heart, liver, kidney, lungs, and intestinal tissues of laboratory mice and naturally infected birds. Demyelination of neurons and formation of autophagosome are unique features of USUV infection. The process of autophagy helps in incorporation of host cellular components in viral replication.

2.2.4.6 Clinical Symptoms

USUV infection in birds produces non specific clinical symptoms such as apathy, depression, anorexia, dehydration, ruffled feathers and moulting. In complicated cases, neurological signs, for instance, convulsions, ataxia, abnormal head postures and movements, tremors, paresis, torticollis and nystagmus are detected. Neurological symptoms are often followed by death.

2.2.4.7 Lesion

The disease in birds is characterized by encephalitis, myocardial degeneration, and necrosis in liver and spleen. Degeneration of Purkinje cells, accumulation of glial nodules surrounding the degenerated Purkinje cells ('glial shrubbery'), perivascular cuffing are characteristic findings in brain of affected birds. Hepatomegaly, enlargement and discoloration of the kidneys, necrosis on the sheathed arteries of spleen are detected in great grey owls and boreal owls. In blackbirds, affected liver and spleen contains myriads of small (up to 1 mm) yellowish foci. Enlarged gallbladder and intestine, hyperaemic meninges and brain are also found in blackbirds.

2.2.4.8 Diagnosis

Clinical Specimens

Blood or serum (paired sera collected in two weeks interval) from live bird and liver, spleen, lung, kidney, gizzard, and intestines are collected and fixed in 10% buffered formalin as post mortem specimens.

Diagnostic Techniques

(a) *Isolation of virus from clinical samples*: USUV can be isolated in Vero, PK-15 and goose embryo fibroblast cells. The laboratories should have containment level 3 to handle the samples suspected for USUV infection.
(b) *Immunological tests*: The tissues collected from the suspected birds and fixed with formalin can be stained by immunohistochemical (IHC) staining for identification of USUV antigen.
(e) *Serological tests*: Detection of four fold rise in antibody titer and seroconversion from IgM to IgG in paired sera sample collected in two weeks interval indicates USUV infection in birds. USUV-IgM appears 5 days after onset of clinical symptoms. USUV antibody titer is detected by haemagglutination inhibition (HI) and plaque reduction neutralization test (PRNT). USUV-specific IgG-capture ELISA is developed for human use. However, the serological tests often produce cross reactivity with other *Flavivirus* infections such as WNV. PRNT is more specific than other serological tests but requires specialized laboratory which can handle the virus. Detection of acute infection is not possible by serological tests as the birds die before development of antibody titer.
(f) *Molecular biology*: Reverse transcriptase-PCR can specifically detect USUV in tissues of suspected birds. Recently real-time PCR is also developed for detection of USUV in human blood and cerebrospinal fluid samples which can be adapted in avian diagnostics.

2.2.4.9 Zoonosis

In 1981, in Central African Republic, a man with fever and rashes was diagnosed as a first human patient of USUV infection (CAR-1981). In Italy (2009), USUV infection was detected in two different patients of meningoencephalitis and orthotropic liver transplantation. Both the patients were immunosuppressed and received blood transfusion before the infection. The common clinical symptoms were persistent fever, headache and neurological disorders. In 2012–13, a sero-surveillance program in Germany and Croatia detected low prevalence of USUV antibodies among the human population.

2.2.4.10 Treatment and Control Strategy

No effective treatment for pet birds against USUV infection is documented. To control the infection in pet birds, exposure to mosquitoes should be restricted. Water should not be stagnant in the vicinity of the aviaries or bird owner's houses.

Culex pipiens, the potent vector of USUV do not prefer to fly a long distance and lack of breeding site will significantly reduce their numbers. Use of mosquito net (window and door) and repellants help to reduce the mosquito population. N, N-diethyl-meta-toluamide (DDET) is most effective repellant against *Culex pipiens*.

No specific vaccine against USUV is available to use in birds. Vaccines against *Flavivirus* (Japanese encephalitis, yellow fever) are available for human use but no cross protection against USUV infection is reported.

2.2.5 Avian Borna Virus Infection

2.2.5.1 History

Borna disease was first detected among animals (horse, sheep) in Southeast Germany during 19th century. It was named after the German district of Borna around the town of Borna in Saxony where the infection remained endemic for prolonged period. The etiological correlation of Borna disease with a virus was established in 1920.

Proventricular dilatation disease (PDD) was first reported from macaws and conures in USA during 1977. At that time, PDD was known as 'macaw wasting or fading syndrome' and 'gastric distension of macaws' as mostly macaws were associated with the syndrome. Actual etiology of PDD remained uncertain for a considerable period. Two independent research groups from Israel and USA (Honkavuori et al. 2008; Kistler et al. 2008) identified a novel genus of the family *Bornaviridae*, provisionally named as avian bornavirus (ABV) as etiological agent of PDD. The virus was identified by molecular techniques such as panviral DNA microarray and high throughput sequencing.

2.2.5.2 Etiology

Avian bornavirus (ABV) belongs to the family *Bornaviridae* and order *Mononegavirales*. The virion is enveloped, spherical, and 80–100 nm in diameter. The virions replicate in the nucleus of the host cells and use the host cellular splicing machineries for generation of mRNAs. The genome is non-segmented, single stranded RNA which encodes six major viral proteins such as nucleoprotein (N), regulatory protein (X), phosphoprotein (P), matrix protein (M), membrane-bound glycoprotein (G), and RNA-dependent RNA polymerase (L).

After the discovery of Avian Bornavirus in psittacine birds, several types of Bornavirus were detected in both psittacine and non-psittacine birds. Recently it is proposed that the genus should include five species such as *Mammalian 1 bornavirus*, *Psittaciform 1 bornavirus* (avian/psittacine bornaviruses 1, 2, 3, 4, 7), *Passeriform 1 bornavirus* (canary bornaviruses C1, C2, C3, LS), *Passeriform 2 bornavirus* (estrildid finch bornavirus EF) and *Waterbird 1 bornavirus* (avian bornavirus 062CG).

Till date, 14 ABV genotypes have been detected in psittacine (ABV-1, 2, 3, 4, 5, 6, 7) and non psittacine birds (ABV-C1, ABV-C2, ABV-C3 in canaries; ABV-CG in Canada geese; ABV-EF in estrildid finch; ABV-BF in Bengalese finch).

2.2.5.3 Host Susceptibility

Avian borna viruses are detected in over 80 species of birds of which more than 70 species belongs to psittaciformes. The members of psittaciformes commonly infected with ABV include *Cacatuidae* (cockatoos, cockatiels) and *Psittacidae* (lovebirds, macaws, parakeets, parrots, Amazon parrots, conures) (Table 2.8). Among non-psittacine birds, canary, long-wattled umbrella bird, weaver finch, red-tailed hawk, falcon, Canada geese, swan, duck, bald eagle are found naturally infected with ABV (Table 2.8). PDD is reported from United States, Australia, Middle East, South America, South Africa and Japan.

2.2.5.4 Transmission

The studies suggest about faecal-oral or faecal-oronasal transmission of ABV between the captive birds. In a bird with PDD, ABV infected cells are detected in the intestinal villi from where they are excreated through the faeces. Sometimes ABV infected birds remain healthy and act as source of infection for other birds kept in the same aviary. It is observed that birds with high serum antibody titer against ABV or viral RNA load are prone to become clinically infected with PDD. Overcrowding in aviaries, hand-feeding of parrot chicks are detected as predisposing factors for PDD. No gender based predisposition of PDD is observed. Transmission route of ABV in wild birds is still unexplored.

2.2.5.5 Pathogenesis

Irrespective of transmission route, classical borna disease virus enters central nervous system. In experimentally infected rats, centrifugal spread of virus into peripheral nerves and autonomic nerve fibers and ganglia are detected. In avian borna virus infection in birds, autonomous nervous system of the upper and middle digestive tract, including the esophagus, crop, proventriculus, ventriculus, and duodenum is chiefly infected. Further spread of ABV in extraneural tissues such as smooth or heart muscle fibers, liver, kidney, spleen, pancreas, lung, gonads, thyroid, and skin is observed. In mammalian borna virus infection, extraneural spread in hepatocytes, kidney epithelial cells, and myocytes of the intestine and heart is associated with immunosuppression of the host.

2.2.5.6 Clinical Symptoms

The incubation period of ABV infection in birds is highly variable (10 days–years). Clinically the birds show gastrointestinal dysfunction or neurological signs or both. The symptoms of gastrointestinal dysfunction are impaction of proventriculus, dysphagia, polyuria, regurgitation, diarrhoea, presence of undigested food (seeds) in faeces, and crop stasis which leads to starvation and death. Death due to circulatory collapse or food aspiration is also found. Neurological signs are ataxia, seizure, blindness, tremor, abnormal gait, reduced proprioceptive skills, motor deficit and peripheral neuritis in sciatic, brachial and vagal nerves.

2.2.5.7 Lesion

No gross lesion is observed in sudden death of birds due to PDD. In majority of the birds suffering with PDD (70%), proventriculus is thin walled and distended with seeds. Rupture of proventriculus wall releases the food particles and causes

Table 2.8 Distribution and susceptible hosts of Avian Bornavirus

Psittacine birds	Continent/Country	Non-psittacine birds	Continent/Country
Cockatiel (*Nymphicus hollandicus*)	Asia	Canary (*Serinus canaria*)	Spain, Germany
White cockatoo (*Cacatua alba*), Ducorps's Cockatoo (*Cacatua* ducorpsii), sulphur-crested cockatoo (*Cacatua galerita*), red-vented cockatoo (*Cacatua haematuropygia*), Salmon-crested cockatoo (*Cacatua moluccensis*), Little corella (*Cacatua sanguinea*)	Asia, Europe	Long-wattled umbrella bird (*Cephalopterus penduliger*)	Spain
Galah (*Eolophus roseicapillus*)	Asia	Bearded barbet (*Lybius dubius*)	Spain
Red tailed black Cockatoo (*Calyptorhynchus magnificus*)	Asia	Honey creeper (*Chlorophanes spiza*)	Spain
Palm Cockatoo (*Probosciger atterimus*)	Asia	Weaver finch	Spain
Red-breasted parakeet (*Psittacula alexandri*)	Asia	Red-tailed hawk (*Buteo jamancensis*)	–
Eclectus parrot (*Eclectus roratus*)	Asia, Europe	Falcon (*Falco peregrinus*)	–
Rainbow lorikeet (*Trichoglossus moluccanus*)	Asia	Canada geese (*Branta canadensis*)	USA
Blue-and-yellow macaw (*Ara ararauna*), Golden collared macaw (*Ara auricollis*), green-winged macaw (*Ara chloropterus*), Blue-throated macaws (*Ara glaucogularis*), scarlet macaw (*Ara macao*), military macaw (*Ara militaris*), red shouldered macaw (*Ara nobilis*), red-fronted macaw (*Ara rubrogenys*), chestnut-fronted macaw (*Ara severus*)	USA	Roseate spoonbills (*Ajaja ajaja*)	–
Hyacinth macaw (*Anodorhynchus hyacinthinus*)	USA, Europe	Toucans (*Ramphastos sp.*)	–
Spix's macaw (*Cyanopsitta spixii*)	USA	Bald eagle (*Haliaeetus leucocephalus*)	–

(continued)

Table 2.8 (continued)

Psittacine birds	Continent/Country	Non-psittacine birds	Continent/Country
Blue-crowned Conure (*Aratinga acuticaudata*), Peach-fronted Parakeet (*Aratinga aurea*), golden-capped parakeet (*Aratinga auricapillus*), red-masked parakeet (*Psittacara erythrogenys*), Finsch's Conure (*Aratinga finschi*), Golden parakeet (*Aratinga guarouba*), jenday conure (*Aratinga jandaya*), sun conure (*Aratinga solstitialis*), dusky-headed parakeet (*Aratinga weddellii*)	USA	American geese, swans, ducks	–
Nanday Conure (*Nandayus nenday*)	USA, Europe	Estrildid finches (*Estrildidae*)	Germany
Burrowing parrot (*Cyanoliseus patagonus*)	USA, Europe	Bengalese finch (*Lonchura striata*)	–
Grey cheeked parakeet (*Brotogeris pyrrhoptera*)	USA, Europe	Trumpeter swans (*Cygnus buccinator*), feral mute swans (*Cygnus olor*)	USA
Green-cheeked conure (*Pyrrhura molinae*), black-capped parakeet (*Pyrrhura rupicola*)	USA, Europe		
Thick-billed parrot (*Rhynchopsitta pachyrhyncha*)	USA		
Blue-fronted amazon (*Amazona aestiva*), white-fronted amazon (*Amazona albifrons*), orange-winged amazon (*Amazona amazonica*), yellow-naped parrot (*Amazona auropalliata*), red-lored parrot (*Amazona autumnalis*), Cuban amazon (*Amazona leucocephala*), ellow-crowned parrot (*Amazona ochrocephala*), Yucatan amazon (Amazona xantholora)	USA		
Pileated parrot (*Pionopsitta pileata*)	USA		
Bronze-winged parrot (*Pionus chalcopterus*)	USA, Europe		

(continued)

Table 2.8 (continued)

Psittacine birds	Continent/Country	Non-psittacine birds	Continent/Country
Green-thighed parrot (*Pionites leucogaster*)	USA		
Red-fan parrot (*Deroptyus accipitrinus*)	USA		
Pacific parrotlet (*Forpus coelestis*)	USA		
Congo African grey parrot (*Psittacus erithacus*)	Africa, Europe		
Red-fronted parrot (*Poicephalus gulielmi*)	Africa, Europe		
Greater vasa parrot (*Coracopsis vasa*)	Africa, Europe		
Yellow-collared Lovebird (*Agapornis Personata*)	Africa, Europe		
Canindae macaw (*Ara glaucogularis*), Vinaceous Amazon (*Amazona vinacea*)	USA		
Umbrella cockatoo (*Cacatua alba*), Solomons cockatoo (*Cacatua ducorpsii*), Sulfur-crested cockatoo (*Cacatua galerita*)	Japan		
Barred parakeet (*Bolborhynchus* spp.), Golden parakeet (*Guaruba* spp.), superb parrot (*Polytelis* spp.)	Europe		

peritonitis. Sometimes, enlargement of duodenum and adrenal glands, pale area on the epicardium is observed. Microscopic lesions consist of lymphocytic infiltration along with plasma cells in the ganglia and nerve plexus (specially myenteric plexus supplying the digestive tract) of proventriculus, intestine, crop, esophagus, adrenal gland, conduction fibers of heart, central nervous system and spinal cord. Perivascular cuffing by lymphocytes are detected in cerebral cortex, cerebellum, spinal cord and in peripheral nerves such as sciatic, brachial and vagus nerves.

2.2.5.8 Diagnosis

Clinical Specimens

For *intra vitam* (ante-mortem) diagnosis of ABV infection, faeces, blood, swabs of crop and cloaca, tissue biopsies from crops can be collected. Left lateral sac of the crop (cranial portion) is preferred site for biopsy collection. The biopsy should be elliptical and it should contain a blood vessel so that the nerve sections can be visualized. After post-mortem, brain, crop, intestine and adrenal glands are collected.

Diagnostic Techniques

(a) *Clinical signs, haematology*: Clinical signs are mostly associated with gastrointestinal upset and/or neurological signs. Non-regenerative anemia, leukocytosis, heterophilia, decreased total protein and albumin, increased level of muscle enzymes such as lactate dehydrogenase, creatine kinase, aspartate amino-transferase are detected in PDD infected birds.

(b) *Radiography*: Distended proventriculus, ventriculus, crop and small intestine with ingesta and gas and prolonged gastrointestinal transit time are observed in infected birds by contrast radiography, contrast fluoroscopy and ultrasonography. Spontaneous ruptures of the dilated proventriculus are rarely observed. Although these findings are not specific for PDD. In healthy neonatal birds, distension of proventriculus and crop is also found. Contrast radiography is performed in birds by introducing barium sulfate or iodine-based contrast media (@ 10–15 ml/kg) into the crop by gavage. Barium sulfate produces better contrast but causes airway irritation. In psittacine birds, normal gastrointestinal transit time is 90 min–3 h.

(c) *Isolation of virus from clinical samples*: Isolation of ABV can provide confirmatory diagnosis for PDD. It can be done in quail cell lines (CEC-32, QM7) and other avian cell lines. Incorporation of mammalian cell line for virus isolation is not always successful.

(d) *Histopathology*: Histopathological investigation is another way of confirmatory diagnosis. Lymphocytic infiltration in the ganglia and nerve plexus of proventriculus, ventriculus, intestine and crop is considered as diagnostic for PDD.

(e) *Serology*: ELISA, indirect immunofluorescence assay and Western blot are developed for detection of anti-ABV antibodies. The serological tests cannot differentiate between PDD infected birds from asymptomatic carrier of ABV.

(f) *Molecular biology*: Reverse transcriptase-PCR (RT-PCR) for detection of L, M and N genes of ABV is developed. Quantitative real time-PCR for detection of P gene is recently developed. Brain, crop, intestine and adrenal glands collected after post-mortem and crop tissue, blood, cloacal swabs, and faeces can be used for ABV-RNA extraction. However, both false-positive (from asymptomatic bird) and false-negative results can be obtained by RT-PCR.

2.2.5.9 Zoonosis

Zoonotic potentiality of ABV is not established.

2.2.5.10 Treatment and Control Strategy

PDD is a highly contagious infection and it spreads rapidly from one bird to another within a flock. Decision to offer long term treatment or euthanasia of the affected bird is crucial. Euthanasia is the best policy for management of PDD, although not preferred by most of the owners. Management of inflammation, indigestion and secondary bacterial infections are currently considered as line of treatment for PDD. Use of nonsteroidal anti-inflammatory drugs (NSAIDs, e.g. celicoxib, 20 mg/kg body weight, orally) along with antivirals (amantadine hydrochloride, 10 mg/kg po or 20 mg/kg with food) is recommended to treat PDD in birds. Use of surfactants (for reduction of gas production), metoclopramide (0.5 mg/kg body weight, intramuscularly) and B complex vitamins are suitable supportive therapy.

Diet of the PDD infected birds should be easily digestible (preferably formulated diets), and in liquid or pelleted forms because the proventriculus and ventriculus function is adversely affected in PDD. Addition of vegetables in the diet will increase intestinal motility. Toys and cage accessories should be provided to the birds to avoid ingestion of foreign bodies.

For prevention of PDD, new birds should be quarantined and checked for PDD before introduction into aviaries. Maintenance of strict biosecurity and hygienic measures should be followed. Overcrowding should be avoided in the aviaries.

2.2.6 Beak and Feather Disease

2.2.6.1 History

Fist description of Beak and feather disease (BFD) was observed in 1907 in an Australian journal ('The Emu') and the author described about wild red-rumped parrots (*Psephotus haematonotus*) in the Adelaide hills being unable to fly due to loss of feathers (Ashby 1907). In 1916, death of a captive sulphur-crested cockatoo (popular by its name 'Cocky Bennett') at the age of 120 years in Sydney was published in local news paper. The bird was suspected for BFD due to loss of feather and presence of elongated beak. Psittacine beak and feather disease (PBFD) was first scientifically documented in 1975 in sulfur-crested cockatoos, lovebirds, budgerigars and galahs in Australia (Pass and Perry 1984).

2.2.6.2 Etiology

Beak and feather disease is caused by beak and feather disease virus (BFDV) which belonged to the genus *Circovirus* and family *Circoviridae*. Circoviruses are icosahedral, non-enveloped and the smallest known autonomously replicating animal virus, measuring 15–26 nm in diameter. The viruses have an ambisense, circular, single-stranded DNA genome (2000 nt) which can encode a replicase enzyme and capsid protein. The virus possesses highest mutation rate and genetic diversity although it is antigenically conserved. No serotype variation of *Circovirus* is detected. The virus is considered as a model to study host parasite interaction due to its simple genome structure. It is also the representative of ancient viral form and *Circovirus* sequences are detected in fossils of vertebrates, invertebrates, protozoa, plant, fungi, algae and bacteria.

2.2.6.3 Host Susceptibility

BFD virus mostly infects psittacine birds (more than 60 species) besides other bird families such as Passeriformes, Columbiformes and Anseriformes (Table 2.9). A few susceptible bird species are enlisted as endangered or threatened by International Union for Conservation of Nature (http://www.iucnredlist.org). BFDV infection is considered as a significant conservation threat. The infection is more fatal in young birds (0–3 years) due to poor development of immune system and the viral load is more prevalent in parental bird species than the hybrids. Cockatoos mostly show chronic viral infection with excreation of virus through the faeces and dystrophic feathers. Sometimes, cockatoos do not show any visible symptoms. Occasionally, birds other than the common susceptible species are also infected with BFDV ('host switch over'). Recently, BFDV infection is detected in rainbow bee-eaters (*Merops ornatus*), a species of Coraciiformes, unrelated to psittacine birds.

Mostly the BFD infection is detected in Australia, New Zealand, Europe (Poland, UK, Denmark, Portugal, Germany, Italy), United States, Africa (Zambia, Zimbabwe) and Asia (Japan, China, Taiwan, Thailand). Circumstantial evidence indicated that the infection was originated in Australia. The virus was disseminated from Australia in early 1970s to European countries with the imported parrots. From Europe, the virus is further distributed to Africa, New Zealand, Japan and United States during unregulated parrot trafficking.

2.2.6.4 Transmission

Transmission of BFDV can take place by both horizontal and vertical means. Horizontal route via direct contact with infected birds is the major mode of transmission in both wild and captive birds. The virus is excreated in high titers in the environment through feather dust, crop secreations, and faeces of infected birds. In aviaries, the virus once spread is difficult to control due to its high infectious and persistence nature. In the fomites, BFDV can persist for several years and after a long period, the fomites may act as a source of ancestral viral genotype in the host population.

In forests, 'host switch over' is facilitated by horizontal transmission. The switch over mostly takes place in the unoccupied nests in the trees where competition exists between Psittaciformes and other birds for reproductive opportunities.

2.2.6.5 Pathogenesis

BFDV depends on host cell machineries for their replication and prefers the actively dividing cells such as basal follicular epithelium, lymphoid tissues, and intestinal epithelium. The virus causes necrosis of basal epithelium in the birds. The necrosis is found to be responsible for feather dystrophy and beak and claw deformities. BFDV also causes lymphoid depletion and associated immunosuppression. Secondary infection with bacteria, fungi, parasite takes place due to immunosuppression and it is responsible for 70% bird mortality as observed in ducks, pigeons, geese, black-backed gull and other avian species. More experiments are needed to explore the pathogenesis of BFDV in pet birds.

Table 2.9 Common susceptible hosts of BFDV

Bird order	Bird species	Common name
Psittaciformes	*Amazona aestiva*	Blue-fronted Amazon
	Amazona albifrons	White-fronted Amazon
	Amazona amazonica	Orange-winged Amazon
	Amazona auropalliata	Yellow naped Amazon
	Amazona autumnalis	Red lored Amazon
	Amazona vinacea	Vinaceous-breasted Amazon
	Amazona aestiva	Turquoise-fronted Amazon
	Poicephalus robustus	South African Cape Parrot
	Alisterus scapularis	Australian king parrot
	Aprosmictus erythropterus	Red-winged parrot
	Poicephalus senegalus	Senegal Parrot
	Pionites leucogaster	Green-thighed parrot
	Eclectus roratus	Eclectus Parrot
	Poicephalus rufiventris	African red-bellied parrot
	Poicephalus gulielmi massaicus	Jardine parrot
	Poicephalus rueppellii	Ruppell's parrot
	Pionites melanocephalus	Black headed parrot
	Pionus chalcopterus	Bronze winged parrot
	Guarouba guarouba	Golden parakeet
	Psittacula echo	Echo Parakeet
	Psittacula krameri	Ring-necked parakeet
	Psittacula eupatria	Alexandrine parakeet
	Cyanoramphus novaezelandiae	Red-crowned parakeet
	Psittacula krameri	Rose-ringed parakeet
	Cyanoramphus unicolor	Antipodes parakeet
	Coracopsis vasa	Vasa parrot
	Psittacara finschi	Crimson-fronted parakeet
	Cacatua alba	White cockatoo
	Cacatua leabeateri	Major Mitchell's cockatoo
	Cacatua galerita	Sulphur-crested cockatoo
	Calyptorhynchus banksii	Red-tailed black cockatoo
	Calyptorhynchus lathami	Glossy black cockatoo
	Cacatua ducorpsii	Solomon's corella
	Cacatua galerita triton	Triton cockatoo
	Cacatua goffiniana	Tanimbar corella
	Cacatua haematuropygia	Philippine cockatoo
	Cacatua moluccensis	Moluccan cockatoo
	Cacatua ophthalmica	Blue-eyed cockatoo
	Cacatua sulphurea	Yellow-crested cockatoo
	Cacatua tenuirostris	Eastern long-billed corella
	Nymphicus hollandicus	Cockatiel

(continued)

Table 2.9 (continued)

Bird order	Bird species	Common name
	Melopsittacus undulatus	Budgerigar
	Psephotus haematogaster	Blue Bonnet
	Trichogolossus haematodus	Rainbow lorikeet
	Lorius chlorocercus	Yellow bib lorikeet
	Psitteuteles goldiei	Goldie's lorikeet
	Eos reticulate	Blue-streak lorikeet
	Agapornis roseicollis	Lovebird
	Agapornis lilianae	Nyasa lovebird
	Agapornis nigrigenis	Black-cheeked lovebird
	Agapornis roseicollis	Peach-faced lovebird
	Eolophus roseicappillus	Galah
	Cacatua tenuirostris	Eastern long billed corella
	Platycercus elegans	Crimson rosella
	Platycercus eximius	Eastern rosella
	Cacatua alba	White cockatoo
	Forpus coelestis	Pacific parrotlet
	Ara macao	Scarlet Macaw
	Ara ararauna	Blue-and-yellow macaw
	Ara severa	Chestnut-fronted macaw
	Ara nobilis	Red-shouldered macaw
	Ara chloropterus	Red-and-green macaw
	Ara militaris	Military macaw
	Aratinga solstitialis	Conure
Passeriformes	*Serinus canaria*	Canary
Columbiformes	*Columbia livia*	Pigeon
Anseriformes	*Ariser sp.*	Goose
	Hybrid of Pekin (*Anas platyrhynchos domestica*) and Muscovy duck (*Cairina moschata*)	Mulard duck

2.2.6.6 Clinical Symptoms

BFDV infection has three clinical forms in birds such as per acute, acute and chronic. Sudden death without any symptom or mild symptom (feather dystrophy) occurs in per acute and acute forms, respectively. These forms are common in juvenile birds. In chronic form of the infection, weight loss, lethargy, anaemia, diarrhoea, shedding of developing feathers, abnormal development of new feathers is observed (Figs. 2.14, 2.15 and 2.16). Mostly, contour, tail and down feathers are lost symmetrically and they are replaced with dystrophic feathers that fail to grow. Deformities of beak and claws are not a constant feature. It depends on species of the bird and other predisposing factors. Cockatoos are more susceptible to beak and claw deformities than other psittacine birds. Beak elongation, transverse or longitudinal fractures, palatine necrosis are possible beak deformities observed (Fig. 2.17). Chronic infection is not always fatal, the birds may survive for several years. Sometimes death occurs due to secondary infection.

Fig. 2.14 Young Budgerigar affected with Beak and feather disease (*Courtesy* M. Scott Echols, Medical Center for Birds, California)

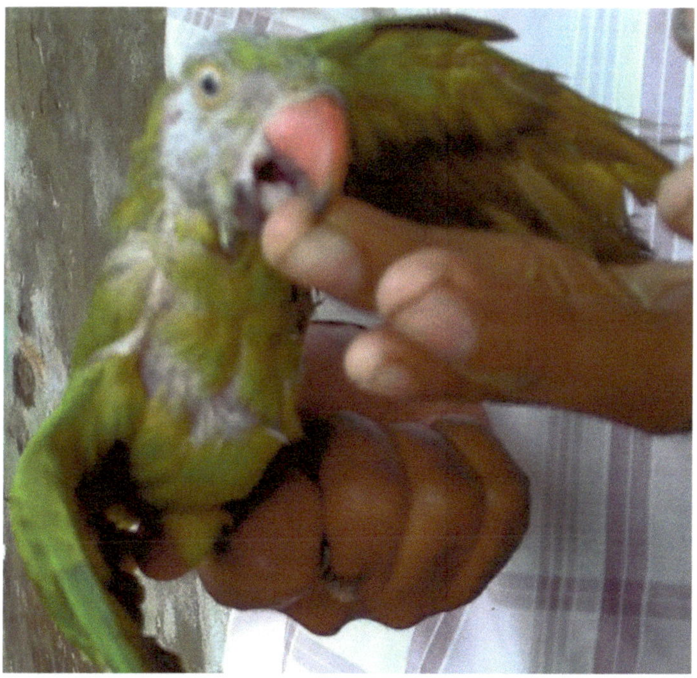

Fig. 2.15 Parrot affected with Beak and feather disease (*Courtesy* Barun Dev Das, Animal Resources Development Department, Government of West Bengal, India)

Fig. 2.16 Gouldian finches affected with Beak and feather disease (*Courtesy* Prof. Elena Circella, Avian Diseases Unit, Department of Veterinary Medicine, University of Bari, Italy)

2.2.6.7 Lesion

Gross lesions in feathers of BFDV infected birds are retention of sheaths, fracture of the proximal rachis, haemorrhage in pulp cavity, short clubbed feathers, curled feathers and circumferential constrictions. In naturally infected cockatoos, vane of feather is ragged with multiple fractures. Hooklets, barbules and barbs are poorly developed and fractured (Fig. 2.18). Hyperkeratotic sheaths are found in affected feathers which results terminal clubbing and mid-shaft constriction. In beaks of infected birds, abnormal elongation, palantine necrosis, transverse to longitudinal fractures is detected.

Histopathological examinations revealed basophilic intranuclear and/or intracytoplasmic inclusion bodies in feather epithelial cells, follicular epidermal cells and macrophages.

2.2.6.8 Diagnosis

Clinical Specimens

Feathers (newly grown quill portion is the best specimen), blood or serum, cloacal swabs, pharyngeal swabs can be collected as clinical specimens from the suspected birds. Although, feathers produce low amount of viral DNA because most of the fully grown feathers are separated from the blood supply. The blood should be collected from vein (not toenail) to avoid environmental contamination of BFDV.

Fig. 2.17 Fracture at tip of the beak (*Courtesy* LafeberVet)

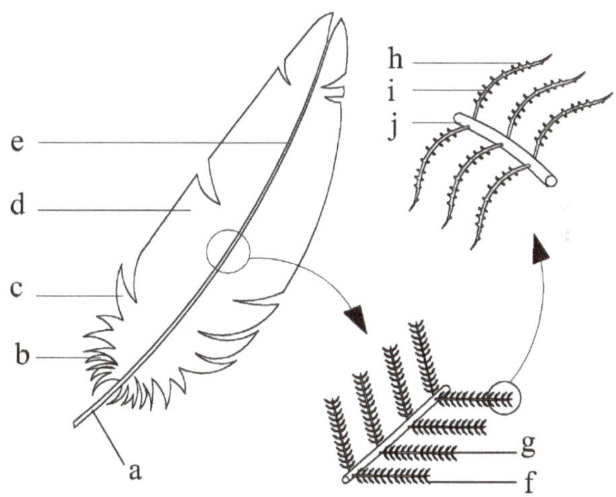

Fig. 2.18 Structure of a bird feather (schematic). *a* Shaft *b* barbs *c* afterfeather *d* vane *e* rachis *f* barb *g* barbules *h* hooklets *i* barbules *j* barb

Diagnostic Techniques

(a) *Clinical signs*: Progressive feather loss preliminarily suggests BFDV infection. Feather loss is also associated with Polyomavirus infection, trauma, bacterial folliculitis, malnutrition, endocrine abnormalities, and adverse drug reactions to penicillins and cephalosporins. Specific laboratory tests should be performed to confirm the BFDV infection.

(b) *Histopathology*: Histopathology with light or electron microscopy is a reliable and frequently used technique for confirmation of BFD. Presence of intranuclear or intracytoplasmic or both type of inclusion body acts as primary indicator of BFDV infection. The virus is confirmed by electron microscopy or immunohistochemistry.

(c) *Serology*: Haemagglutination inhibition (HI) and blocking ELISA are developed for detection of Avian Circovirus antibodies. The serological tests cannot confirm the active BFDV infection as the antibodies might be present due to latent infection. Agglutination of red blood cells by BFDV varies with the source. Red blood cells collected from South American birds (Amazon, Macaw) are generally not agglutinated with BFDV.

(d) *Molecular biology*: PCR is one of the sensitive tools for rapid detection of BFDV infection in pet birds which should be coupled with clinical signs, lesion and histopathology. Conventional PCR targeting ORF C1 or ORF V1 section of the viral genome, nested-PCR, duplex shuttle PCR, and real-time PCR are developed for detection of BFDV. Whole genome sequencing is a recent progress for diagnosis of BFDV which is a cost effective, rapid and sensitive technique due to small genome size of the virus.

2.2.6.9 Zoonosis

Zoonotic potentiality of BFDV is not established.

2.2.6.10 Treatment and Control Strategy

Currently there is no known treatment for BFDV infection. Secondary infection should be diagnosed and treated properly. Avian gamma interferon injection (intramuscular) along with quaternary ammonium compound (as nebulizer) has shown success in treatment of BFD.

No vaccine is commercially available to control BFDV infection. Studies revealed that maternal antibodies against BFDV can protect the young birds. Surviving birds sometimes develop long lasting immunity. The only way to control the disease is through maintenance of hygiene, strict isolation or culling of infected birds.

2.2.7 Other Viral Infection

2.2.7.1 Psittacid Herpesvirus Infection (Pacheco's Disease)

Pacheco and Bier (1930), a veterinarian from Brazil first described an outbreak of acute, fatal hepatitis in psittacine birds. This syndrome became known as 'Pacheco's disease'. Later in 1975, psittacid herpesvirus type 1 (PsHV-1) was confirmed as etiological agent. Psittacid herpesvirus type 1 (PsHV-1) is closely related with Gallid herpesvirus-1 (infectious laryngotracheitis of chicken). PsHV-1 has been classified into 4 genotypes (1–4) on the basis of variations in UL16 gene sequence. The genotypes of the virus have preference for different hosts (Table 2.10). Amazon parrots and conures (Patagonian conures) most commonly suffer with

Table 2.10 Susceptible hosts of Psittacid herpesvirus-1 genotypes

Virus	Genotype	Susceptible hosts
Psittacid herpesvirus-1	1	Amazon parrot, cockatiels, cockatoos
	2	Amazon parrot, cockatiels, cockatoos, African grey parrots
	3	Amazon parrot, cockatiels, cockatoos, African grey parrots, macaws
	4	Amazon parrot, cockatiels, cockatoos, African grey parrots, macaws

Pacheco's disease. The infection is common in United States, United Kingdom, Spain, South Africa, Kenya and Japan.

PsHV-1 is transmitted by direct contact with the infected birds. Persistently infected birds can shed the virus through faeces and pharyngeal secreation. In most of the cases, the infected birds die suddenly without showing any syndrome. If the birds survive, non-specific clinical signs such as depression, anorexia, diarrhoea, tremor and instability are observed. Neurological disorder is followed by death. Gross lesions are not distinct. Intranuclear inclusion bodies (Cowdry type A) are observed in liver, kidney, spleen, pancreas and small intestine.

Cloacal swabs, pharyngeal swabs, faeces, newly emerged feathers (blood/pin feather) can be collected from live bird as clinical specimens. After post-mortem, liver, spleen, kidney, lung, cerebellum can be used for detection of the virus. Laboratory confirmation of PsHV-1 infection depends on isolation of virus in cell lines, demonstration of virus in clinical samples by electron microscopy, detection of viral DNA by PCR or real-time PCR. Presence of intranuclear inclusion body in tissues is inconclusive because many other viruses (*Avian Polyoma*, *Psittacine Adenovirus*) also produce the same.

Use of antiviral (Acyclovir, oral or intramuscular, intravenous injection) in early stage of infection can prevent the outbreak. An inactivated virus vaccine adjuvinated with oil is available for selected psittacine birds against Pacheco's disease. The vaccine is recommended to use subcutaneously in smaller psittacines and subcutaneously or intramuscularly in larger psittacines (more than 100 g body weight). Maintenance of strict hygiene, quarantine (30 days) of newly procured birds, and regular use of disinfectants in cages can prevent the PsHV-1 infection.

2.2.7.2 *Psittacine adenovirus* Infection (PsAdV)

Psittacine adenovirus (PsAdV) was first reported from Senegal parrots (*Poicephalus senegalus*) with acute infection. The virus was confirmed by amplification of hexon gene (L1 variable loop) by PCR (Raue et al. 2005). *Psittacine adenovirus* belongs to the family *Adenoviridae* and genus *Aviadenovirus*. *Adenovirus* is non-enveloped and has an icosahedral capsid with a diameter of 70 nm. The hexon protein is the major capsid protein and it has conserved pedestal regions (P1, P2) and the variable loops (L1–L4). *Adenovirus* infection is reported from pet birds such as budgerigars, macaws, Amazon parrots and cockatoos.

Aviadenoviruses are present in faeces, urine, tracheal and nasal secretions of infected birds. The virus is readily transmitted by horizontal mode. Direct faecal contact and aerial spread are major ways of horizontal transmission. Fomites, personnels and transport also contribute in horizontal transmission of *Aviadenovirus*. In most of the cases, the infected birds die suddenly without showing any syndrome. If the birds survive, non-specific clinical signs such as depression, anorexia, diarrhoea, ruffled feathers are observed. Gross lesions include hepatomegaly, splenomegaly, nephromegaly, dilatation of duodenum and , and congestion of lungs. The livers become enlarged, friable, haemorrhagic, pale or mottled. Basophilic intranuclear inclusion bodies are observed.

Faeces from live birds and organs (kidney, liver, intestine) after post mortem are collected from the suspected birds for confirmatory diagnosis. PsAdV can be isolated in primary chicken embryo kidney (CEK) or chicken embryo liver (CEL) cell lines. Embryonated chicken eggs can be inoculated by yolk sac, chorioallantoic membrane and allantoic cavity route. The eggs should be free of pathogen (SPF) and antibodies against Aviadenovirus. Replication of the virus can be confirmed by death of embryos and gross microscopic lesions observed in hepatocytes. Psittacine embryonated eggs are very expensive and are not readily available for diagnostic purpose. Other diagnostic methods for detection of PsAdV in clinical samples are electron microscopy due to typical morphology of the virus and demonstration of basophilic or eosinophilic, intranuclear inclusion bodies in liver and intestinal epithelium by haematoxylin and eosin staining. PsAdV in clinical samples can be confirmed by PCR.

No specific treatment for PsAdV infection is documented. Secondary bacterial infection can be treated by broad spectrum antibiotics. Aviadenoviruses are extremely resistant to the environment (heat, pH 3–9) and common disinfectants (ether, chloroform, trypsin, 50% alcohol) and they can persist for prolonged period in the cages or aviaries. Treatment with formalin, aldehydes and iodophors for more than 1 h can inactivate the virus.

2.2.7.3 Psittacine Poxvirus Infection

Poxvirus infection was reported for first time from lovebirds (*Agapornis personata*, *A. roseicollis*) in Germany (Kraft and Teufel 1971). Psittacine poxvirus belongs to the genus *Avipoxvirus*, family *Poxviridae* and subfamily *Chordopoxvirinae*. The viral genome consists of double stranded DNA (230–300 kbp). Several psittacine and passerine birds, specially Amazon and pionus parrots, lovebirds and canaries are susceptible to Psittacine Poxvirus infection. Transmission of Poxvirus from infected owners or caretakers is possible into the birds having skin injuries. Mechanical transmission is also possible by mosquitoes and mites. In a contaminated cage or aviary, infected aerosols generated from dried scabs and feathers may act as source of infection.

The infection in psittacine birds is characterized by ocular discharge, rhinitis, conjunctivitis and ulcerations on the eyelid. Typical crusty lesions develop on the eyelid margins, lateral and medial canthi of the eyes and occasionally on face and feet (Figs. 2.19 and 2.20). Persistent cutaneous lesion for a period of 13 months is

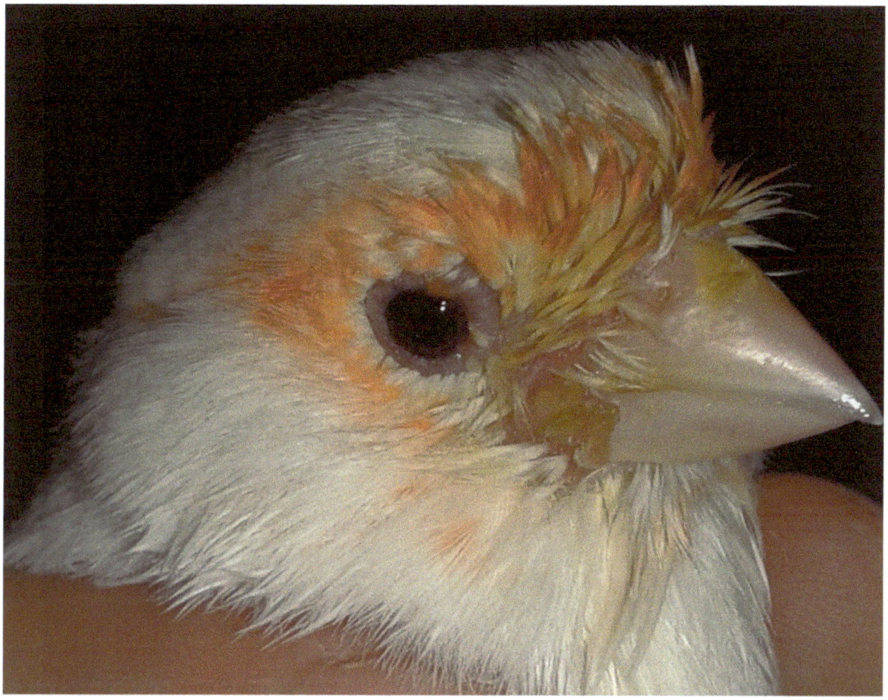

Fig. 2.19 Conjunctivitis and scaby lesion in a canary infected with Poxvirus (*Courtesy* Prof. Elena Circella, Avian Diseases Unit, Department of Veterinary Medicine, University of Bari, Italy)

Fig. 2.20 Poxvirus lesions in a pigeon (*Courtesy* Prof. Elena Circella, Avian Diseases Unit, Department of Veterinary Medicine, University of Bari, Italy)

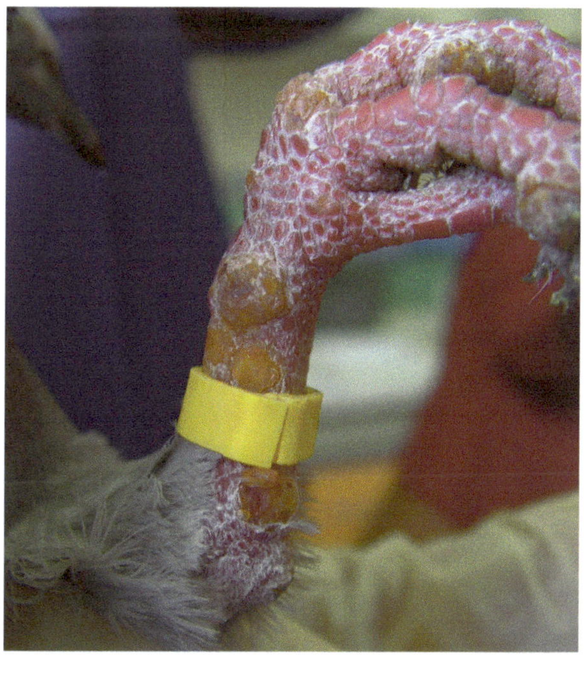

detected in yellow-shafted flicker. Diptheretic lesions develop in some birds which is associated with dyspnea and higher mortality rate. Diphtheritic form produces necrotic lesions in the trachea, larynx and oral cavity. Gross lesions include necrosis of heart, liver, air sac, lungs, peritoneum, and accumulation of necrotic debris on the surface of the alimentary tract. Intracytoplasmic inclusion bodies (Bollinger bodies) are observed in the mucosa of sinus, trachea, crop, esophagus and throat.

Presence of typical cutaneous lesions primarily suggests about Poxvirus infection. Vesicular fluid from cutaneous lesions, faeces and pharyngeal swabs can be collected as clinical specimens. The infection is confirmed by electron microscopy, detection of Bollinger body in tissue samples and Psittacine Poxvirus specific-PCR.

No specific treatment and vaccine is available to control Psittacine Poxvirus infection. Infected birds should be separated from the healthy group. Cages, fomites and utensils should be properly disinfected because the virus persists for prolonged period in the dried scabs and the aerosols generated from the infected scabs.

2.2.7.4 Avian Polyoma Virus Infection

Among the pet birds, *Polyomavirus* was detected for first time in young budgerigars (*Melopsittacus undulatus*) in 1980. The virus was named as budgerigar fledgling disease polyomavirus which was renamed later as Avian Polyomavirus. The virus belongs to the family *Papovaviridae*. The virion is icosahedral, non-enveloped with a diameter of 45–50 nm. The viral genome is a circular double-stranded DNA and

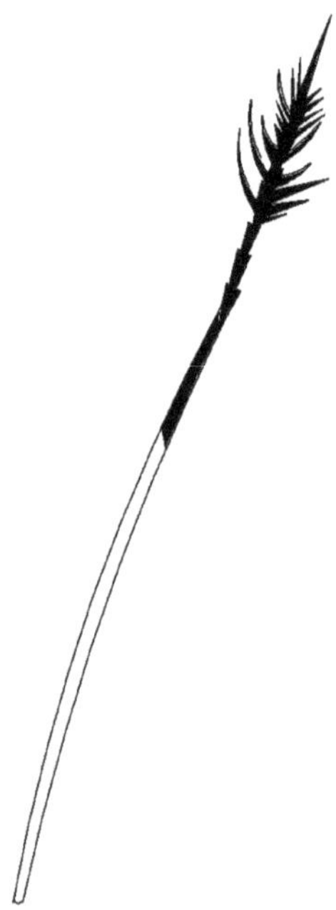

Fig. 2.21 Filoplume feather (schematic)

Fig. 2.22 Feather disorders in society finches due to Avian Polyomavirus infection (*Courtesy* Prof. Elena Circella, Avian Diseases Unit, Department of Veterinary Medicine, University of Bari, Italy)

Table 2.11 Age related syndrome of Avian Polyoma virus infection

Age of the bird	Clinical symptom
3–6 weeks	Death without symptom
5–16 weeks	Depression, anorexia, crop stasis, regurgitation and ecchymosis in subcutaneous tissues
16 and 21 weeks	Feather deformities followed by death
More than 24 weeks	Viraemia without clinical symptoms

has two regions-early and late. The early region encodes tumour protein and the late region encodes four structural proteins (VP1, VP2, VP3, VP4).

Other than budgerigars, lovebirds, canaries, finches, macaws, eclectus parrots, conures, cockatoos and Indian ringneck parakeets are also susceptible to Avian Polyoma virus infection. The infection is common in Canada, China, Australia, Germany, Slovakia and Italy. Direct contact with infected birds is the major way of transmission. Contaminated cages, fomites, utensils, nestboxes, egg incubator may also act as source of infection as the virus is highly stable in the environment.

The infection is more fatal in young birds of less than 16 weeks age. In fledgling and young budgerigars, death without symptom, or brief illness showing feather dystrophy, loss of down feathers, presence of 'filoplumes' (feather like projection with a thin rachis and few barbs, Figs. 2.21 and 2.22) on head and neck, abdominal distension followed by death, are observed. It is known as 'French molt' or 'Budgerigar and fledgling disease'. In cockatoos, two clinical forms i.e. acute death with haemorrhages in the feather shaft and pneumonia with gasping (in young

Fig. 2.23 Subcutaneous haemorrhages in African grey parrot due to Avian Polyomavirus infection (*Courtesy* Prof. Elena Circella, Avian Diseases Unit, Department of Veterinary Medicine, University of Bari, Italy)

birds) are detected. In other susceptible birds age related syndrome is prevalent (Table 2.11). Gross lesions include distension of heart with hydropericardium, swollen liver, congested kidneys, and hemorrhage in the body cavity (Fig. 2.23). Intranuclear and basophilic inclusion bodies are detected in spleen, liver and kidney.

Blood, cloacal swabs from live animals and tissues from spleen, liver and kidney can be collected as clinical specimens. The presence of Avian Polyomavirus in the clinical samples is confirmed by electron microscopy, virus neutralization test, immunofluorescent antibody staining, in situ hybridization, PCR and real-time PCR.

No specific treatment for Avian Polyomavirus is documented for pet birds. An inactivated and oil adjuvinated vaccine is commercially available. A dose of 0.25 ml is recommended for birds below 200 g of body weight and larger dose (0.5 ml) is given to the birds with more than 200 g of body weight. Primary vaccination is done at 5 weeks of age which should be followed by a booster after 2–3 weeks. Annual vaccination is recommended. Sometimes thickened skin at the vaccination site is observed as adverse reaction.

2.2.7.5 Psittacine Papillomavirus Infection

Papovavirus infection was detected in budgerigars (*Melopsittacus undulatus*) and splendid parakeets (*Neophema splendida*) which died suddenly without distinct clinical symptoms (Graham and Calnek 1987; Pass et al. 1987). Psittacine Papillomavirus belongs to *Papovaviridae* family. Pet birds such as Amazon parrots, African grey parrots, macaws, finches, budgerigars, canaries and parakeets are most susceptible to Psittacine Papillomavirus infection. Direct contact with infected bird is the major way of transmission. Introduction of infected bird into aviary or cage rapidly transmit Papillomavirus infection into other birds.

The infection is characterized by reddish cauliflower like growth (papilloma) in oral (larynx, crop and upper gastrointestinal tract) and cloacal mucosa. Presence of a large mass in larynx causes wheezing and change of voice. Papillomas in oral mucosa hinder swallowing and digestion which causes anorexia, chronic weight loss and vomition. In clocal papillomatosis, raised, coalescing mass appears at the cloaca (Fig. 2.24). Presence of fresh blood is noted in the droppings. Sometimes cloacal prolapse is detected. Gross lesion is characterized by proliferation of epithelial cells on thin fibrovascular stalks. Neoplastic growth of bile duct, pancreas and liver is also observed specially in Amazon parrots and macaws.

Presence of typical cauliflower like growth in cloaca or oral mucosa primarily suggests about Psittacine Papillomavirus infection. Tissue biopsy samples from

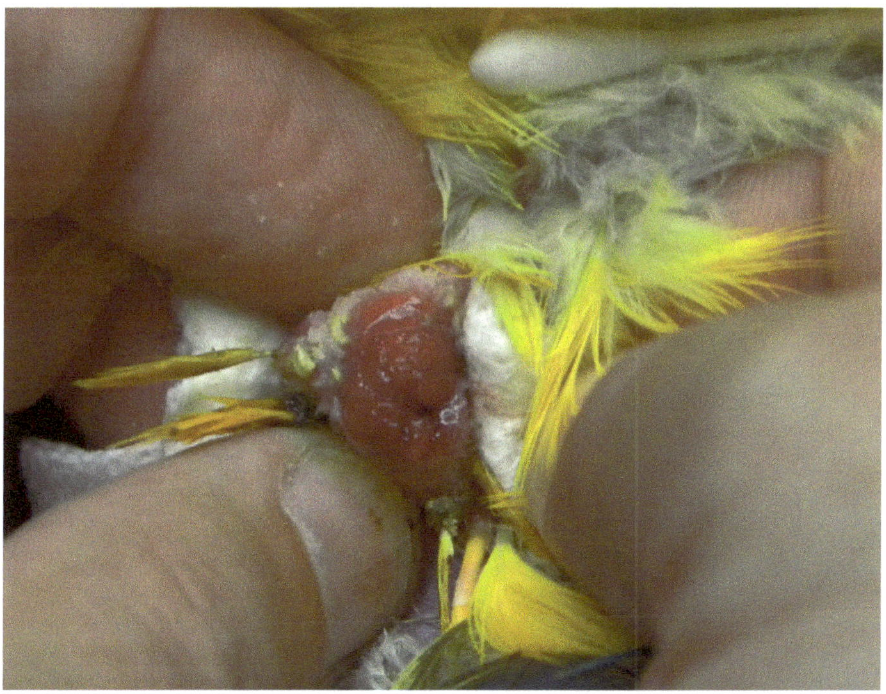

Fig. 2.24 Cloacal papilloma in a blue and green macaw (*Courtesy* Kenneth R. Welle)

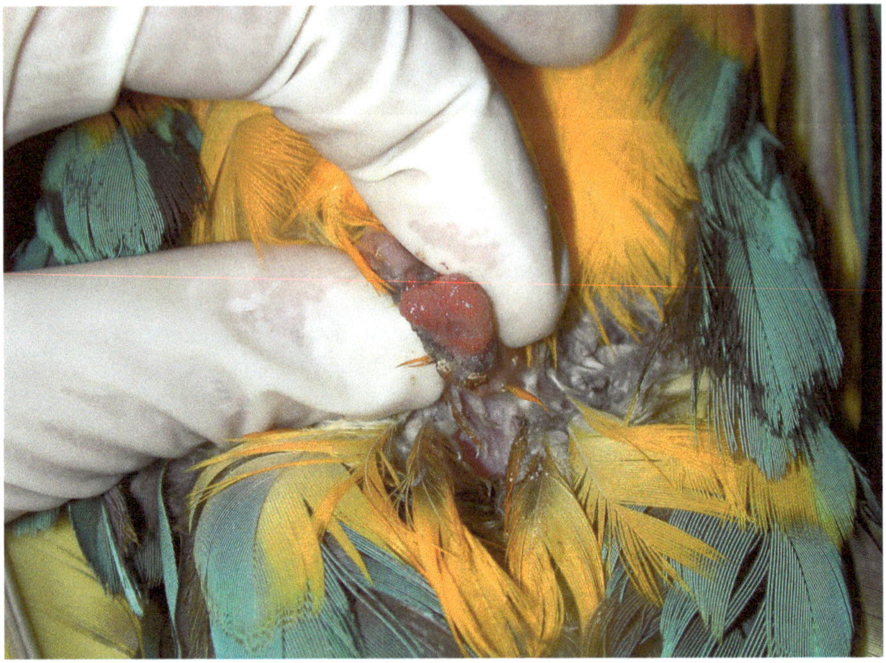

Fig. 2.25 Collection of biopsy by gloved hand from cloacal papilloma in a macaw (*Courtesy* Ashley Zehnder)

cloaca can be collected by a sterile speculum, moistened cotton-tipped applicator or a gloved hand (Fig. 2.25). Histological examination of biopsy samples, in situ hybridization and PCR can confirm the infection.

Treatment of papillomas includes chemical cauterization with silver nitrate or surgical removal of the mass. In chemical cauterization, possibility of re-appearance of the papillomas is detected. Use of interferon (50,000 IU/kg, intramuscular) in some species of birds can prevent the recurrence of growth. In a few countries, autogenous vaccine is used to prevent the infection.

2.2.7.6 Avian Reovirus Infection

Among the members of family *Reoviridae*, *Orbivirus* mostly causes pet bird infection. Parrots, budgerigar, cockatiel, duck and American woodcock are susceptible to *Orbivirus* infection. Pheasants, pigeons and raptors mostly act as carriers. The carrier birds can shed the virus through faeces and contaminate the environment that can act as a source of infection. Biting insects sometimes also help in transmission of infection. In psittacine birds, conjunctivitis, swollen eyelids, enteritis, emaciation, incoordination and other neurological signs are observed. In budgerigars and cockatiels, sudden death without clinical symptom is detected. Stunted growth and feather deformity is observed in Muscovy ducks. Gross lesions include swollen liver, kidney and spleen with necrotic areas. Accumulation of fluid in lungs and pericardium and myocarditis are often detected as gross lesions. Cloacal swabs and tissues from liver, spleen, and kidney can be collected as clinical specimens. The virus can be isolated in chick embryo liver cells or chick kidney cells. The virus is confirmed by electron microscopy, immunofluorescence staining and PCR.

2.2.7.7 Coronavirus Infection (Infectious Bronchitis)

Psittacine birds (budgerigar, Amazon parrot), pigeons, ostrich, rhea are sometimes infected with Infectious Bronchitis virus of *Coronaviridae* family. Respiratory signs, mucopharyngitis, ulcerated crop and esophagus, swollen kidney, egg peritonitis are commonly observed. In ostriches, thin walled and blood filled proventriculus is detected. Nasal swabs, pharyngeal swabs can be collected as clinical specimens in 50% glycerol. The virus can be isolated in chicken embryo liver cells and primary embryo liver cells derived from blue and yellow macaw embryos. Presence of virus can be confirmed by electron microscopy and pan-coronavirus reverse transcriptase-PCR.

2.3 Parasitic Diseases

2.3.1 Toxoplasmosis

2.3.1.1 History

Splendore (1913, Brazil) first observed *Toxoplasma* like organisms in blood smear prepared from an infected rabbit. In the same year, Nicole and Manceaux detected the same organisms at Gondi (Tunisia) and described them as *Toxoplasma gondii*. In pet birds (pigeons), Carini (1911) first observed *Toxoplasma* like parasites in smears prepared from liver and spleen in São Paulo, Brazil. Based on phenotypical detection in smears prepared from blood and tissues, *Toxoplasma* was described subsequently in different species of birds. Catar (1974) detected the organisms in mistle thrush (*Turdus viscivorus*), song thrush (*Turdus philomelos*), robin

Table 2.12 *Toxoplasma gondii* genotypes common in animals, birds and human

Conventional genotype	ToxoDB PCR-RFLP genotypes
Type I	#10
Type II	#1
Type III	#3
Type 12 (atypical)	#4
Type BrI (atypical)	#6
Type BrII (atypical)	#11
Type BrIII (atypical)	#8
Type BrIV (atypical)	#17
Chinese 1 (atypical)	#9

(*Erithacus rubecula*), house sparrow (*Passer domesticus*) and pheasants (*Phasianus colchicus*) in Slovakia.

2.3.1.2 Etiology

Toxoplasma gondii is an obligate intracellular protozoa and a member of suborder *Eimeriina*, phylum Apicomplexa. Feral and domestic cats (*Felidae* family) are the definitive host of the parasite. It has a wide range of intermediate hosts including different species of birds. *T. gondii* has three conventional clonal lineages (type I, II, III) which has different host preference and virulence pattern. Most of the animal strains belong to type III, whereas, type I and II strains commonly infect human. Occasionally, all the typical genotypes (I, II and III) and their variants are isolated from birds such as free-range chickens, raptors etc. Progress in this area identified several other genotypes ('atypical') of *T. gondii* such as Type BrI, BrII, BrIII, BrIV, Type 12, Africa 1 and Chinese 1. A classification scheme is adopted to designate each genotype as 'Toxo DB-PCR-RFLP genotype' followed by a specific number (Table 2.12). A total of 189 ToxoDB PCR-RFLP genotypes is identified so far (2012).

2.3.1.3 Host Susceptibility

Clinical toxoplasmosis is detected in birds belonged to *Passeriformes*, *Psittaciformes*, *Columbiformes*, *Strigiformes*, *Galliformes* and *Anseriformes* orders (Table 2.13). Among the pet birds, passerines (canaries, finches, mynahs), pigeons and partridges are commonly infected with *Toxoplasma*. Clinical toxoplasmosis in pet birds is reported from different countries of Europe, North America, South America and Oceania (Table 2.13).

Apparently healthy wild and domestic birds belonged to different families (*Accipitriformes*, *Anseriformes*, *Galliformes*, *Gruiformes*, *Charadriiformes*, *Columbiformes*, *Strigiformes*, *Passeriformes*) can also harbour *T. gondii* without showing clinical symptoms. Examples of carrier birds include goshawk (*Accipiter gentilis*), common buzzard (*Buteo buteo*), kestrel (*Falco tinnunculus*), pallid harrier (*Circus macrourus*), black vulture (*Aegypius monachus*), red-tailed hawk (*Buteo jamaicensis*), pheasant (*Phasianus colchicus*), turkey (*Meleagris gallopavo*), mallard duck (*Anas platyrhynchos*), pintail duck (*Anas acuta*), coot (*Fulica atra*), blackheaded gull (*Larus ridibundus*), common tern (*Sterna hirundo*), collared dove (*Streptopelia decaocto*), woodpigeon (*Columba palumbus*), common pigeon (*Columba livia*), ferruginous pygmy owl (*Glaucidium brasilianum*), little owl (*Athene noctua*), chaffinch (*Fringilla coelebs*), house sparrow (*Passer domesticus*), tree sparrow (*Passer montanus*), yellowhammer (*Emberiza citrinella*), starling (*Sturnus vulgaris*), black bird (*Turdus merula*), mistle thrush (*Turdus viscivorus*), song thrush (*Turdus philomelos*), robin (*Erithacus rubecula*), great tit (*Parus major*), treecreeper (*Certhia familiaris*), jackdaw (*Corvus monedula*), rook (*Corvus frugilegus*).

Table 2.13 Clinical toxoplasmosis in pet birds in different countries

Order	Birds	Country/continent	Gross lesions
Passeriformes	Canary (*Serinus canarius*), greenfinch (*Carduelis chloris*), goldfinches (*C. carduelis*), sirkins (*C. spinus*), bullfinches (*Pyrrhula pyrrhula*), linnets (*C. cannabina*)	United States, United Kingdom, Uruguay, Australia, Italy, New Zealand	Non-suppurative inflammation of optic nerve, uveitis, choroiditis, focal necrosis and detachment of retina, atrophy of periocular tissue
	Mynah (*Acridotheres* spp.), gold crested mynah (*Ampieliceps coronatus*)	United States (Indiana), Netherlands	–
	Hawaiian crow (*Corvus hawaiiensis*)	Hawaii, United states	–
	Satin bowerbird (*Ptilornorhyncus violaceus*), bowerbird (*Sericulus chrysocephalus*), red-whiskered bulbul (*Pycnonotus jocosus*)	Australia	–
Psittaciformes	Budgerigars (*Melopsittacus undulatus*)	Switzerland	Enlarged spleen, necrotizing myocarditis, hepatitis, intestinal pneumonia are observed in naturally infected budgerigars. Hyperemia and hemorrhagic to fibrino-necrotic enteritis, multifocal areas of necrosis in intestine and brain (with tachyzoites) are detected in experimentally inoculated budgerigars
	Regent parrot (*Polytelis anthopeplus*), superb parrot (*P. swansonii*), crimson rosella (*Platycercus elegans*)	Australia	–
	Vinaceous Amazon parrot (*Amazona vinacea*)	Brazil	Edema and congestion of lungs, cloudy air sacs, mild hepatomegaly, necrosis in myocardium
	Red lory (*Eos bornea*), Black-winged lory (*Eos cyanogenia*)	United States	Enlarged spleen, necrotizing myocarditis, hepatitis, intestinal pneumonia
	Swainson's lorikeet (*Trichologlossus moluccanus*)	Netherland	–

(continued)

Table 2.13 (continued)

Order	Birds	Country/continent	Gross lesions
	North Island kaka (*Nestor meridionalis*)	New Zealand	Hepatospleenomegaly and swollen lungs
Columbiformes	Rock dove (*Columba livia*)	Brazil, United States	Mild encephalitis and neuritis
	Crown pigeons (*Goura crista, G. Victoria, G. scheepmaker*)	Belgium, Netherlands, United states	–
	Strait pigeons (*Ducula spilorrhoa*), Wonga pigeon (*Leucosarcia melanoleuca*)	Australia	–
	Bleeding heart dove (*Gallicolumba luzonica*)	Netherland	–
	Nicobar pigeons (*Caloenas nicobarica*), luzon bleeding-heart pigeons (*Gallicoluba luzonica*), orange-breasted green pigeon (*Treron bicinta*)	United States	–
	Kereru (*Hemiphaga novaeseelandiae*)	New Zealand	–
Strigiformes	Barred owl (*Strix varia*)	Canada	Multifocal necrotic areas (1 mm diameter) containing numerous tachyzoites and surrounded by inflammatory cells
Galliformes	Domestic turkey	Germany	–
	Partridges (*Perdix perdix*)	Czech Republic	–
	Erckel's francolin (*Francolinus erckelii*)	Hawaii, United States	Focal discoloration of liver and heart, edematous lungs
Anseriformes	Hawaiian goose nene goslings (*Nesochen sandicensis*)	Hawaii, United States	Edematous, consolidated lungs and necrosis in liver, brain, heart and muscles
Apterygiformes	North Island brown kiwi (*Apteryx mantelli*)	New Zealand	Hepatospleenomegaly and swollen lungs

2.3.1.4 Transmission

Toxoplasma gondii has three stages in life cycle—oocyst (with sporozoites), tissue cyst with bradyzoites and rapidly multiplying form or tachyzoites. Oocysts are infective stage of the protozoa present in cat faeces (definitive host) . The infected cats can shed millions of oocysts within 3–10 days of infection, regardless of the presence of clinical signs. The oocysts are activated within 1–5 days after faecal excreation and can survive in soil and water for prolonged period (up to 1 year). The feral cats bury the faeces into soil but earthworms and other soil associated insects bring them into the top layer of soil. The oocysts are ingested by intermediate hosts (birds, rodents, sheep, marine mammals and human) from the soil or water. The parasite invades tissues of intermediate hosts and produce bradyzoites within tissue cyst.

2.3.1.5 Pathogenesis

After ingestion of oocysts by the intermediate hosts, sporozoites prefer to invade most of the vital organs, with predilection for the reticuloendothelial and central nervous systems. The parasite enters the host cell by an active process. After intracellular multiplication, the progeny parasites go into the blood circulation by lysis of the infected cells and finally reach vital organs through the blood circulation. Onset of clinical symptoms depends on type of organs invaded by the parasite.

The parasitemia (presence of parasite in blood) declines after development of host immunity. The parasites localize and persist in the form of 'tissue cysts'. Most of these tissue cysts are benign in nature and do not produce any clinical symptom during persistence. In immunosuppression due to stress, concurrent viral infection or immunosuppressive therapy, the tissue cysts are ruptured. Granulomatous lesions develop with invading inflammatory cells into the surrounding tissues of the site where the cyst persisted.

2.3.1.6 Clinical Symptoms

In canaries infected with toxoplasmosis, weight loss, diarrhoea, dyspnoea, crusty exudates around eye lids, collapsed eyeballs, inflammation of choroid with or without retinal involvement, cataracts and blindness, head twitch, walking in circles (due to encephalitis) are commonly observed. Dull, visionless, closed eyes and encephalitis are consistent features in passerine birds (Fig. 2.26). In psittacine birds, non-specific clinical symptoms are observed. In lories, respiratory distress is distinct. In pigeons infected with *T. gondii*, anorexia, emaciation, high fever, weakness, conjunctivitis and convulsions are detected.

Fig. 2.26 Finches with closed eye (*Courtesy* LafeberVet)

2.3.1.7 Lesion

In passerines (canaries), gross lesions include osseous replacement of eye globe, non-suppurative inflammation of optic nerve, anterior and posterior uveitis, swelling of the lenticular fibers in the lens, choroiditis, focal necrosis and detachment of retina and atrophy of periocular tissue (sunken appearance of eye). *T. gondii* tachyzoites are detected in choroid, retina (nerve fiber layer), vitreous and lens of the affected birds. Histological evidence of pneumonia and non-suppurative encephalitis associated with tissue cysts is also observed. Gross lesions observed in affected psittacine and other birds are enlisted in Table 2.13.

2.3.1.8 Diagnosis

Clinical Specimens

Blood, serum, eye suspensions from live birds and organs (spleen, liver, intestine, eye, brain, gizzard, proventriculus) collected after post mortem in buffered neutral 10% formalin can be used as clinical specimens. For bioassay (isolation), collected tissues are homogenized in 0.85% normal saline solution and brain tissue homogenates are digested with acidic pepsin before bioassay.

Diagnostic Techniques

(a) *Direct examination*: Preliminary diagnosis is made by demonstration of *T. gondii* tachyzoites in Giemsa stained impression smears prepared from collected organs. In smears, tachyzoites are crescentic to globular in shape (Fig. 2.27).
(b) *Histological examination*: In formalin fixed tissues, globular to oval shaped *T. gondii* tachyzoites are detected which are smaller in size than their appearance in impression smear (Fig. 2.28). *T. gondii* 'tissue cysts' appear as a globular structure with a thin cyst wall and small, slender bradyzoites are present within it (50–500). The bradyzoites can be visualized by periodic acid Schiff (PAS) staining. Immunohistochemical staining with polyclonal antibody raised against whole parasite can confirm the presence of *T. gondii* in formalin fixed tissues.
(c) *Serological tests*: Modified agglutination test (MAT) is a sensitive, specific and easy to do serological test for detection of *T. gondii* antibodies in different species of birds. ELISA, indirect FAT in tissues can also be performed.
(d) *Bioassay*: *T. gondii* can be isolated in laboratory mice by subcutaneous inoculation of eye suspensions or tissue homogenates (liver, brain) collected from suspected birds. In positive cases, inoculated mice will die and *T. gondii* tachyzoites or tissue cysts are detected from dead mice.
(e) *Molecular biology*: The parasite can be confirmed by *T. gondii*-specific PCR. Nested PCR analysis based on *T. gondii pppk-dhps* gene is recently developed for confirmation. Genotyping of *T. gondii* depends on restriction fragment length polymorphism of surface antigen 2 gene (SAG2).

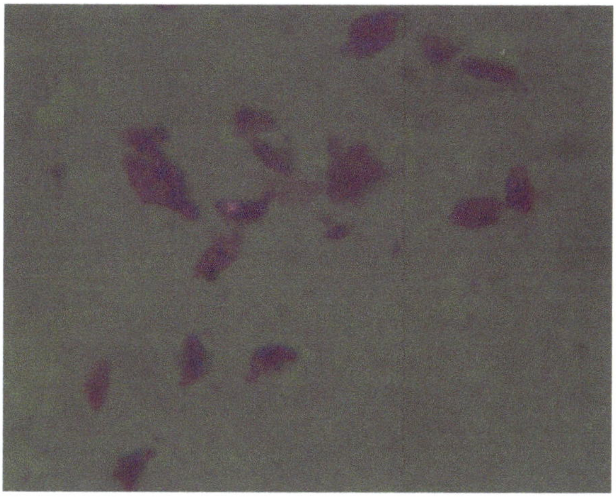

Fig. 2.27 *Toxoplasma gondii* tachyzoites from infected mice and stained with Giemsa (×100, *Courtesy* Surajit Baidya, Department of Veterinary Parasitology, West Bengal University of Animal and Fishery Sciences, India)

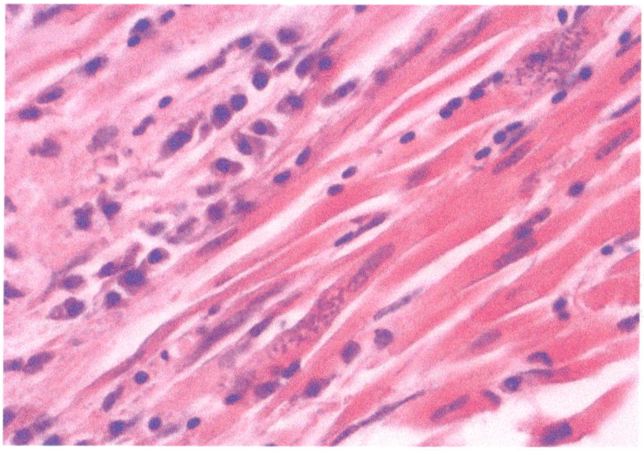

Fig. 2.28 Necrosed myodcardium of a Vinaceous Amazon parrot associated with intracellular tachyzoites and lymphocytes, macrophages and plasma cell infiltration (×40, H & E stain, *Courtesy* Prof. Roselene Ecco, Veterinary School, Universidade Federal de Minas Gerais, Brazil)

2.3.1.9 Zoonosis

Human get the infection by ingestion of vegetables, fruits and other foods or water contaminated with *T. gondii* oocysts. Direct transmission of *T. gondii* from pet birds to human is not documented. Feral or domestic cats may ingest the *T. gondii* infected carcasses of pet birds disposed into the surroundings without proper measures. The cats may become infected with toxoplasmosis and start to shed oocysts into the environment.

Person to person transmission is rarely possible during organ transplantation or blood transfusion. Most of the human infection is aymptomatic but sometimes cervical or occipital lymphadenopathy is observed. In immunosuppressed persons, encephalitis, chorioretinitis and pneumonitis are detected. In pregnant women, abortion and fetal infection causing congenital hydrocephalus, intracranial calcifications and mental retardation is documented.

2.3.1.10 Treatment and Control Strategy

Pyrimethamine (0.5 mg/kg body weight, orally, 12 h interval) is recommended for treatment of clinical toxoplasmosis in birds. Use of Tiamulin fumarate (300–400 mg/kg feed for 7 days in different species of birds; 225–250 mg/l drinking water for 3–7 days in pigeons and poultry) and clindamycin (25 mg/kg body weight, oral, 48 h interval) is also observed. In canaries, treatment with trimethoprim (80 mg/ml) and sulfadiazine (400 mg/ml) in drinking water for 14 days was apparently successful.

Prophylactic use of coccidiostats (monensin, decoquinate) is recommended in calves and lambs to prevent toxoplasmosis. Low acute toxicity of decoquinate is observed in avian species and no regulation or evidence is available regarding its prophylactic use. No vaccine is also available for birds to prevent toxoplasmosis.

2.3.2 Giardiasis

2.3.2.1 History

Antony van Leeuwenhoek (Delft, Netherlands, 1681) first observed *Giardia* under his self made simple microscope during investigation of his own diarrhoeic stool. He described it as 'animalcule' with 'flattish belly' and 'sundry little paws' (flagella). Under microscope, the organisms showed a slow and helical motion with occasional rapid movement by 'paws'. Later (Dobell 1932) it was concluded as trophozoite stage of *Giardia* spp. In 1859, Vilem Dusan Lambl (Prague, Czech Republic) described the organism in more details and named it as *Cercomonas intestinalis*. In 1915, Charles Wardell Stiles coined the name *Giardia lamblia* in honour of Professor Alfred Mathieu Giard (Zoologist, France) and Dr. Vilem Dusan Lambl (Czech) for their contribution in progress of *Giardia* associated knowledge.

Leibovitz (1962) first described *Giardia* infection in a budgerigar in United States with a history of chronic diarrhoea and debility. In 1977, Jones and Carroll also observed presence of *Giardia* in the intestine of budgerigars in United Kingdom. In 1978, *Giardia* infection causing high mortality in parakeets was reported from United States which was successfully treated by dimetridazole (Panigrahy et al. 1978).

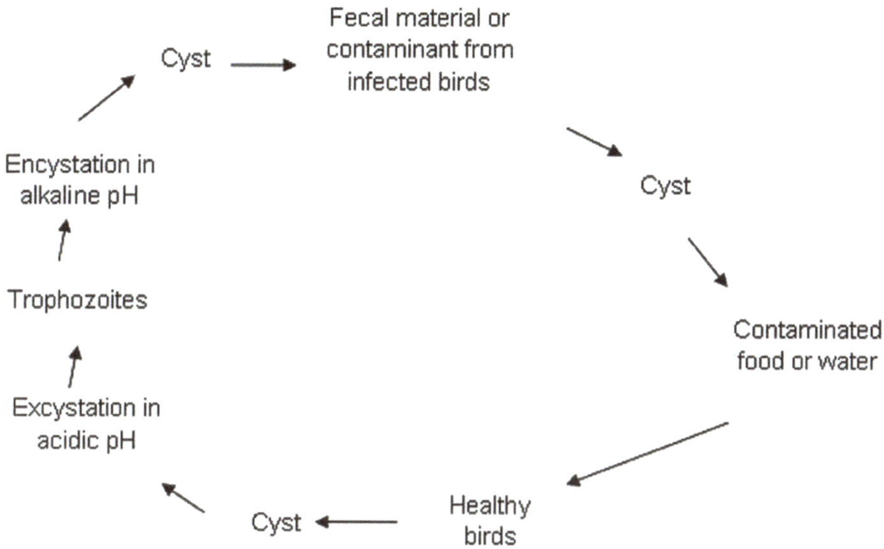

Fig. 2.29 Life cycle of *Giardia* (schematic)

Fig. 2.30 Trophozoites of *Giardia* under confocal microscopy using fluorescent tagged antibody (*Courtesy* Sandipan Ganguly, National Institute of Cholera and Enteric Diseases, Kolkata, India)

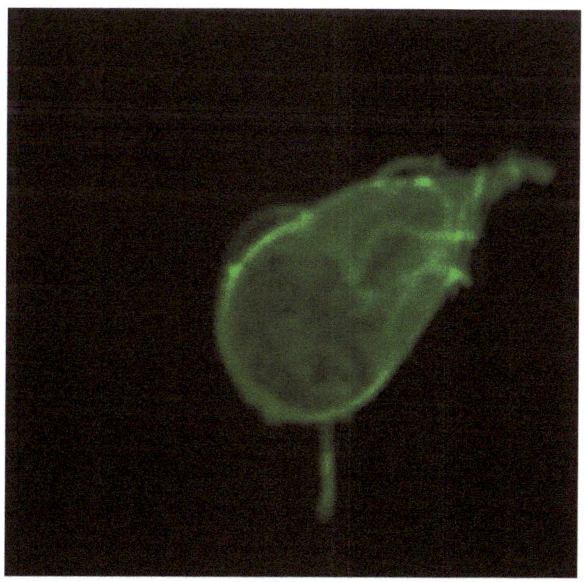

2.3.2.2 Etiology

Avian Giardiasis is caused by *Giardia* spp., an eukaryotic, multicellular, binucleate, flagellated protozoan belonged to Sarcomastigophora phylum. Six valid species of *Giardia* are recognized based on morphological observations using light and electron microscopy. The valid species are *Giardia psittaci* and *G. ardeae* in birds; *G. duodenalis* (syn. *G. lamblia, G. intestinalis*) in human, livestock and wildlife; *G. microti* and *G. muris* in rodents; and *G. agilis* in amphibians. *G. duodenalis* has eight identified genotypes (assemblages, A–H). Assemblage A and B are common in both human and animals and assemblage C–H are restricted within animals only.

Giardia spp. has two stages in life cycle-motile 'trophozoite' and 'cyst' (Fig. 2.29). The cysts are smaller than trophozoites (10 μm × 8 μm), dormant, resistant to adverse environmental condition (like bacterial spore) and infectious form of the protozoa. They are common in streams, lakes and ponds. They can survive for months in cold water also (8 °C). After transmission of the cysts into the host, the cysts pass through acidic pH, increased CO_2 level and slight alkaline pH (proximal small intestine) consecutively and excystation takes place. One trophozoite from each cyst emerges which undergoes cytoplasmic division to produce two trophozoites. Trophozoites are pear shaped (pyriform) structure which measures 12–18 μm in length, 10 μm in breadth and 2–4 μm in thickness (Fig. 2.30). The trophozoites have a concave disc with a raised ridge at the ventral surface of anterior site (broad portion) and eight flagella arranged bilaterally. The trophozoites attach with enterocytes at duodenum and jejunum with the ventral disc ('sucker') for feeding on mucosal secreations. The colonization is followed by binary fission.

Some trophozoites detach from the enterocytes and move forward (tumbling and skipping) with their flagella to re-attach with a new enterocyte. A few of the trophozoites instead of re-attachment prefer to be excreated through the faeces as cyst. During the process of encystment, the trophozoites stop their active motility, become rounded in shape and covered with a cyst wall. Nuclear division takes place and a quadrinucleate, matured cyst is excreated into the environment.

2.3.2.3 Host Susceptibility

Giardia is identified in faecal samples of more than 20 species of birds specially in psittacines. *Giardia psittaci* associated clinical infection is fatal in young budgerigars. Avian giardiasis is also reported in cockatiels, lovebirds, finches, great blue herons, raptors and gray-cheeked parakeets throughout the world. Some psittacines such as blue-fronted Amazon (*Amazona aestiva*), blue and yellow macaw (*Ara ararauna*), scarlet macaw (*Ara macao*) act as carrier (transport host) of *Giardia duodenalis* (assemblage A) without showing clinical symptoms.

2.3.2.4 Transmission

The infection is transmitted by faeco-oral route through ingestion of the food or water contaminated with infective cysts. Feeding or watering trough, cage materials, toys or other inanimate objects contaminated or soiled with faeces of the infected birds (captive or wild) may serve as a source of infection. Asymptomatic birds generally shed the cyst intermittently and thus serve as a potential source of infection.

2.3.2.5 Pathogenesis

Giardia psittaci mostly reside in the duodenum with manifestation of diarrhoea and malabsorption syndrome in birds. The trophozoites adhere to the intestinal villi with ventral suckers. The adherence results inflammatory cell infiltration, villous atrophy, reduced villous to crypt ratio and reduction in disaccharidase enzymes (e.g. lactase). Food absorption is hampered and the food particles are accumulated in the lumen which increases the osmotic pressure and causes diarrhoea. With inhibition of food absorption, chronic weight loss may be noticed in affected birds. There may be deficiency of vitamins and minerals due to absorption failure. Unlike mammals where immunity has a direct relation with occurrence of giardiasis, avian immunity rarely shows such association. Some risk factors are also noticed in birds for giardiasis such as overcrowding, unhygienic cage condition and inadequate nutrition. The studies revealed that giardiasis may be more common among heavily inbred population of birds.

2.3.2.6 Clinical Symptoms

Mucoid and persistent diarrhoea with loose, brown or pale coloured and foul smelling faeces, anorexia, depression, hypoproteinemia, weight loss, ruffled feathers are common clinical symptoms in birds infected with *Giardia* spp. Stunted growth and high mortality are observed in young budgerigars and cockatiels. In cockatiels, feather picking and pruritus are also detected. Feather picking from wings, flanks and legs along with screaming is common. Feather damage is sometimes observed in non-cockatiel birds also. Secondary bacterial, viral or yeast infection are identified in birds with avian giardiasis. Concurrent infection of *G. psittaci* and *Polyomavirus, Cryptosporidium, Chlamydia* and *Macrorhabdus* spp. (Megabacterium) are observed in budgerigars.

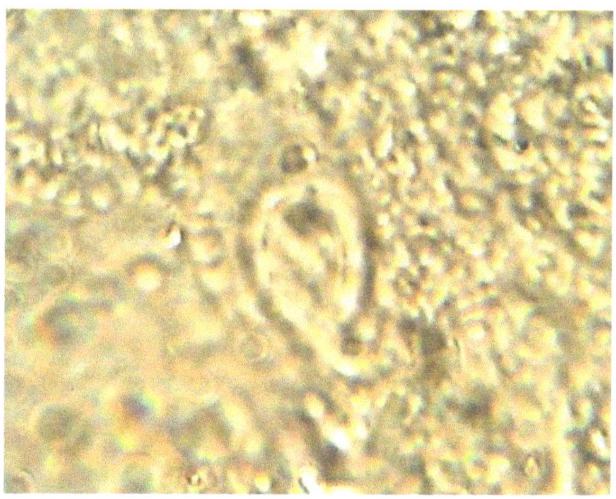

Fig. 2.31 Trophozoites of *Giardia* present in stool smear under wet mount preparation (*Courtesy* Sandipan Ganguly, National Institute of Cholera and Enteric Diseases, Kolkata, India)

2.3.2.7 Lesion

No gross lesion or sometimes distended small intestine with mucous, yellowish or creamy fluids is detected in most species of the birds. Atrophy of villous, infiltration of inflammatory cells, hyperemia of intestinal mucosal layer and presence of numerous trophozoites throughout the entire length of villi are observed.

In cockatiels, thickened skin, haemorrhage and areas of feather loss are observed in patagium (membranous structure that helps in flight) and axillary area. It may progress to squamous cell carcinoma.

2.3.2.8 Diagnosis

Clinical Specimens

Fresh faeces collected in saline from the suspected birds, blood or serum, intestine (duodenum) in 10% formalin can be used as clinical specimen.

Diagnostic Techniques

(a) *Direct examination*: Wet mount examination of collected faecal sample (mixed with warm saline, not tap water) or the content of duodenum can be performed for direct visualization of motile trophozoites (pear shaped) or cysts (Fig. 2.31). The faecal sample can be concentrated with formalin-ethyl acetate or SAF (sodium acetate, acetic acid, formaldehyde) and zinc sulphate. Multiple fresh faecal samples (three consecutive samples) should be tested from a single suspected bird due to *intermittent shedding* of trophozoites and cysts. If the faecal sample is more than 10 min old, possibility of trophozoite visualization is low. If immediate processing is not possible, faecal samples may be preserved in polyvinyl alcohol for trichrome staining. The trophozoites are also destroyed in salt or sugar flotation solutions for faecal sample observation. Faecal sample can be stained with Lugol's iodine for visualization of *Giardia* cysts.

(b) *Molecular biology*: For extraction of DNA from *Giardia*, oocysts present in faecal samples are concentrated. Faecal suspension prepared with sterile distilled water is kept over sucrose solution (1 M) and is centrifuged at 400 g for 15 min at room temperature. The water-sucrose interface is removed with a Pasteur pipette, washed in normal saline and centrifuged at 600 g for 10 min. DNA is extracted from the sediment by a standard nucleic acid extraction kit. PCR targeting *β-giardin* gene, *Giardia* elongation factor 1 alpha gene (*ef1α*) or *Giardia* glutamate dehydrogenase gene (*gdh*) can be performed for confirmation.

(c) *Antigen detection tests*: ELISA (antigen capture) and immunofluorescence tests are available for detection of *Giardia* antigen in clinical samples.

2.3.2.9 Zoonosis

Human giardiasis is caused by *G. duodenalis* and it has two phases i.e. acute and chronic. Flatulence, belching, abdominal distension with cramps, frequent watery diarrhoea with offensive smell occurs in acute phase. In chronic phase, malabsorption syndrome takes place with chronic weight loss. The stools are usually pale or yellow, frequent and of small volume. The prevalence of giardiasis in developing countries is approximately 20% compared to about 5% in the developed world. Transmission of *Giardia* to humans can occur through ingestion of food and water contaminated with infectious cysts. Waterborne transmission is associated with contaminated community water systems of municipality or corporation in urban area, and ponds, rivers and streams in rural area. Contaminated swimming pool also plays a role in transmission of giardiasis. Giardiasis is not specific for any human race and sex but it is more prevalent in children below 4 years of age.

The cysts are excreated in the environment through the faeces of infected human, animals and birds. The studies revealed that psittacine birds (blue fronted Amazon, blue and yellow macaw, scarlet macaw) can carry *Giardia duodenalis* (Assemblage A) cysts which may be disseminated into human.

2.3.2.10 Treatment and Control Strategy

Metronidazole (10–20 mg/kg body weight, oral, 12–24 h interval for two days for psittacines and pigeons) is the drug of choice in confirmed cases of avian giardiasis. Necrosis at site of injection in all species of birds and toxicity in finches are adverse drug reactions of metronidazole. Fenbendazole (50 mg/kg body weight, oral, 24 h interval for 3 days) is an alternative choice for effective treatment. Other drugs such as tinidazole (20 mg/25 ml of drinking water for 7–14 days) or paromomycin (100 mg/kg twice daily for 7 days) are also used successfully against avian giardiasis in budgerigars and barred parakeets. In budgerigars, amphotericin B and metronidazole combination is indicated for synergistic infection of *Giardia*, *Candida*, Megabacterium, *Trichomonas* and other bacterial infections.

With the discontinuation of therapy the infestation may relapse. It is necessary to properly rinse and dry the cage, feeding, watering and other in contact inanimate objects. Cleaning and drying reduce both the number of cysts and their viability and helps to prevent reinfestation. Feeding of boiled water in clean water bottles also helps to reduce the risk of infestation.

2.3.3 Cryptosporidiosis

2.3.3.1 History

Dr. Ernest Edward Tyzzer (1929, United States) first described cryptosporidiosis in bird (chicken) although *Cryptosporidium* species was not identified. It resembled *Cryptosporidium muris* as described in mice earlier by same worker (Tyzzer 1907, 1910). *C. meleagridis* was first identified in turkeys by Slavin (1955). In 1986, Current, Upton and Haynes isolated another species from chickens and coined the name *C. baileyi*.

2.3.3.2 Etiology

Cryptosporidium is a protozoon under phylum Apicomplexa, class Sporozoasida, order Eucoccidiorida, and family Cryptosporidiidae. Three *Cryptosporidium* species such as *C. meleagridis*, *C. baileyi*, and *C. galli* are commonly associated with avian infection. Recent molecular epidemiological studies have also identified several genotypes of *Cryptosporidium* causing infection in birds. The genotypes are: avian genotypes (I–V), goose genotypes (I–IV), black duck genotype, and Eurasian woodcock genotype. In addition, *C. hominis*, *C. parvum*, *C. serpentis*, *C. muris* and *C. andersoni* are detected in a number of birds due to accidental ingestion of Cryptosporidial oocysts. Among all these species and genotypes, *C. baileyi* is considered as the most common cause of avian infection.

Cryptosporidium has a direct life cycle which requires a single host. 'Oocyst' with four sporozoites is the infectious form and it is shed in faeces and cough of infected birds, animals and human. The oocyst can enter a new host and after excystation the sporozoites penetrate to epithelial cells of the gastrointestinal or respiratory tract. The sporozoites undergo six developmental stages such as merogony (asexual multiplication), gametogony (gamete formation), fertilization, oocyst formation, and sporogonyv (sporozoite formation). The sporozoites of *C. baileyi* produce 3 types of meronts (type I, II, III) during merogony to generate merozoites. Two types of oocysts (thin and thick walled) are observed in *C. baileyi* infection. Thin-walled oocysts are not excreated and they excyst in situ within the host to begin auto-infection. Whereas, thick walled (multilayered) oocysts are more resistant to environment and are excreated out to enter new hosts. Sometimes, auto-infection takes place after excreation and the excreated oocysts may infect the same host.

2.3.3.3 Host Susceptibility

Cryptosporidium is detected in more than 30 bird species (Anseriformes, Charadriiformes, Columbiformes, Galliformes, Passeriformes, Psittaciformes and Struthiniformes) in different countries such as Australia, Argentina, Canada, China, Czech Republic, Denmark, Egypt, Germany, Greece, Hungary, Japan, Korea, The Netherlands, Romania, Scotland, Spain, South Africa, Taiwan, Turkey, and United States (Table 2.14).

2.3.3.4 Transmission

Avian cryptosporidiosis is transmitted by ingestion of contaminated food and water (faeco-oral) or inhalation of sporulated oocysts. Feeding or watering trough, cage materials, toys or other inanimate objects contaminated with faeces of the infected birds may serve as a source of infection. In the same aviary or premises, transmission is possible from one avian species to another through contaminated equipments or personnel. Once the oocyst enters the premises possiblity of large scale outbreak increases in the aviaries or zoos.

Table 2.14 Susceptible avian hosts of *Cryptosporidium* spp. in different geographical locations

Parasite	Susceptible host	Organs affected	Geographical location
Cryptosporidium baileyi	Black-headed gulls (*Chroicocephalus ridibundus*)	Bursa of Fabricius, conjunctiva, kidney, respiratory tract, cloaca, rectum	Africa, Asia, Europe, North America, South America
	Great cormorant (*Phalacrocorax carbo*)		
	Cranes (*Gruidae*)		
	Channel-billed toucan (*Ramphastos vitellinus*)		
	Eastern golden-backed weaver (*Ploceus jacksoni*)		
	Cockatiels (*Nymphicus hollandicus*)		
	Grey-bellied bulbul (*Pycnonotus cyaniventris*)		
	Red-rumped cacique (*Cacicus haemorrhous*)		
	Crested oropendola (*Psarocolius decumanus*)		
	Red crowned amazon (*Amazona viridigenalis*)		
	Rose-ringed parakeet (*Psittacula krameri*)		
	Grey partridge (*Perdix perdix*)		
	Mixed-bred falcons		
	Black vulture (*Coragyps atratus*)		
	Saffron finch (*Sicalis flaveola*)		
	Ruddy Shelducks (*Tadorna ferruginea*)		
	Chickens (*Gallus gallus domesticus*)		
	Turkey (*Meleagris gallopavo*)		
	Brown quail (*Coturnix ypsilophora*)		
	Duck (*Anas* spp.)		
	Goose (*Branta canadensis*)		
	Black-billed magpie (*Pica pica*)		
	Bohemian waxwing (*Bombycilla garrulus*)		
	Common myna (*Acridotheres tristis*)		
	Crested Lark (*Galerida cristata*)		
	Gouldian finch (*Chloebia gouldiae*)		
	Red-billed leiothrix (*Leiothrix lutea*)		
	White Java sparrow (*Padda oryzivora*)		
	Zebra finch (*Taeniopygia guttata*)		
	Ostriches (*Struthio camelus*)		

(continued)

Table 2.14 (continued)

Parasite	Susceptible host	Organs affected	Geographical location
Cryptosporidium meleagridis	Parrots	Small intestine, large intestine	Africa, Asia, Europe, Oceania, North America, South America
	Cockatiel (*Nymphicus hollandicus*)		
	Red-legged partridge (*Alectoris rufa*)		
	Rose-ringed parakeet (*Psittacula krameri*)		
	Chickens (*Gallus gallus domesticus*)		
	Turkey (*Meleagris gallopavo*)		
	Bohemian waxwing (*Bombycilla anaries*)		
	Fan-tailed pigeon (*Columba livia*)		
	Rufous turtle dove (*Streptopelia orientalis*)		
	Indian ring-necked parrot (*Psittacula krameri*)		
Cryptosporidium galli	Vinaceous-breasted amazon parrot (*Amazona vinacea*)	Proventriculus	Asia, Europe, Oceania, South America
	Plain parakeet (*Brotogeris tirica*)		
	Chopi blackbird (*Gnorimopsar chopi*)		
	Cockatiel (*Nymphicus hollandicus*)		
	Green-winged saltator (*Saltator similis*)		
	Double-collared seedeater (*Sporophila caerulescens*)		
	Atlantic Canary (*Serinus anaries*)		
	Saffron finch (*Sicalis flaveola*)		
	Rufous-collared sparrow (*Zonotrichia capensis*)		
	Silver-eared Mesia (*Leiothrix argentauris*)		
	Turquoise parrots (*Neophema pulchella*)		
	Flamingo (*Phoenicopterus ruber*)		
	Hornbill (*Buceros rhinoceros*)		
Avian genotype I	Canaries (*Serinus canaria*)	–	Oceania, South America
	Indian peafowl (*Pavo cristatus*)		

(continued)

Table 2.14 (continued)

Parasite	Susceptible host	Organs affected	Geographical location
Avian Genotype II	Cockatiel (*Nymphicus hollandicus*)	Cloaca, rectum, bursa of Fabricius	Asia, Oceania, South America
	Major Mitchell cockatoo (*Cacatua leadbeateri*)		
	Eclectus (*Eclectus roratus*)		
	Galah (*Eolophus roseicapilla*)		
	Sun conure (*Aratinga solstitialis*)		
	Princess parrot (*Polytelis alexandrae*)		
	Alexandrine parrot (*Psittacula eupatria*)		
	White-eyed parakeet (*Aratinga leucophthalma*)		
Avian Genotype III	Cockatiel (*Nymphicus hollandicus*)	Proventriculus	Asia, Oceania, North America, South America
	Red-billed blue magpie (*Urocissa erythrorhyncha*)		
	Peach-faced lovebird (*Agapornis roseicollis*)		
	Galah (*Eolophus roseicapilla*)		
	Sun conure (*Aratinga solstitialis*)		
Avian genotype IV	Japanese white-eye (*Zosterops japonicus*)	–	Europe
Avian Genotype V	Blue-fronted amazon (*Amazona aestiva*)	–	Asia, South America
	Cockatiel (*Nymphicus hollandicus*)		
	Mitchell's cockatoo (*Lophochroa leadbeateri*)		
Goose genotypes I–V	Anseriformes	–	North America
Black duck geno-type	Anseriformes	–	Oceania
Eurasian Wood-cock genotype	Charadriiformes	Proventriculus	Europe
Cryptosporidium andersoni	Galliformes	–	Europe
Cryptosporidium muris	Ostriches (*Struthio camelus*)	–	Asia, Europe
Cryptosporidium parvum	Cockatiel (*Nymphicus hollandicus*)	Small intestine, caecum	Asia, Europe, North America, South America

(continued)

Table 2.14 (continued)

Parasite	Susceptible host	Organs affected	Geographical location
Cryptosporidium blagburni	Gouldian Finch (*Erythrura gouldiae*)	–	
	Red-faced Auroura Finch (*Pytilia hypogrammica*)		
	Plum-headed Finch (*Aidemosyne modesta*)		
Cryptosporidium spp.	Budgerigar (*Melopsittacus undulatus*) Lovebirds (*Agapornis* sp.)	–	

2.3.3.5 Pathogenesis

Following ingestion of oocysts by susceptible birds, it can enter salivary and esophageal glands, proventriculus, small intestine, caecum, colon, cloaca, and bursa of Fabricius and developmental stages of *Cryptosporidium* begins. The intracellular stages of the organism occur within a parasitophorous vacuole. Detachment of enterocytes and villous atrophy in small intestine are common lesions which hamper the absorption of nutrients and causes osmotic diarrhoea and malabsorption syndrome.

Following inhalation of oocysts, primary colonization in upper respiratory tract (sinus) is followed by colonization in lower respiratory tract of the birds (trachea, bronchi, air sacs, lungs). Mucoid exudates are detected in sinus, nasal passage, and trachea. The air sacs and lungs become cloudy and mottled grey-red, respectively.

The site of predilection also varies with species of *Cryptosporidium* infecting the birds. *C. baileyi* prefers to invade mucosal epithelium of a number of organs (small and large intestines, caeca, cloaca, trachea, air sacs, urinary system, bursa and conjunctiva), whereas *C. meleagridis* and *C. galli* are restricted within the intestine and proventriculus, respectively (Table 2.14).

Concurrent viral, bacterial or parasitic infections are observed in birds suffering with both clinical forms of cryptosporidiosis due to bursitis and impairment of

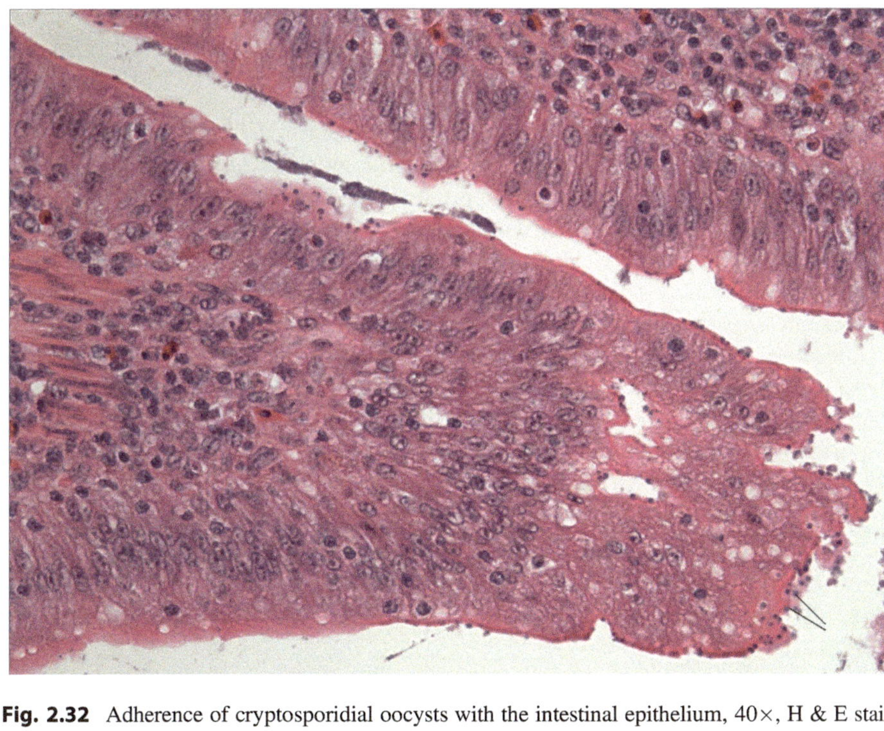

Fig. 2.32 Adherence of cryptosporidial oocysts with the intestinal epithelium, 40×, H & E stain (*Courtesy* Prof. Oscar Fletcher, North Carolina State University, United States)

immunity. *Escherichia coli* and *Isospora* sp. infection are common with *C. galli* infestation.

2.3.3.6 Clinical Symptoms

Two clinical forms such as respiratory and gastrointestinal cryptosporidiosis are observed in birds. In respiratory form, increased mortality, depression, lethargy, anorexia, unthriftines, coughing, sneezing, gurgling, dyspnoea, conjunctivitis and sinusitis are the common clinical symptoms. In gastrointestinal form, lethargy, decreased bodyweight gain, lower pigmentation, and diarrhoea with lime-green stool, chronic or intermittent regurgitation, lockjaw, aspiration pneumonia, seizure, egg binding are the common clinical observations.

C. galli infection in passerine birds is characterized by chronic apathy and weight loss. In Brazil in a cockatiel infected with *C. galli* lethargy for approximately 1 year before death was observed. Lethargy and slow crop emptying was also detected in an Indian ring-necked parrot infected with *C. meleagridis.* In peach-faced lovebirds infected with avian genotype III, chronic vomiting and weight loss was detected.

2.3.3.7 Lesion

Presence of too much mucoid exudates in conjunctival sacs, nasal passage, sinuses and trachea, chemosis, hyperemia, mucus gland distension or cystic hypertrophy/hyperplasia, mottled grey-red lungs, cloudy air sacs, bursal atrophy, hepatospleenomegaly are common gross lesions in respiratory form of cryptosporidiosis in birds.

In gastrointestinal form, distended intestine filled with mucoid contents and gas, detachment of enterocytes, villous atrophy, bursal epithelial cell hypertrophy and hyperplasia and necrosis in the bursa are common gross and microscopic lesions in birds. Histopathology demonstrates the presence of cryptosporidial oocysts adhered to the surface of intestinal brush border (Fig. 2.32). In finches, purulent nephritis with enlarged and pale kidneys is detected. Recently, visceral gout with renal and cloacal lesions is detected in Mitchell's cockatoo infected with *Cryptosporidium* sp.

Dilated proventriculus with thickened proventricular wall is common in lovebirds, cockatiels, red-faced Aurora finch, Australian diamond firetail finch and bronze mannikin finches. Histopathology demonstrates the presence of cryptosporidial oocysts adhered to the surface of proventricular glandular epithelial cells along with hyperplasia and necrosis of epithelial cells.

The organs affected by different *Cryptosporidium* spp. are enlisted in Table 2.14.

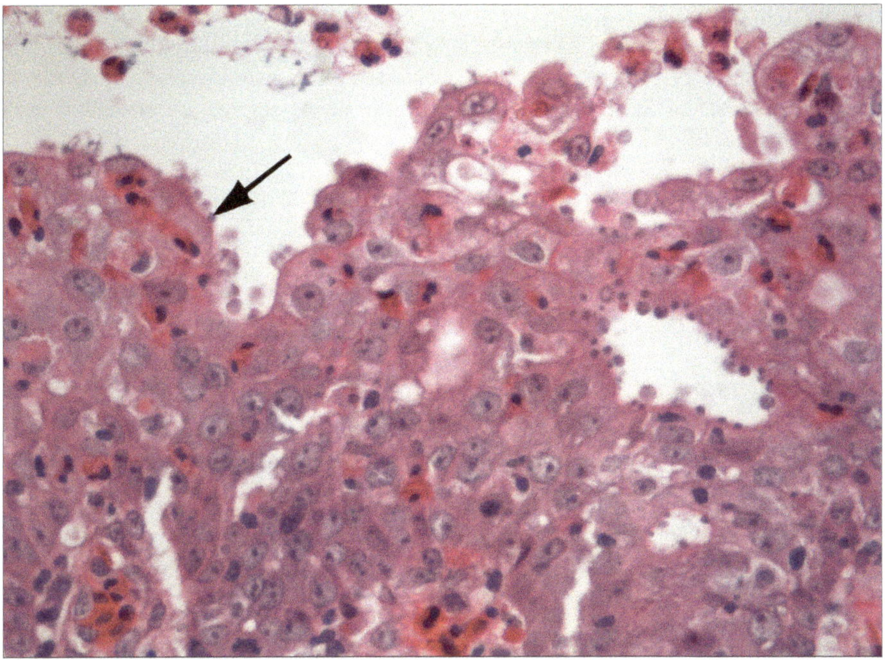

Fig. 2.33 Histological section of trachea of a turkey showing cryptosporidial oocysts, 40×, H & E stain (*Courtesy* Prof. Oscar Fletcher, North Carolina State University, United States)

2.3.3.8 Diagnosis

Clinical Specimens

Fresh faeces or cloacal swabs collected in potassium dichromate solution (2.5–5%) from the suspected birds, blood (collected from wing vein) or serum, and formalin fixed intestine, proventriculus can be used as clinical specimen. The cloacal swabs are filtered through wire mesh (0.3 mm) and the filtrate is centrifuged at room temperature at 1000 g for 10 min. After discarding the supernatant, the concentrated faecal sample is used for further analysis. The faecal samples can be preserved at 4 °C in phosphate buffered saline solution (pH 7.2) with antibiotics and antifungals (streptomycin @ 100 mg/ml, penicillin G @100 IU/ml and amphotericin B @ 0.25 mg/ml).

Diagnostic Techniques

(a) *Direct examination*: Detection of cryptosporidial oocysts in clinical samples is the most common diagnostic technique used in laboratories. It is a low sensitive method which requires minimum 5000–50000 oocysts/g of faeces for detection. The oocysts are non-refractile, spherical (5–6 µm in diameter) in shape and are often confused with yeast cells. Gram's stain, Kinyoun acid-fast stains are used in light microscopy. In Kinyoun stained smear, oocysts appear as red spherical bodies against blue background whereas yeast cells take blue stain. Phase contrast microscopy provides better resolution and the oocysts appear as bright bodies with 1–4 dark granules (granules are absent in yeast cells).

Faecal samples can be concentrated by centrifugal flotation in high specific-gravity salt or sugar solutions (Sheather's sugar flotation technique, discontinuous density sucrose gradient) for detection of oocysts. The sample after concentration should be used rapidly because in presence of sugar solutions oocysts are distorted or collapsed.

In histologic sections stained with hematoxilin and eosin (H & E), cryptosporidia appear as basophilic bodies (Fig. 2.33).

(b) *Molecular biology*: *Cryptosporidium* is detected by means of PCR, followed by either restriction fragment length polymorphism (RFLP) or sequencing of the amplified fragments. The gene commonly used for determining the species or genotype is 18S rRNA. Other target genes are actin, heat shock protein (HSP-70) and *Cryptosporidium* oocyst wall protein (COWP). Subtyping of *C. meleagridis* is based on 60-kDa glycoprotein (GP60) gene. Duplex-real-time PCR is recently developed for confirmation of *C. galli* and avian genotype III.
(c) *Serological tests*: ELISA can be used for detection of anti-*Cryptosporidium* antibodies in serum samples.
(d) *Antigen detection assays*: Capture enzyme-linked immunoassays (ELISA), direct fluorescent antibody (DFA) assay using commercially available antibodies are used for detection of Cryptosporidial antigen which is more sensitive and specific than staining. The detection of cryptopsordial species is not possible due to cross reactivity.

2.3.3.9 Zoonosis

Cryptosporidial oocysts are excreated in the faeces of infected birds and animals which can be transmitted to human by faeco-oral route. Human cryptosporidiosis is mostly caused by *C. hominis*, *C. parvum* and *C. meleagridis*. The pet birds and domestic chicken may act as source of *C. parvum* and *C. meleagridis* oocysts in the environment. Watery diarrhoea, abdominal cramping, and increased gas production are common clinical signs of *Cryptosporidium* infection in human. Severe complications such as pancreatitis, cholangitis, respiratory distress and death are observed in immunosuppressed individuals. Person to person transmission is possible.

2.3.3.10 Treatment and Control Strategy

Macrolide antibiotic (erythromycin) is used successfully against cryptosporidiosis in cliff swallow and owls. Azithromycin (40–45 mg/kg body weight, 24 h interval, oral) is recommended for treatment of avian cryptosporidiosis in most of the bird species. United States Food and Drug Administration (FDA) have approved nitazoxanide for use in humans suffering with cryptosporidiosis.

The cryptosporidial oocysts are highly resistant to common disinfectants used in aviaries and environmental stress such as temperature variation, desiccation, humidity and change of pH. Maintenance of strict hygiene and biosecurity in the aviaries to reduce the possibility of oocyst contamination is the control strategy for avian cryptosporidiosis.

2.3.4 Other Parasitic Infections

Other parasites affecting pet birds include protozoa, helminths (nematodes, cestodes, trematodes), and arthropods. Captive birds kept in small and dirty cages are more susceptible to parasitic infestations. Capability of a parasite to survive and reproduce in pet birds ('parasite fitness') depends on host immunity and nutrients for parasite available in the host body. The parasites may avoid the birds with good health condition due to strong immune response, but the parasites also avoid the birds with very poor health due to less possibility to obtain nutrients. The balance between immunity and nutrient availability is crucial for selection of host by the parasites.

2.3.4.1 *Sarcocystis falcatula* (Avian Sarcocysticosis, Sarcosporidiosis)

Sarcocystis belongs to the phylum Apicomplexa, suborder Eimeriorina and the family Sarcocystidae. It is a protozoon with two different hosts in their life cycle. Opossum (*Didelphis virginiana*) acts as definitive host in which infectious oocysts are generated and are excreated with faeces. Sporulated oocyst contains two sporocysts and each of them contains 4 sporozoites. Insects (flies, cockroaches), cowbirds (*Molothrus* sp.), and grackles (*Quiscalus* sp., *Hypopyrrhus* sp., *Lampropsar* sp., *Macroagelaius* sp.) act as natural intermediate hosts. Raptors (owls, eagles) may act as both definitive and intermediate hosts. The sporocysts after transmission into intermediate hosts (via ingestion) undergo schizogony or merogony and produce sarcocysts in the cardiac or striated muscles. When the definitive hosts further ingest the intermediate hosts with sarcocysts in the muscles, the life cycle is completed.

Cockatoos, cockatiels, grey parrots can get the protozoon by the ingestion of insects with sarcocysts. Three clinical forms such as acute pulmonary form, muscular form, and neurologic forms are observed in birds. In acute form, death without development of sarcocysts in the muscle occurs. In other two forms, clinical signs for instance dullness, weakness, respiratory signs, yellow urates are detected. Pulmonary hemorrhage and edema are the cause of death. Presence of sarcocysts in muscles of heart, breast, thigh, neck, and esophagus is the major lesion. Hepatosplenomegaly, lung consolidation, pulmonary edema are also observed.

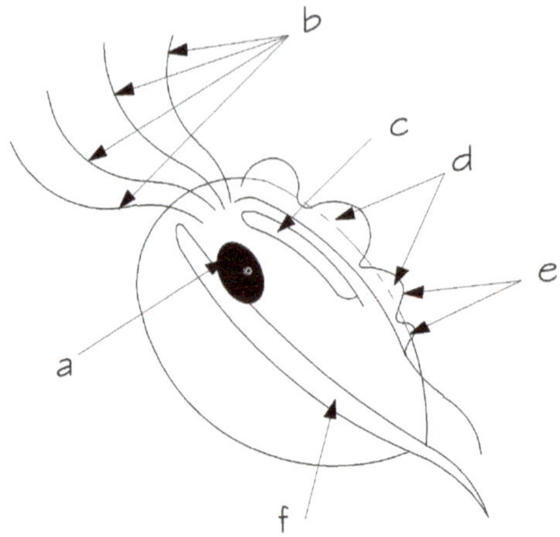

Fig. 2.34 Morphology of *Trichomonas gallinae* (schematic). *a* Nucleus *b* anterior flagella *c* parabasal body *d* undulating membrane *e* posterior flagella *f* axostyle

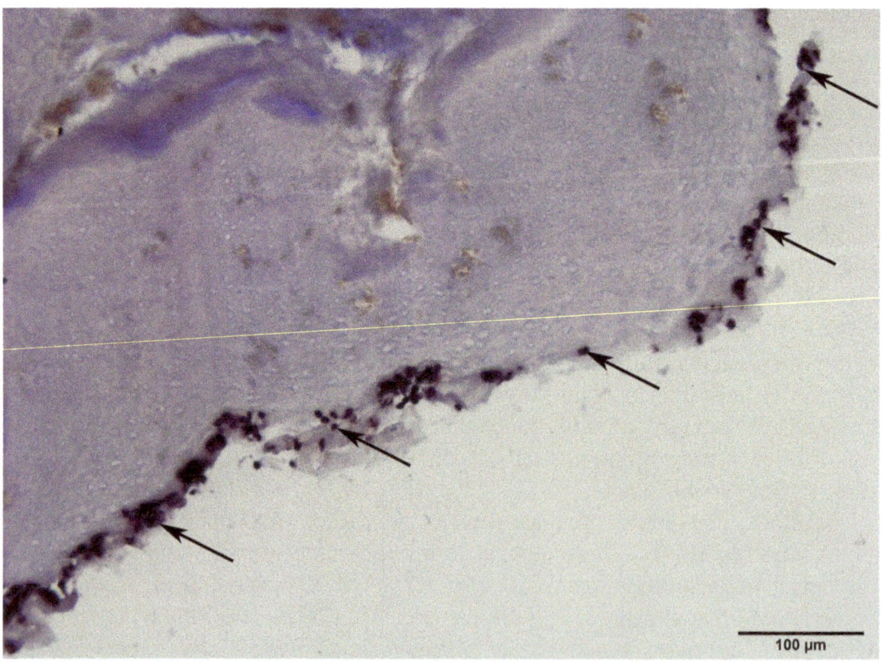

Fig. 2.35 Section of a pigeon crop infected with *Trichomonas gallinae* (Haematoxylin and eosin). Arrows indicate some of the parasites in the epithelium (*Courtesy* Clinic for Poultry and Fish Medicine, Department for Farm Animals and Veterinary Public Health, University of Veterinary Medicine Vienna, Austria)

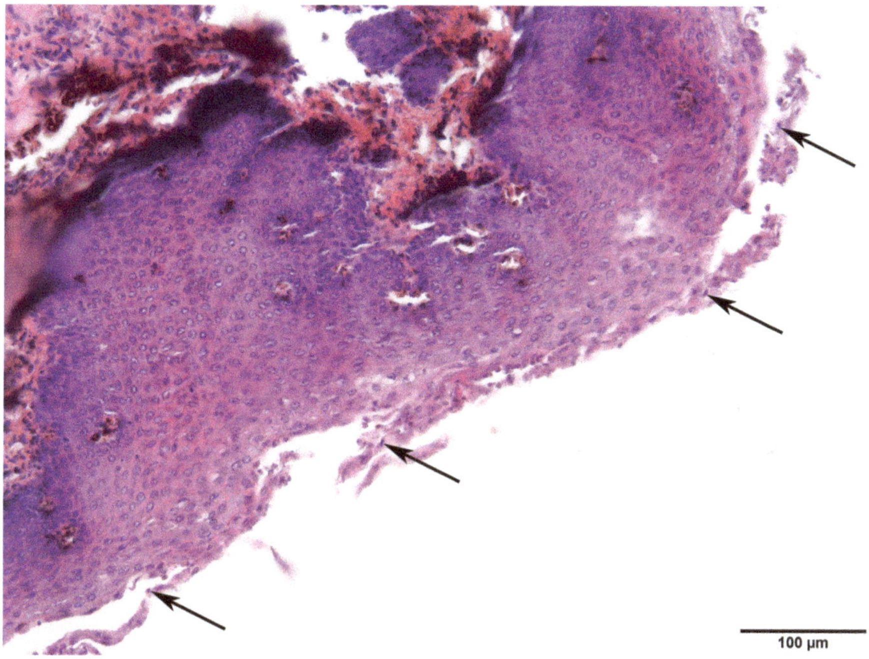

Fig. 2.36 In situ hybridization (ISH) confirmed the presence of *Trichomonas gallinae* in crop epithelium of a pigeon (*Courtesy* Clinic for Poultry and Fish Medicine, Department for Farm Animals and Veterinary Public Health, University of Veterinary Medicine Vienna, Austria)

Diagnosis is usually based on visualization of sarcocysts in the muscles after post-mortem. Ante-mortem diagnosis is possible by muscle biopsy, indirect immunofluorescent assay and PCR. Successful treatment of birds can be done with pyrimethamine (0.5 mg/kg body weight, 12 h interval for 14–28 days) and sulphadiazine-trimethoprim (30 mg/kg body weight, 8–12 h interval). Removal of insects from cages, maintenance of cleanliness, fencing of outdoor aviaries to prevent opossum access can prevent the occurrence of sarcocysticosis in birds. Avian sarcocystosis is usually not considered to be a public health hazard.

2.3.4.2 *Trichomonas gallinae* (Avian Trichomoniasis)

Trichomonas sp. is a motile pear shaped protozoon which contains a central, longitudinal rod up to the posterior end (axostyle) and four flagella attached to the anterior end. An undulating membrane extending from the anterior to posterior end is present (Fig. 2.34). The membrane encloses a flagellum which does not have any free end. *T. gallinae* has direct life cycle without any intermediate hosts or/and vectors. They multiply by longitudinal binary fission.

In budgerigars, finches, and cockatiels, transmission of *T. gallinae* occurs through ingestion of contaminated food and water. Young domestic pigeons (squabs) get the infection through ingestion of contaminated 'pigeon milk'. Adult pigeons act as carrier of *T. gallinae*.

After entry through the oral route *T. gallinae* can invade the mucosal surface of the buccal cavity, sinuses, pharynx, esophagus, crop, proventriculus and conjunctiva (rare invasion) depending on the species of affected birds (Figs. 2.35 and 2.36). Clinical signs include weight loss, hypersalivation, vomiting, and diarrhoea. Death due to starvation occurs with progression of the infection. Gross lesion includes formation of diphtheritic membrane or white plaques on gastrointestinal mucosa (oropharynx to esophagus), blockage of esophagus lumen with cheesy material and ingluvitis (inflammation of crop). Solid, white to yellow circular masses appears in liver of affected pigeons (Fig. 2.37).

Buccal cavity and crop can be collected as clinical specimens after post-mortem of the bird. Wet mount prepared from fresh crop wash with saline can be observed under microscope for detection of typical pear shaped *Trichomonas* sp. In live birds, wet mount prepared from fresh faeces or crop swabs can be observed under microscope for detection of *Trichomonas* sp. However, diagnosis of avian trichomoniasis is difficult in ante-mortem samples because the parasite is unstable in the environment due to lack of cyst formation capacity.

Metronidazole (10–20 mg/kg body weight, 12–24 h interval, for 2 days), ronidazole (6–10 mg/kg body weight, 24 h interval, for 6–10 days), carnidazole (100–200 mg/kg body weight, oral, once), dimetridazole [1 tea spoonful/gallon

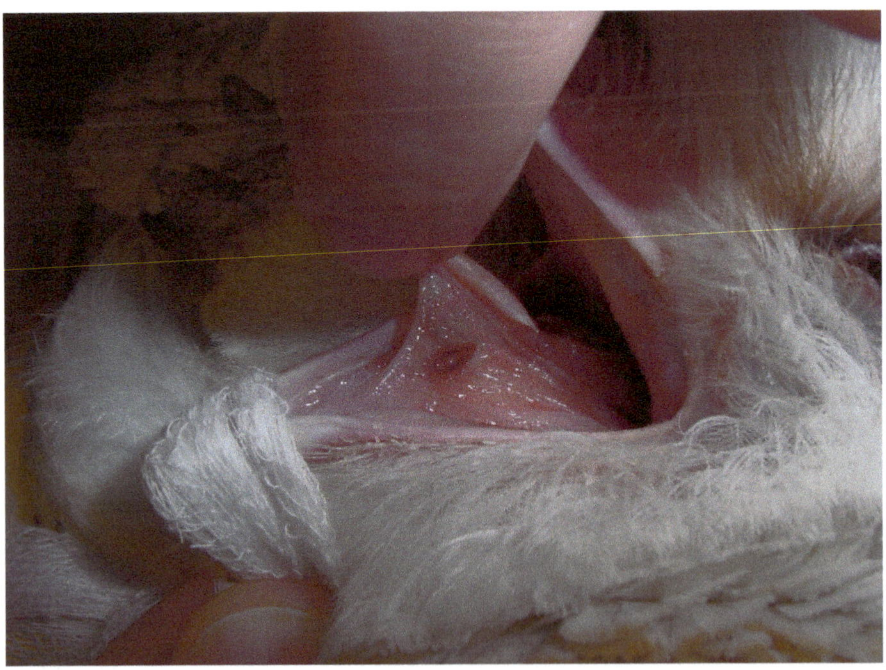

Fig. 2.37 Oral fluke in barn owl (*Courtesy* Dr. M. Scott Echols, The Medical Center for Birds, California)

(4.5 L) drinking water] and ipronidazole (500 mg/galon drinking water for 7–30 days) are recommended for avian trichomoniasis.

2.3.4.3 *Coccidia* (*Eimeria* spp., *Isospora* spp.)

Eimeria spp. and *Isospora* spp. belongs to phylum Apicomplexa and they have a direct life cycle without any intermediate host or vector. Sporulated oocysts are the infectious form and ingestion of food and water contaminated with oocysts is the major way of transmission. The oocyst wall is crushed in gizzard and the sporozoites are released. The sporozoites enter intestinal mucosa and undergo schizogony or merogony to produce progeny oocysts. The new oocysts leave intestinal mucosa and are excreated through the faeces. The direct life cycle requires 6–8 days time to complete.

Coccidiosis is common in mynahs, toucans, pigeons, lories, finches, budgerigars and canaries. Clinical symptoms do not appear until the birds are immunosuppressed or infected with enormous numbers of coccidial oocysts. During schizogony in intestine, mucosa is damaged and enteritis is produced. Lethargy, weight loss, severe haemorrhagic diarrhoea is detected as clinical symptoms within 4–6 days after infection. Nonsporulated oocysts are shed with the faeces.

Diagnosis is made through examination of faeces and smears taken from the suspected lesion. The oocysts or macrogametes are detected in the smears in positive cases. The oocysts of *Eimeria* contain four sporocysts each with two sporozoites, whereas, two sporocysts each with four sporozoites are present in oocysts of *Isospora*.

Treatment with amprolium (50–100 mg/l drinking water for 5–7 days), trimethoprim and sulphamethoxazole (25 mg/kg body weight, 24 h interval, oral) are recommended for coccidiosis in pet birds. Amprolium produces toxicity in falcons (@22 mg/kg body weight, 24 h interval) and trimethoprim-sulphamethoxazole is recommended for toucans and mynahs.

2.3.4.4 Atoxoplasmosis

Atoxoplasmosis is a dreadful disease caused by the protozoa *Atoxoplasma* sp. causing significant mortality among the fledgling birds. Although atoxoplasmosis is mostly asymptomatic among the adult birds, fatal infection with hepato-spleenomegaly may be noticed among the young birds of canaries and Passeriformes.

The infected birds usually shed the oocysts via faeces and the healthy one picks up the infection by consuming the oocysts. Asymptomatic adults may also shed such oocysts and can be a potent but silent source of infection for the in-contact and prone young birds. Interestingly, the *Atoxoplasma* sp. from one host may not be infectious to other species of birds exhibiting certain degree of host specificity. Infected birds may periodically shed large number of oocysts. In general, the birds were detected to shed the oocysts up to eight months of infection. The oocysts are environmentally stable and can persist for along period.

Birds of less than 1 year of age, usually exhibit clinical symptom and adults are usually asymptomatic. Clinical findings are usually non-specific and include anorexia, depression, weight-loss, chronic diarrhoea, lethargy and depression. In few cases mortality may reach up to 80%. The enlarged liver, spleen and dilated intestinal loops can be palpated in few cases. During the period of parasitaemia, the protozoa *A serini* undergoes scizogony in the polymorphomononuclear cells (PMNs) and spread throughout the body circulation. Thereafter, the parasites reach out to their site of predilection reticuloendothelial cells (RE cells) of vital internal organs like liver, spleen, pancreas and intestinal epithelial cells. During this process of multiplication and propagation, the parasite causes heavy damage to these organs resulting in hepatospleenomegaly and enteritis in young birds.

When any adult bird is suspected as an asymptomatic carrier, microscopic examination of the fecal samples may confirm the presence of the oocysts (20.1 × 19.2 μm). In the height of severe infection, zoite form of the parasite may be detected in the lymphocytes following processing the buffy coat smear with Romanowsky stain. Reddish intracytoplasmic inclusion bodies are detected in the PMN cells when stained with Giemsa. Impression smear prepared from the enlarged liver, spleen or other affected organ may be helpful. PCR or nucleic acid detection method may give more confirmatory diagnosis.

There is no effective therapy available for the disease. The birds may remain infected and continues to show periodic clinical symptoms even after therapy up to 4 months. Primaquine has been recommended to suppress the tissue form of the parasite where as sulfachlorpyridazine can be used to decrease the faecal shedding of the oocysts. Generally, toltrazuril is employed @ 12.5 mg/kg/day, orally for 14 days and sulfachlorpyridazine @ of 150–300 mg in one liter of drinking water. This sulfachlorpyridazine should be given for 5 days in a week for at least 2–3 weeks.

2.3.4.5 Ascaridiasis (Roundworm Infestation)

Ascaridia hermaphrodita, *A. columbae*, *A. nymphii* and *A. galli* have detected in hyacinth macaw, pigeon, Australian parrots, budgerigar, cockatiel, and princess parrot. In some ground feeding species of birds (e.g. dove), *Baylisascaris procyonis* (raccoon roundworm) and *Baylisascaris columnaris* (skunk roundworm) infestation is observed due to occasional consumption of food, water and faeces contaminated with *Baylisascaris* eggs. Ingestion is the major route of transmission in pet birds for other Ascarids also.

Clinical signs include lethargy, diarrhoea and death. Engorgement of duodenal loop with roundworms is major necropsy finding. In *Baylisascaris procyonis* infestation, considerable damage of central nervous system and subsequent ataxia, depression, head tilt, stumbling, recumbancy, torticollis, wing paralysis and death is common (Fig. 2.38).

Detection of thick shelled Ascarid eggs in faecal sample is the major diagnostic approach. During necropsy, presence of roundworm in the intestine and larval stage in tissue section can aid in diagnosis. Confirmation can be done by PCR targeting ribosomal DNA, internal spacers (ITS-1, ITS-2), and mitochondrial gene (mtDNA-cox2) of Ascarids.

Treatment with fenbendazole (20–100 mg/kg body weight, once) is recommended for ascaridiasis in most of the birds except finches, marabou storks and vultures due to acute toxicity. The drug is contraindicated during growth of feathers

Fig. 2.38 Sternal recumbancy of a lovebird (*Courtesy* LafeberVet)

also. Other common anthelmentics such as piperazine dihydrochloride and mebendazole are also contraindicated in psittacines, pigeons, cormorants, pelicans, raptors, finches and ostriches due to toxicity and reported death. Regular deworming of birds kept in crowded and unclean cages is needed to avoid ascaridiasis.

Other partasitic infections of pet birds are described in Table 2.15 (Figs. 2.39, 2.40, 2.41, 2.42, 2.43 and 2.44).

2.4 Fungal Diseases

2.4.1 Cryptococcosis

2.4.1.1 History

Cryptococcus was first isolated from peach juice in Italy and it was named as *Saccharomyces neoformans* (Sanfelice 1894) and subsequently, it was also isolated from a German patient (Busse 1894). Two years later, an encapsulated bacilliform yeast, which was named *Saccharomyces subcutaneous tumefaciens*, was detected from an apparently healthy man in France, later identified as *Cryptococcus gattii* (Curtis 1896). Vuillemin (1901) examined several of these cultures and due to lack of *Saccharomyces* specific characteristics he placed these species in the genus *Cryptococcus*. In 1905, von Hansemann reported the first case of meningoencephalic cryptococcosis. In 1951, Emmons isolated *C. neoformans* from pigeon nests and their droppings. Disseminated cryptococcosis in a macaw was described later (Clipsham and Britt 1983).

Table 2.15 Other important parasitic infection of pet birds

Parasite	Susceptible hosts	Clinical signs/gross lesions	Treatment
Thelazia spp.	Parrots, cockatoos, other companion birds	Eye spasm, hyperaemia, eye itching and irritation, conjunctivitis, chemosis	The worms can be manually removed after they are incapacitated with 0.125% demecarium bromide
Ceratospira spp.			Ivermectin (0.2 mg/kg, PO, SC, or IM, repeated in 10–14 days) may be used to kill the worms followed by manual removal or flushing
Capillaria	Budgerigars, canaries, macaws, pigeons etc.	Anorexia, chronic haemorrhagic diarrhoea, weight loss, regurgigation, anemia	–
Encephalitozoon hellum (protozoa)	Peach-faced, masked and Fischer's lovebirds (*Agapornis roseicollis*, *A. personata*, *A. fischeri*), European goldfinch (*Carduelis carduelis*), canary (*Serinus canaria*), budgerigar (*Melopsittacus undulatus*), eclectus parrots (*Eclectus roratus*), double yellow-headed Amazon parrots (*Amazona ochrocephala*), yellow-streaked lory	Enteritis, hepatitis, nephritis, keratoconjunctivitis, sinusitis, and lower respiratory tract infections. Infection is associated concurrent viral or bacterial infection	Fenbendazole (20–100 mg/kg body weight, once). It is toxic for finches, marabou storks and some vultures. The anthelmentic should not be used when birds are actively growing feathers
Cochlosoma spp. (protozoa)	Cockatiel (*Nymphicus hollandicus*), finch (subclinical infection in adults)	Diarrhoea, weight loss, pruritis and feather picking, respiratory signs	Metronidazole, ronidazole
Spironucleus meleagridis (protozoa)	Australian king parrots (*Alisterus scapularis*, cockatiel (*Nymphicus hollandicus*), splendid grass parakeet (*Neophema splendid*)	Chronic diarrhoea, weight loss and death.	Nitroimidazoles (ronidazole, carnidazole)
Haemoproteus spp. (haemoprotozoa)	Cockatoos (*Cacatuidae*), parrot (carrier). It is transmitted by flies (*Hippoboscid*) and midges (*Culicoides*)	Pathogenicity uncertain. Splenomegaly, hepatomegaly, pulmonary edema are detected in heavily infested birds	Chloroquine and primaquine (not used in asymptomatic birds)

(continued)

Table 2.15 (continued)

Parasite	Susceptible hosts	Clinical signs/gross lesions	Treatment
Plasmodium spp. (Avian malaria)	Passerine birds are the definitive host and act as reservoir. It is transmitted by *Culex* and *Aedes* mosquitoes to other birds. Documented in canaries, waterfowl, raptors, pigeons and parrots	Depression, anorexia, vomiting, dyspnoea, hemoglobinuria, bright green faeces due to increased biliverdin, pale mucus membranes, conjunctival edema and death	Chloroquine and primaquine (Therapeutic use: 25 mg/kg body weight, followed by 15 mg/kg body weight at 12, 24, and 48 h in conjunction with 0.75–1.0 mg/kg body weight primaquine at 0 h)
			Quinacrin HCl (@7.5 mg/kg body weight, 24 h interval for 7–10 days)
			(Prophylactic use: weekly dosing during the season when mosquito population increases)
			Pyrimethamine (0.5 mg/kg body weight, 12 h interval for 14–28 days)
			Azithromycin (45 mg/kg body weight, 24 h interval)
Leucocytozoon (protozoa)	Waterfowl, turkeys, young raptors and some passerines. Relatively uncommon in parrots. Black flies (family Simuliidae) act as vector	Anorexia, depression, dehydration, hemolytic anemia and associated hemoglobinuria	Chloroquine and primaquine
Trematoda (Fluke)	Cockatoos, owls, falcons, Gouldian finches.	Anorexia, Depression, anaemia, diarrhoea, hepatomegaly, increased firmness of liver, striation and mottling of liver. Flukes are present in any location such as blood vessel, mouth etc.	Fenbendazole (20–50 mg/kg body weight, 24 h interval for 3 days)
			Praziquantel (10–20 mg/kg body weight, oral, repeat after 10–14 days). For intramuscular injection: 9 mg/kg body weight, 24 h interval for 3 days, then oral administration for 11 days. It is toxic for finches in higher dosage.
			Chlorsulon (20 mg/kg body weight, oral, three times, 2 weeks apart)
	Birds get the infection by ingestion of insects or mollusc (second intermediate host)		Mebendazole (25 mg/kg body weight, 24 h interval for 5 days)

(continued)

Table 2.15 (continued)

Parasite	Susceptible hosts	Clinical signs/gross lesions	Treatment
Capillaria (nematode, hairworm or threadworm)	Parrots, macaws, budgerigars, canaries, pigeons, and raptors. Direct or indirect transmission takes place. Insects and earthworms act as intermediate hosts	Anorexia, dysphagia, diarrhoea, weight loss, hyperemic streaks and diphtheritic lesions in organs	Anthelmentics are recommended
Raillietaenia, Choanataenia, Gastronemia, Idiogenes, Amoebataenia (tapeworm)	Finches, African grays, cockatoos, and eclectus parrots. Tapeworms have indirect life cycles and require an intermediate host (grasshoppers, beetles, ants, horse flies)	Anorexia, weight loss, diarrhoea. Tapeworm segments are observed in droppings or hanging from the vent	Praziquantel
Knemidocoptes pilae (scaly face and leg mite)	Budgerigars, New Zealand parakeets, grass parrots, Polytelis parrots	Crusty lesions on beak, cere, face, legs, margins of vent and wing tips	Ivermectin (0.2 mg/kg body weight, repeat after14 day. Frequent use can cause ataxia and depression. In budgerigars, finches, kingfishers and woodpeckers, very small dose (0.02 ml) may be toxic. In small birds one drop of the drug is directly applied on the lesion) Moxidectin
Dermanyssus gallinae (red or roost mite)	Reported in canaries. Common in birds having direct contact with wild birds	These mites feed on the birds during night and leave them at morning. Skin irritation, anaemia in young birds is observed.	Ivermectin, Moxidectin, carbaryl dusting powder (toxicity in higher dose)
Cytodites nudus, Sternostoma tracheacolum (air sac mite)	Canaries, Gouldian finches, parrots, budgerigars, cockatiels	Lethargy, change of voice, production of sucking or clicking noise due to irritation caused by the mite. Visualization of mites in the trachea with powerful light source, or detection of eggs in faeces or tracheal wash can diagnose it	Ivermectin with antibiotic (in Gouldian finches, a drop of ivermectin is applied onto the skin between the scapulae to avoid toxicity)

(continued)

Table 2.15 (continued)

Parasite	Susceptible hosts	Clinical signs/gross lesions	Treatment
Neopsittaconirmus spp., *Psittaconirmus* spp., *Eomenopon* spp., *Pacifimenopon* spp., *Ciconiphilus* spp., *Menacanthus* spp. (Chewing lice)	Budgerigars, cockatiels, lovebirds, raptors, falcon, cattle egret, snowy egret, rusty-marginated guan, Indian peafowl,	The entire life cycle of chewing lice (egg, three nymph, adult) is developed on a single host. Severe bite-induced pruritus, anorexia, weight loss are observed. Self-mutilation activities of the bird increase the risk of secondary bacterial infection	Carbaryl dusting powder, pyrethrin sprays
Hippoboscid fly (Louse fly/parrot fly)	Raptors, pigeon, wild parrots, red crowned parakeets, ostriches, swift, swallows	Blood sucking parasite which lays eggs on the host body (abdomen). Non-pathogenic but it can transmit blood parasite (*Haemoproteus* spp.). Clinical myiasis and anaemia is detected in young birds. It can be transmitted into human handlers also	Carbaryl dusting powder, pyrethrin sprays
Echidnophaga gallinacea (Stickfast flea)	Psittacines, raptors, pigeons	Stickfast flea attaches firmly around the head and ear of the bird. In severe cases hyperkeratinisation, irritation and anaemia may occur	Pyrethrin sprays

2.4.1.2 Etiology

Cryptococcus belongs to the Filobasidiella clade of the Tremellales, under the order Tremellomycetes, phylum Basidiomycota. The genus *Cryptococcus* includes over 37 species majority of which do not cause any infection in mammals. Important pathogenic species are *C. neoformans*-*C. gattii* species complex, which includes *C. neoformans* var. *neoformans*, *C. neoformans* var. *grubii*, and *C. gattii* (*C. bacillisporus*). Among them, *C. neoformans* var. *neoformans* and *C. gattii* are considered as primary pathogens in immunocompetent avian hosts. Other cryptococcal species such as *C. uniguttulatus*, *C. albidus* and *C. laurentii* are occasionally detected in droppings and crops of pigeons and psittacine birds although clinical significance is uncertain.

Cryptococcus life cycle is predominantly divided into two phases i.e. vegetative and sexual growth phase. Two major morphological forms (yeast and pseudohyphae) exist in the vegetative growth phase. The predominant form found in the environment and avian and animal hosts is unicellular budding yeast. The yeast

Fig. 2.39 Bird feather infected with lice (*Courtesy* Dr. Sychra Oldřich, University of Veterinary and Pharmaceutical Sciences Brno, Czech Republic)

Fig. 2.40 Layer birds infested with *Dermanyssus gallinae* (*Courtesy* Prof. Elena Circella, Avian Diseases Unit, Department of Veterinary Medicine, University of Bari, Italy)

Fig. 2.41 Canaries infested with *Dermanyssus gallinae* (*Courtesy* Prof. Elena Circella, Avian Diseases Unit, Department of Veterinary Medicine, University of Bari, Italy)

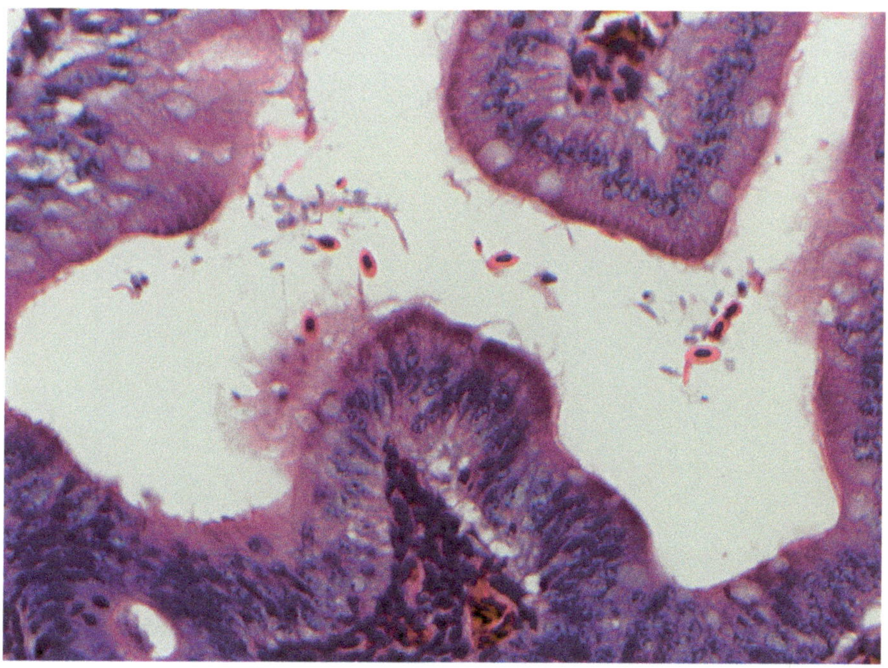

Fig. 2.42 Section of a female lutino cockatiel (*Nymphicus hollandicus*) infested with *Spironucleus meleagridis* (*Courtesy* Dr. Lauren V Powers, Carolina Veterinary Specialists, United States; Prof. John Barnes, NC State University, North Carolina, United States)

Fig. 2.43 Budgerigar infected with mite (*Courtesy* Prof. Elena Circella, Avian Diseases Unit, Department of Veterinary Medicine, University of Bari, Italy)

Fig. 2.44 Bird infected with scaly leg mite (*Knemidokoptes*) infestation of a bird (*Courtesy* Petra Maria Burgmann, Canada)

cells are thin walled, spherical to oval with varying diameter (2–20 μm) and they reproduce by mitotic division. The buds are present at the narrow base (Fig. 2.45).

The morphological transition from the yeast phase to hyphal phase (pseudohyphae) is noticed during sexual mating. They are not considered as true dimorphic fungi probably due to their predominant existence as yeast form in the environment and hosts and the lack of involvement of this transition in the infection process. Further, both at 25 and 37 °C they can produce yeast like colonies in the isolation media. Recently unusually large yeast like morphological form (30–100 μm) is also detected in clinical samples, known as 'giant' or 'titan' cells.

Another unique morphological feature of *Cryptococcus* is the presence of capsule. The capsule can be best observed in fresh preparations by staining with diluted India ink or phase contrast microscopy (Fig. 2.46). Giemsa can also partially stain

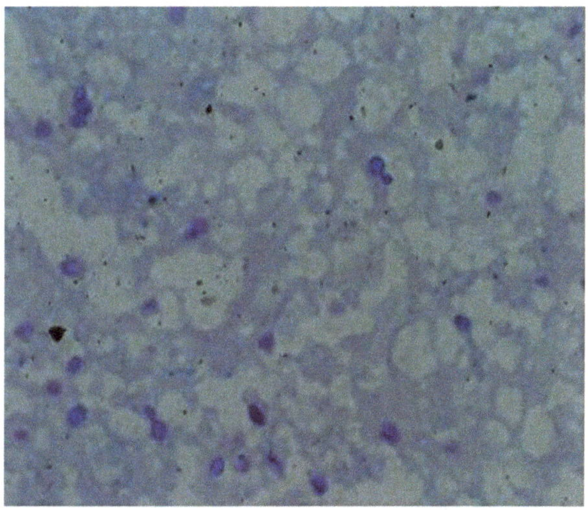

Fig. 2.45 Budding *Cryptococcus* spp. (*Courtesy* Prof. P.P. Gupta, Ex-Director of Veterinary Research, Punjab Agricultural University, Punjab, India)

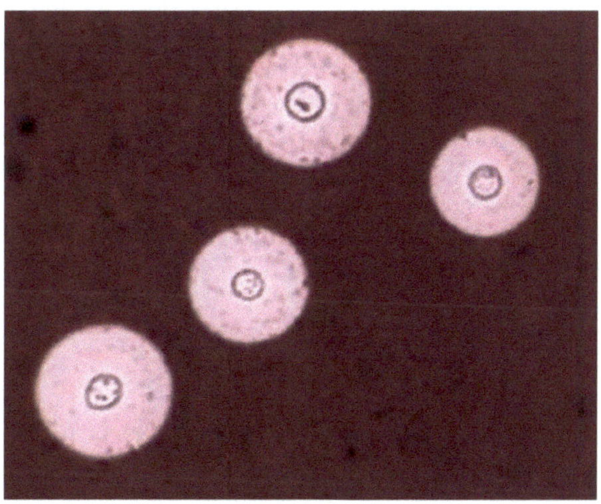

Fig. 2.46 Encapsulated *Cryptococcus* spp. in India ink preparation, 400× (*Courtesy* Prof. P.P. Gupta, Ex-Director of Veterinary Research, Punjab Agricultural University, Punjab, India)

the capsule. Like other eukaryotic organism *Cryptococcus* also possesses mitochondria which serves as source of energy, and is involved in processes of aging, calcium homeostasis, apoptosis, and regulation of virulence.

2.4.1.3 Host Susceptibility

Presence of *C. neoformans* in the faeces of different avian species except raptors (Table 2.16.) is a saprobiotic phenomenon. Avian faeces rich in creatine, urea, uric acid and protected from sunlight and ultraviolet light, high flock density and poor sanitary conditions create a microenvironment for *C. neoformans*. Pigeons as mechanical carrier of *C. neoformans* in their feathers, feet and crop are mostly studied avian species.

Description of clinical cryptococcosis in birds is relatively rare although reported in pigeons, kiwis, major Mitchell's cockatoo, moluccan cockatoos, thick-billed parrot, African grey parrot, green-winged macaw, Papua lories, blackcapped lories, Goldie's lorikeet, and ring necked parrot.

2.4.1.4 Transmission

Cryptococcus is transmitted primarily through inhalation route in birds, animals and human. The basidiospores are major infectious particles, small in size (2–3 μm) which can easily invade the lung alveoli than the encapsulated yeast (10–60 μm). Rarely ingestion of large number of organisms may cause the infection as observed during *C. gattii* infection of psittacine birds in Brazil. Psittacine birds have the habit of chewing wooden objects, such as the perches made of *Eucalyptus* spp. In tropical countries, *C. gattii* is commonly associated with *Eucalyptus* trees.

2.4.1.5 Pathogenesis

The basidiospores lodge in the lung alveoli of the birds after inhalation. The capsule of the fungi helps in evasion of immune system and survival within the host. The capsule is constituted with mannan (polysaccharide) which is highly hydrophilic. It makes a gelatinous zone surrounding the yeasts that conceals the pattern recognition receptors (PRR) of the yeast from the immune system. The capsule also prevents antibody binding and phagocytosis of the yeasts. *C. gattii* can produce extracellular fibrils which can prevent the phagocytosis by neutrophils and help to establish the primary pulmonary infection.

After primary colonization in upper respiratory tract, haematogenous spread of yeast into different organs such as heart, liver, spleen, intestine, kidneys, and central nervous system is observed in different psittacine birds. Occasionally, coelomic dissemination occurs between organs within close proximity. Superficial colonization of yeast is detected in choanas, sinus, upper beak, and infraorbital sinus of African grey parrot, Goldie's lorikeet and Beccaris's crowned pigeon. Minimal inflammatory response with epithelioid macrophages, multinucleated giant cells and heterophils or absence of inflammatory response is detected in different organs of psittacine birds.

Table 2.16 *Cryptococcus* spp. detected in avian hosts in different countries

Cryptococcus spp.	Avian hosts	Country
C. neoformans var. *grubii* molecular type VNI	Budgerigars (*Melopsittacus undulatus*)	Brazil, Germany
C. neoformans, *C. gattii*	White eyed parakeet (*Aratinga leucophthalmus*), Peach-fronted parakeet (*Aratinga aurea*), Jandaya parakeet (*Aratinga jandaya*), Scaly-headed parrot (*Pionus maximiliani*), Cockatiel (*Nymphicus hollandicus*), Alexandrine parakeet (*Psittacula eupatria*), Nanday parakeet (*Nandayus nenday*), Festive amazon (*Amazona festiva*), Red-browed amazon (*Amazona rhodocorytha*), Mealy amazon (*Amazona farinose*)	Brazil
C. neoformans	Budgerigars (*Melopsittacus undulates*), monk parakeet (*Myiopsitta monachus*)	Brazil
C. neoformans	Sun parakeet (*Aratinga solstitialis*), blue fronted amazon (*Amazona aestiva*)	Brazil
C. neoformans	Red-cowled cardinal(*Paroaria dominicana*), yellow canary (*Serinus canarius*, *Serinus flaviventris*), saffron finch (*Sicalis flaveola*), double-collared seedeater (*Sporophila caerulescens*)	Brazil
C. neoformans	Pigeons (*Columba livia*)	United States
C. neoformans	Kiwis (*Apteryx australis mantelli*)	New Zealand, Australia
C. neoformans var. gattii	Major Mitchell's Cockatoo (*Cacatua leadbeateri*)	Australia
C. neoformans (serotype AD)	Exotic birds	Chile
C. neoformans var. *neoformans*	White face duck (*Dendrocygna viduata*), eagle owl (*Bubo africanus cinerascene*), peacock (*Pavo cristatus*), spotted eagle owl (*Bubo africanus*)	Nigeria
C. uniguttulatus, *C. laurentii*	Pigeons (*Columba livia*)	Sweden
C. uniguttulatus	Slender billed parakeet (*Enicognathus leptorhynchus*), bluecheeked rosella (*Platycercus adscitus*)	Chile
C. albidus	Bluefronted Amazon parrot (*Amazona aestiva*)	Chile
C. neoformans var. *gattii* (serovar B)	Papua lori (*Charmosyna papou*), blackcapped lories (*Lorius lory*), Goldie's lorikeet (*Trichoglossus goldiei*), ring necked parrot (*Psittacula krameri*), African grey parrot (*Psittacus erithacus*)	Brazil

2.4.1.6 Clinical Symptoms

Clinical cryptococcosis causes respiratory distress, lethargy, emaciation, diarrhoea, anaemia, incoordination, progressive paralysis and development of gelatinous mass in choana, sinus, upper beak, and infraorbital sinus in birds (Figs. 2.47 and 2.48). In psittacines, ocular infections causing conjuntival, corneal and intraocular lesions

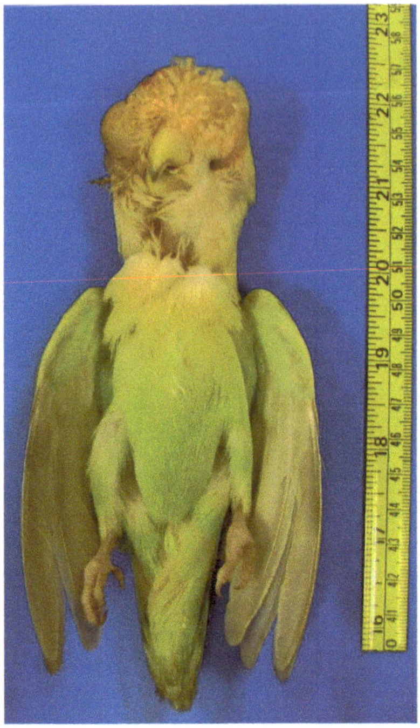

Fig. 2.47 A soft palpable swelling at the head of a lovebird suffering with cryptococcosis (*Courtesy* Prof. Adjunta Karin Werther, Universidade Estadual Paulista, Brazil)

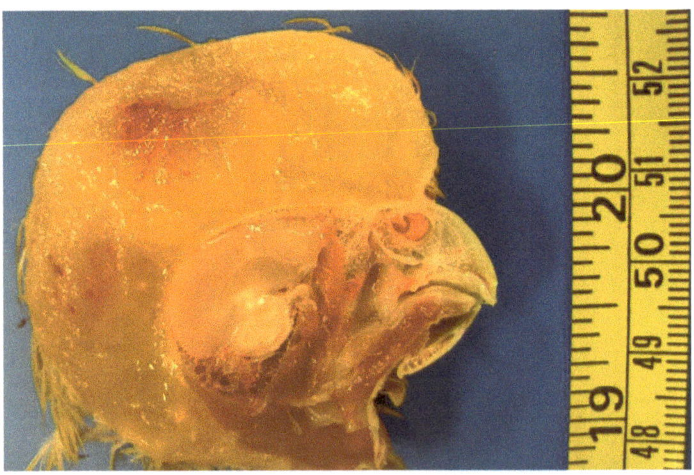

Fig. 2.48 Presence of a gelatinous mass at subcutaneous space of the infected lovebird's head (*Courtesy* Prof. Adjunta Karin Werther, Universidade Estadual Paulista, Brazil)

and blindness are also detected. In domestic pigeons (*Columba livia*), rare report of clinical cases revealed the central nervous system signs, weight loss, dyspnoea, and infraorbital sinus mass. Development of thick crust over the right naris is observed in Major Mitchell's cockatoo suffering with clinical cryptococcosis.

2.4.1.7 Lesion

Gelatinous myxomatous mass (granulomata) in respiratory tract, abdominal cavity, sinuses and brain is the most common lesion observed in birds. In psittacines, hepatomegaly, multifocal hepatitis, yellow areas in the capsule and parenchyma of liver, congestion of lung, replacement of pulmonary parenchyma with yellow gelatinous material, thickened air sacs, sinuses filled with yellow coloured substance, nephromegaly, tubular and glomerular degeneration, splenomegaly, congested intestine with black coloured material in lumen are observed.

2.4.1.8 Diagnosis

Clinical Specimens

Cloacal/choanal swabs, dry faeces, blood/serum, aspiration biopsy of the gelatinous mass present in upper beak, infraorbital sinus and choana, visceras collected in 10% formalin after post mortem are considered as clinical specimens of avian cryptococcosis. Dry faecal samples (2.5 gm) are suspended in 15 ml sodium chloride solution in tubes containing chloramphenicol (0.1 mg/ml).

Diagnostic Techniques

(a) *Direct examination*: The smear can be prepared from the aspiration biopsy samples and is stained with India ink, Nigrosin, or Romanowski for demonstration of capsule. The Romanowski stain produces clearer capsule against the lightly stained background. The tissue samples can be stained with Periodic Acid Schiff base (PAS)—Haematoxylin stain which will outline the yeast cell and the capsule will appear as clear zone surrounding the cell (Fig. 2.49).

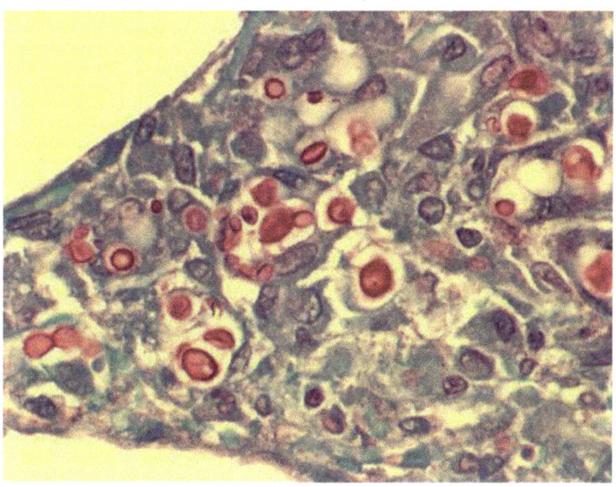

Fig. 2.49 *Cryptococcus* spp. in PAS stained smear (*Courtesy* Prof. P.P. Gupta, Ex-Director of Veterinary Research, Punjab Agricultural University, Punjab, India)

In Mayer's mucicarmine stain, the capsule and cell wall appears as red. Another characteristic feature of *Cryptococcus* is narrow base budding (Fig. 2.45) in comparison to other yeasts (*Blastomyces*) having broad based budding. Sometimes false positive results are produced due to confusion with globules of myelin, lysed cells, lymphocytes and dead yeasts after successful treatment.

(b) *Isolation of Cryptococcus* spp. *from clinical samples*: Faecal suspension (dry faeces mixed with sodium chloride and chloramphenicol) of the suspected birds can be inoculated into corn meal agar, Sabouraud's Dextrose agar, blood agar, honey agar, brain heart infusion agar and malt agar. The specific medium is birdseed agar/Staib medium (Niger seed agar)/sunflower seed extract agar with antibiotics. The plates are incubated at 28–37 °C for 2 days to 2 weeks. The colonies are initially small, convex, mucoid, creamy in colour, increases in diameter up to several centimeters after prolonged incubation. In birdseed agar the colonies appear as brown coloured in the center of the plate due to the production of melanin by the action of phenyl oxidase. The optimum capsule production is detected in chocolate agar after incubation at 37 °C with 5% CO_2 tension. The isolates are confirmed by API system or different biochemical tests such as urea hydrolysis, assimilation of inositol and creatinine, non-fermentation of carbohydrates, and melanin production. The L-Canavanine glycine bromothymol blue media can differentiate *C. neoformans* and *C. gattii* by the formation of distinctive blue coloration with the growth of *C. gattii*.

(c) *Detection of Cryptococcal antigen*: Detection of Cryptococcal antigen in clinical specimens can be performed by latex particles coated with polyclonal serum and antigen-capture ELISA. The detection of antigen is possible from both live and dead organisms. During initial phase of therapy, disintegration of the yeast cells releases the capsule which produces high titer. So, the tests should not be done within 6–8 weeks after initiation of the therapy. The titer can be observed even after successful treatment as the dead organisms also have the intact capsule.

(d) *Molecular biology*: Multiplex PCR and mating type-PCR are used for determination of molecular and mating types of *C. neoformans* var. *neoformans* and *C. gattii* isolates.

2.4.1.9 Zoonosis

In human the patients with suppressed immunity (suffering from AIDS, lymphoma, haematologic malignancy and using corticosteroids for prolonged period) are mostly susceptible to the cryptococcal infection. It causes meningitis, meningoencephalitis and death in human. Inhalation of cryptococcal basidiospores is the major way of transmission in human. Cryptococcal infection in patients after exposure to the birds is observed. It was first reported from United States in a renal transplanted patient who developed cryptococcal meningitis transmitted from her pet cockatoo (Nosanchuk et al. 2000). Report of cryptococcal meningitis is also documented in immunocompetent patient who acquired the infection from her pet magpie, although, her contact with the bird was limited to passing by the cage when entering home (Lagrou et al. 2005).

2.4.1.10 Treatment and Control Strategy

Fluconazole (5–15 mg/kg body weight, oral, 12 h interval for 15–60 days) and itraconazole (10 mg/kg body weight, oral, 24 h interval for 15–90 days with food) are recommended for treatment of avian cryptococcosis in pet birds. Use of itraconazole is contraindicated in African grey parrots.

Proper ventilation, regular cleaning of droppings and organic debris from the cages can prevent the occurrence of avian cryptococcosis. In tropical countries, wooden perches made of *Eucalyptus* tree should be avoided because psittacine birds like to chew the perches.

2.4.2 Aspergillosis

2.4.2.1 History

In 1729, Micheli first described *Aspergillus* who found the similarity between the spore chain of the fungi with the brush ('Aspergillium') used for sprinkling holy water in the churches. Later in 1842, pathogenic *Aspergillus* (*A. candidus*) was detected in air sac lesion of a bullfinch by Rayer and Montagene. Whereas, *A. fumigatus* was first detected in lung of a great bustard (*Otis tardaga*) in 1863 by Fresenius, who was the first to use the term 'aspergillosis' for this respiratory infection. The major members of aflatoxins were first detected and its association with *Aspergillus flavus* was established in 1961 during investigation of mysterious 'turkey-X disease' causing high mortality in turkey poults (Sargeant et al. 1961). In 1962, the name 'aflatoxin', using first letter from 'Aspergillus' and the first 3 letters from 'flavus' was proposed.

2.4.2.2 Etiology

Aspergillus is the fungus which belongs to Trichocomaceae family under the order Eurotiales and phylum Ascomycota. The genus *Aspergillus* contains more than 250 species grouped into sub genera and species complex. There are eight such sub genera i.e. *Aspergillus, Fumigati, Circumdati, Candidi, Terrei, Nidulantes, Warcupi,* and *Ornati*. The important species complexes are *A. fumigatus, A. flavus, A. terreus, A. niger, A. nidulans,* and *A. ustus*. Among them, *A. fumigatus* is the most common cause of avian aspergillosis. Other species complexes such as *A. flavus, A. terreus, A. niger,* and *A. nidulans* are occasionally associated with avian aspergillosis. *A. fumigatus* is the major cause probably due to smaller size of the spores than other species which helps in easy transmission.

2.4.2.3 Host Susceptibility

All species of domestic, pet and wild birds including chicken, duck, quail, and geese are susceptible to avian aspergillosis due to anatomic and physiologic characteristics

Table 2.17 *Aspergillus* spp. detected in avian hosts in different countries

Aspergillus spp.	Avian hosts	Country
Aspergillus spp.	Goliath Heron (*Ardea goliath*), Great horned owl (*Bubo virginianus*), Rhea (*Rhea Americana*), palm cockatoo (*Probosciger aterrimus*), African grey parrot (*Psittacus erithacus*), Moluccan cockatoo (*Cacatua moluccensis*), cape parrot (*Poicephalus robustus*), yellow naped Amazon parrot (*Amazona ochrocephala*), quaker parrot (*Myiopsitta monachus*), blue and gold macaw (*Ara ararauna*), eclectus parrot (*Eclectus roratus*), harlequin macaw (*Ara ararauna, Ara chloroptera*), blue-fronted parrot (*Amazona aestiva*)	United States, Brazil
A. fumigatus	Pigeons (*Columba livia*), common peafowl (*Pavo cristatus*), bearded vulture (*Gypaetos barbatus*), Eurasian black vultures (*Aegypius monachus Linnaeus*), Himalayan Griffon Vulture (*Gyps himalayensis*), herring gulls (*Larus a. argentatus L.*), seagulls (*Larus cachinnans micaellis*), love bird (*Agapornis roseicollis*), red-faced love bird (*Agapornis pullaria*), stitchbird or hihi (*Nofiomystis cincta*), wild geese (*Chloëphaga poliocephala*), pink-footed Geese (*Anser brachyrhynchus Baillon*)	Sudan, Chile, Belgium, United Kingdom, Bulgaria, Italy, Spain, New Zealand
A. flavus	Canada geese (*Branta Canadensis*), king shag (*Phalacocorax albivenier*), juvenile red-crowned crane (*Grus japonensis*), Magellanic penguin (*Spheniscus magellanicus*), gentoo penguin (*Pygoscelis papua*), chinstrap penguin (*Pygoscelis antarctica*), king penguin (*Aptenodytes patagonica*), little blue penguin (*Eudyptula minor*), yellow-eyed penguin (*Megadyptes antipodes*), rockhopper penguin (*Eudyptes chrysocome*), adelie penguin (*Pygoscelis adeliae*), peruvian penguin (*Spheniscus humboldti*),black footed/jackass penguin (*Spheniscus demersus*), budgerigar (*Melopsittacus undulatus*)	Canada, Spain, Scotland, Australia, New Zealand, Antarctica, South Africa
A. terreus	pigeon (*Columba livia*)	India
A. niger	great horned owl (*Bubo virginianus*)	Canada

of avian respiratory system (Table 2.17). Among the wild birds, some species such as goshawks (*Accipiter gentilis*), gyr falcons (*Falco rusticolus*), penguins, and auk (*Alca torda*) are more susceptible to respiratory fungal infections. Diving birds (auk, penguin) are more susceptible due to air re-circulation during diving.

2.4.2.4 Transmission

Inhalation of fungal spores (conidia) is the major way of *Aspergillus* transmission in birds. In addition to inhalation route, ingestion of contaminated feed (seed mixture) with the fungal spores is another way of transmission. Due to small non-expanding lung and presence of air sacs, the birds are more susceptible to *Aspergillus* infection. Primary colonization of the spores takes place in the air sacs because the inhaled air passes through the sacs to reach the lungs. Other predisposing factors for *Aspergillus* infection include higher body temperature which helps in fungal growth, lung injury, stress due to malnutrition and vitamin deficiency, use of immunosuppressive drugs, very young or old age and use of hay or straw contaminated with fungal spores in preparation of aviary litter.

2.4.2.5 Pathogenesis

The conidia of *Aspergillus* spp. reach the lung parenchyma through inhaled air. The conidia are trapped between atrium and infundibulum of parabronchus region of the lungs. Conidial sialic acids act as ligand for adherence with the alveolar epithelial cells specially with fibrinogen and fibronectin, commonly found in the wounded epithelial surfaces. So, lung injury acts as major predisposing factor for causation of invasive aspergillosis. Gliotoxin, fumagillin, and helvolic acid produced by the fungi causes damage in the respiratory mucosa and slow ciliary movement facilitating the attachment of the conidia.

The conidial maturation begins which causes loss of hydrophobic layer and exposure of the inner cell wall. The cell wall is composed of galactomannan, chitin, and β glucan which act as ligand for the soluble and cell associated pattern recognition receptors (PRR). The soluble receptors act as opsonin and can bind with fungal cell wall carbohydrate which enhances phagocytosis by the alveolar macrophages. Most of the conidia are killed by reactive oxygen species (ROS) or acidified phago-lysosome produced within the alveolar macrophages. In immunosuppressed birds having defective alveolar macrophage function the conidia are able to escape the phago-lysosome mediated killing.

The conidia which survive the first line of defense can germinate. The germination involves conidial swelling (isotropic growth) followed by protrusion of germ tube (polar growth). They produce a necrotic focus (plaque) without a structured granuloma which can obstruct the trachea or bronchi and fill up the air sacs. Hyphae with fruiting bodies can penetrate the air sacs or lungs and produce serositis. Fungal growing hyphae also invade the endothelial cell lining of blood vessels (angioinvasion) from the albuminal side to the luminal side. Dissemination of infection into vital organs occurs through the haematogenous route.

2.4.2.6 Clinical Symptoms

Avian aspergillosis is considered as an opportunistic infection in birds with immunosuppression. Primary infection in immunocompetent hosts sometimes takes place when numerous spores are inhaled. Acute and chronic types of aspergillosis may develop in affected birds. Mostly young birds suffer with acute infection in which sudden death of birds occurs without prior symptoms probably due to hepatic damage by aflatoxins

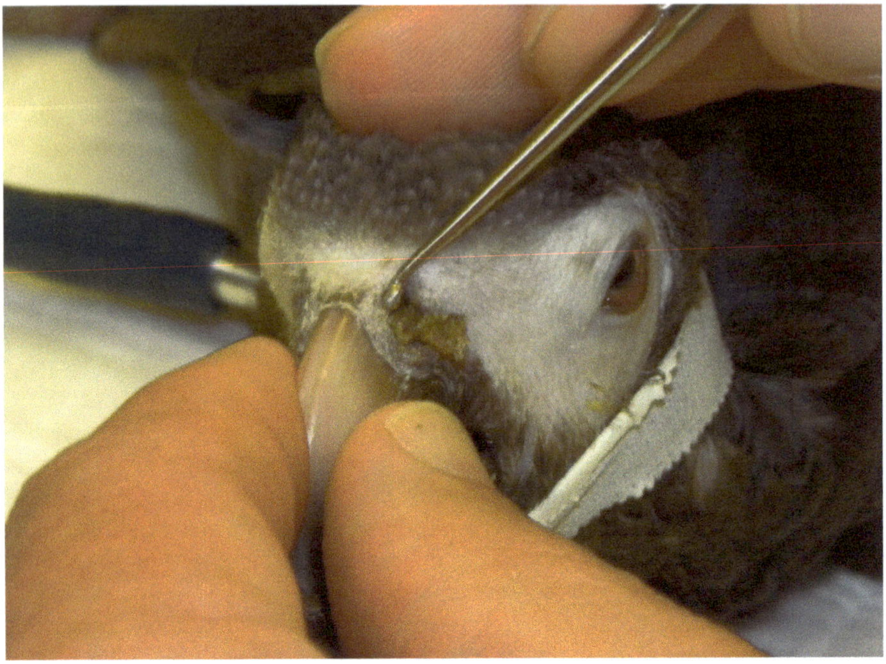

Fig. 2.50 Rhinolith in a Timneh African Grey parrot (*Courtesy* Dr. Kenneth R. Welle)

released by the fungi. If the birds are alive for a few days, general symptoms such as laboured breathing, anorexia, diarrhoea, polydypsia, and cyanosis develop.

Chronic aspergillosis is more common in adult birds. It produces non-specific syndrome such as fever, diarrhoea, respiratory distress, change in voice ('sore throat' sound due to tracheal and syringeal inflammation), change in behaviour, emaciation, cheesy deposit in conjuntival sac, and nares, development of rhinolith ('nose-stone') and facial swelling, neurological disorders, dermatitis, wing droop due to humerous involvement (Fig. 2.50). Death occurs due to obstruction of airways and respiratory failure.

2.4.2.7 Lesion

Congestion and yellow nodules in lungs, thickened air-sacs with small whitish-yellow plaques are detected in acute aspergillosis (Figs. 2.51 and 2.52). In chronic form, granulomatous lesions (nodules and plaques) are observed in periphery of lungs and air sacs which may occlude the trachea and bronchi. Typical granulomas are also detected in kidneys, oviduct, and ovaries.

2.4.2.8 Diagnosis

Clinical Specimens

Whole blood, tracheal washings, air sac fluids, tissue biopsies from respiratory tract granuloma and vital organs such as lungs, air sacs, kidney, liver collected after post

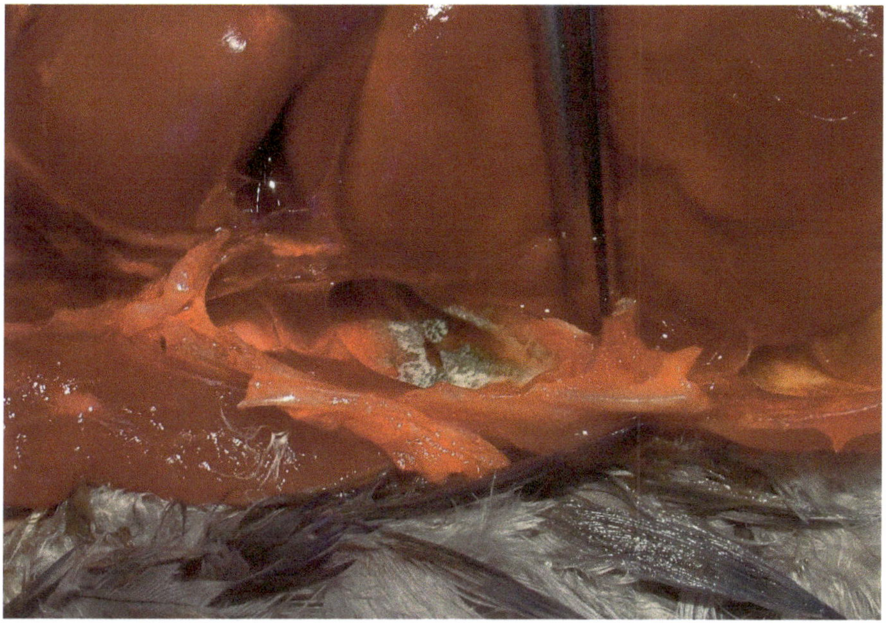

Fig. 2.51 Air sac lesion in African grey parrot infected with *Aspergillus* spp. (*Courtesy* Prof. Elena Circella, Avian Diseases Unit, Department of Veterinary Medicine, University of Bari, Italy)

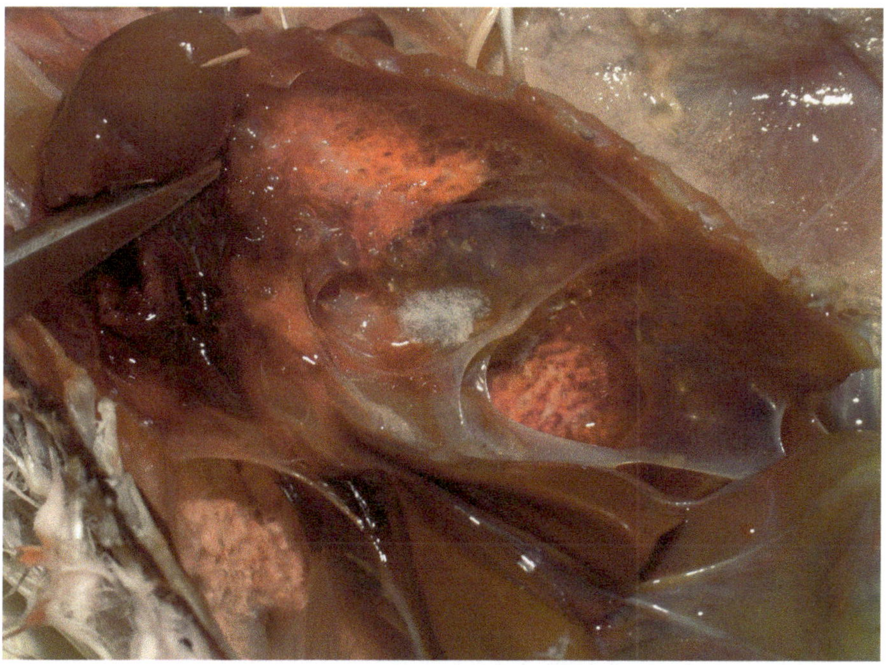

Fig. 2.52 Air sac lesion in cockatoo infected with *Aspergillus* spp. (*Courtesy* Prof. Elena Circella, Avian Diseases Unit, Department of Veterinary Medicine, University of Bari, Italy)

mortem can be used as clinical specimens for laboratory confirmation of avian aspergillosis.

Diagnostic Techniques

(a) *History and clinical symptoms*: History of immunosuppressive treatment and clinical symptoms such as sudden change of voice along with other general symptoms primarily suggests about clinical aspergillosis.
(b) *Direct examination*: The tissues collected as clinical specimens should be cleared with 10% KOH and are observed under microscope. Histopathological staining can be performed with periodic acid Schiff (PAS), Grocott's silver or methenamine silver stain for detection of tissue invasion. In the tissue section, *Aspergillus* hyphae are narrow and septate which are not easily distinguishable from other fungi (Fig. 2.53). Different species of *Aspergillus* has characteristic fruiting body structures which can be identified by an expert. Confirmation of aspergillosis is also possible by immunohistochemistry.
(c) *Isolation of Aspergillus from clinical specimens*: *Aspergillus* can be isolated in Sabouraud dextrose agar (SDA) with or without chloramphenicol (0.05 gm/l) and other common bacteriological media such as blood agar. The cyclohex-imide can inhibit the growth. The plates are incubated at 37 °C for 4–5 days. *A. fumigatus* is thermophilic which is able to grow at 55 °C and can survive at more than 75 °C. However, repeated isolation from the clinical specimen is required, along with correlation with history, clinical signs, histopathological observations for proper diagnosis of clinical aspergillosis.
(d) *Blood biochemical tests and radiography*: Haematological parameters such as leukocytosis (20,000–100,000), heterophilia with a left shift (degenerative shift), monocytosis, lymphopenia, and change in β-globulin concentration

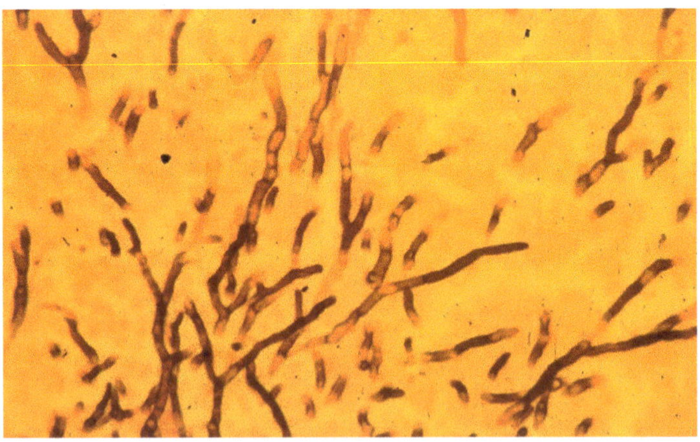

Fig. 2.53 Septate and dichotomously branching *Aspergillus* spp. hyphae in lungs (GMS) (*Courtesy* Prof. P.P. Gupta, Ex-Director of Veterinary Research, Punjab Agricultural University, Punjab, India)

indicate about occurrence of infection such as aspergillosis. Radiographic changes of pneumonia and airsacculitis (lateral and ventrodorsal views) also suggests respiratory tract infection such as aspergillosis. Both these biochemical tests and radiogrpahy cannot confirm the infection.

(e) *Detection of Aspergillus antigen*: Detection of *Aspergillus* antigen is useful in acute infection. Galactomannan (GM) is the predominant antigen released by *A. fumigatus* in the circulation during angioinvasion which can be detected by latex agglutination test, sandwich ELISA in blood samples. The detection limit of both the tests is 15 ng/ml and 1 ng/ml, respectively. However, GM detection assay is not specific for *Aspergillus* as it is cross-reacting with other fungi such as *Penicillium, Fusarium, Alternaria*, and *Histoplasma*. Similarly, β-D-glucan (BDG) can be detected for identification of *Aspergillus*. It is also produced by a lot of other fungi such as *Candida*, *Fusarium*, *Pneumocystis* etc. So the test can predict the general fungal infection rather than specifically aspergillosis.

(f) *Serological tests*: Serological tests such as counter-immunoelectrophoresis, agar gel immunodiffusion and enzyme-linked immunosorbent assays (ELISA) can be employed as supportive diagnostic assays to detect the antibodies in chronic infection. In immunosuppressed birds production of antibody is undetectable.

(g) *Molecular biology*: Confirmation of *Aspergillus* spp. can be done by conventional PCR and real-time PCR from clinical samples. *A. fumigatus* isolates can be characterized by PCR-RFLP using *BccI*, *MspI* and *Sau3AI* restriction enzymes.

2.4.2.9 Zoonosis

A. fumigatus causes invasive aspergillosis in immunocompromised human which involves lung parenchyma, pleura, trachea and bronchi. It is common in the patients with haematological malignancy, prolonged antibiotic users or stem cell transplant recipients. Transmission of clinical aspergillosis from pet birds is not documented so far.

2.4.2.10 Treatment and Control Strategy

Removal of granulomatous lesion is essential for effective treatment. It is difficult due to remote location of granuloma within respiratory tract, risk of surgical trauma and anaesthesia. In most of the confirmed cases, treatment is restricted with antifungals. Nebulization, oral, intravenous, nasal or air sac flushing, surgical irrigation of the abdominal cavities are ways for drug application. Antifungals used in avian aspergillosis with suitable dosage are described in Table 2.18. Sometimes, immunostimulants, for example, levamisole and microbial products (β-glucans) can be used for boosting of the immune system.

No vaccine is currently available against avian aspergillosis. Maintenance of hygiene and nutrition, avoiding mouldy feeds, proper ventilation and spraying of fungistatic agents [nystatin, thiabendazole, copper sulphate (1 g per 2 L of water)] in large aviaries can prevent avian aspergillosis.

Table 2.18 Antifungals used in birds against aspergillosis

Antifungals	Route and Dosage	Comments
Amphotericin-B	Nebulization: 1 mg/kg body weight for 20 min, 3–4 times a day for 10–14 days	It should be slowly injected via intravenous route, otherwise cardiac arrhythmia and mild convulsion occurs. The drug is potentially nephrotoxic for birds (except raptors). It is rapidly excreated by raptors
	Intravenous: 1.5 mg/kg body weight, 8 h interval, for 3–5 days in most species	
	Intravenous: 1 mg/kg body weight, 8–12 h interval, in raptors, psittacines (tracheal granuloma)	
Clotrimazole	Intratracheal: 2 mg/kg body weight, 24 h interval for 5 days in psittacines (tracheal granuloma)	–
	Nebulization: 1% in 2 or 3 ml saline 1.5 h/day for 4–6 weeks	
Fluconazole	Oral: 15 mg/kg body weight, twice a day for 15–60 days	It shows variable activity against *Aspergillus* spp. Use of nephrotoxic drug with fluconazole is contraindicated
Itraconazole	Oral: 10 mg/kg body weight once a day for 7–35 days	It is well absorbed when taken with food. In African Grey parrot, it should be avoided as it causes severe anorexia and depression
Ketoconazole	Oral: 10–30 mg/kg body weight, twice a day for 30–60 days	Some species of *Aspergillus* show resistance. Tablets may be mixed with food or fruit juice for feeding. Hepatotoxic drugs should be avoided along with ketoconazole

2.4.3 Other Fungal Infections of Pet Birds

2.4.3.1 Avian Candidiasis

Candida spp. is a common inhabitant of avian enteric tract, although, it can cause infection in young, immunosuppressed and stressed birds and during prolonged use of antibiotics. *Candida albicans*, *C. krusei*, and *C. tropicalis* are major cause of avian candidiasis. *C. parapsilosis* is rarely isolated from the crops of the birds. Among the pet birds, clinical candidiasis is reported from peach-faced lovebirds, Fisher's lovebirds, pigeons, cockatiels, Amazon parakeets, budgerigars and peacocks. In Eclectus parrot, concomitant infection of histoplasmosis and candidiasis is detected. Healthy cockatiels and some other psittacines are also detected to act as carrier of *Candida* spp. Infected birds show whitish, pseudomembranous lesions in oral cavity, crop and esophagus. Crop is the most suitable organ for fungal growth due to its pouch like structure and availability of nutrients. General symptoms such as anorexia, depression, delayed crop emptying, and regurgitation is sometimes observed. Presence of thickened mucosa (pseudomembranes) and ulcers in oral cavity, esophagus and crop is the major necropsy findings. The smear prepared

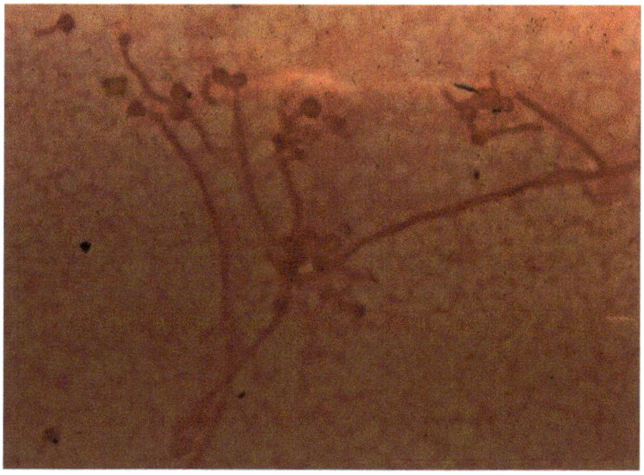

Fig. 2.54 Budding yeast, hyphae and pseudohyphae of *Candida albicans* (PAS) (*Courtesy* Prof. P. P. Gupta, Ex-Director of Veterinary Research, Punjab Agricultural University, Punjab, India)

from the clinical specimens can be observed under microscope either by 10% KOH preparation or by staining with Gram's stain method. In the tissue section *Candida* can be observed by PAS-haematoxylin or methenamine silver stain. They appear as unicellular budding yeast or hyphae and pseudohyphae (Fig. 2.54). *Candida* can be isolated in common fungal or bacteriological media such as Sabouraud dextrose Agar (with penicillin, streptomycin, chloramphenicol to prevent the bacterial growth), potato dextrose agar, blood agar and brain heart infusion agar. The plates are incubated at 25–30 °C for 2–3 days. Immunohistochemistry is also useful in detection of *Candida* spp. in birds. Several types of PCR such as semi nested and nested PCR, real time PCR, multiplex PCR followed by DNA sequencing or pyrosequencing are developed for detection of Candidal DNA. The target gene includes rRNA (5.8S, 18S, 28S) gene, internal transcribed spacer (ITS) and intergenic spacer (IGS) region genes. Treatment with nystatin (3,00,000–600,000 IU/kg body weight, 8–12 h interval for 7–14 days in psittacines; 1,000,000 IU/kg body weight, 12 h interval in raptors and pigeons) is recommended for pet birds in confirmed cases of avian candidiasis. The drug is not well absorbed from the gastro intestinal tract and it should be mixed with food for better absorption. Except vomition in some species the drug is safe and cost-effective.

2.4.3.2 Mycotic Proventriculitis (*Macrorhabdus ornithogaster* Infection)

Proventriculitis was originally described as *Megabacterium* associated infection in aviaries. Due to presence of a eukaryotic nucleus, mycotic staining properties, presence of chitin in cell wall, and finally phylogenetic analysis based on 18S rRNA and 26S rRNA gene, *Megabacterium* is classified as yeast. It is renamed as *Macrorhabdus ornithogaster*. The infection is reported from captive-bred budgerigars (*Melopsitticus undulatus*), parrotlets (*Forpus* spp.), canaries (*Serinus canaria*), pheasants (*Phasianus colchicus*), red-legged partridges (*Alectoris rufa*), helmeted guinea fowl (*Numida meleagris*) and gray partridges (*Perdix perdix*) as primary pathogen or concomitant with other infections. Faecal-oral route is the

Fig. 2.55 Regurgitation of a budgerigar infected with *Macrorhabdus ornithogaster* (*Courtesy* Prof. Elena Circella, Avian Diseases Unit, Department of Veterinary Medicine, University of Bari, Italy)

probable transmission route of *Macrorhabdus* infection in birds. The yeast prefers to colonize isthmus of proventriculus or ventriculus in birds. The infection is chronic and it produces certain non-specific symptoms such as anorexia, continuous loss of body weight, depression, plumage disorder, regurgitation, diarrhoea and dehydration (Fig. 2.55). Sudden death due to rupture of proventriculus is observed in budgerigars and parrots. Swollen and hyperemic proventriculus, covering of isthmus with thick, transparent-to-white mucus, and mucosal erosions in gizzard are specific necropsy findings. Fresh faecal samples from live birds and tissues from vital organs (esophagus, crop, proventriculus, gizzard, pancreas, liver etc.) can be collected in 10% buffered formalin after post-mortem. Smears prepared from faecal samples can be stained by Gram's method. The tissue sections are stained with periodic acid-Schiff (PAS), Grocott's methenamine silver nitrate (GMS) and Mayer's hematoxylin and eosin. Detection of typical gram-positive, large rod shaped organisms (20–60 μm long and 2–3 μm wide) suggests *Macrorhabdus* infection (Fig. 2.56). Isolation of *Macrorhabdus* is difficult because it does not grow in commonly used solid media. It can be isolated only in liquid medium (minimum essential media) with 20% foetal bovine serum (FBS) and 5% sucrose and more than two weeks incubation. Antifungals for instance nystatin and amphotericin-B can be used for the treatment (Figs. 2.57, 2.58 and 2.59).

Other fungal infections of pet birds are described in Table 2.19.

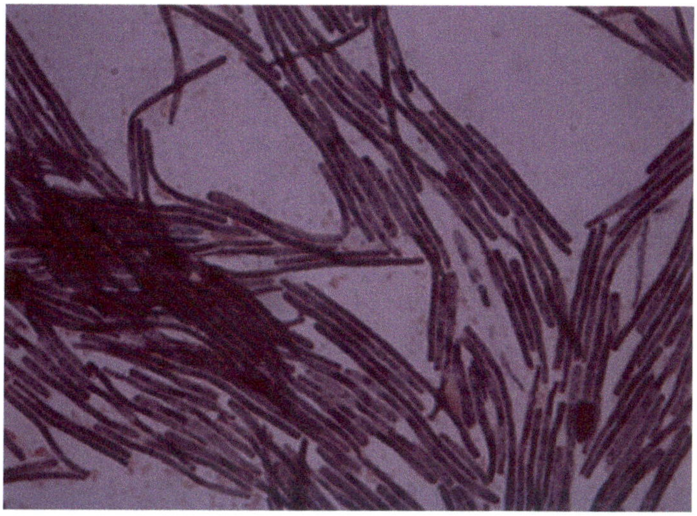

Fig. 2.56 Microscopic appearance of *Macrorhabdus ornithogaster* (Gram's stain) (*Courtesy* Prof. Elena Circella, Avian Diseases Unit, Department of Veterinary Medicine, University of Bari, Italy)

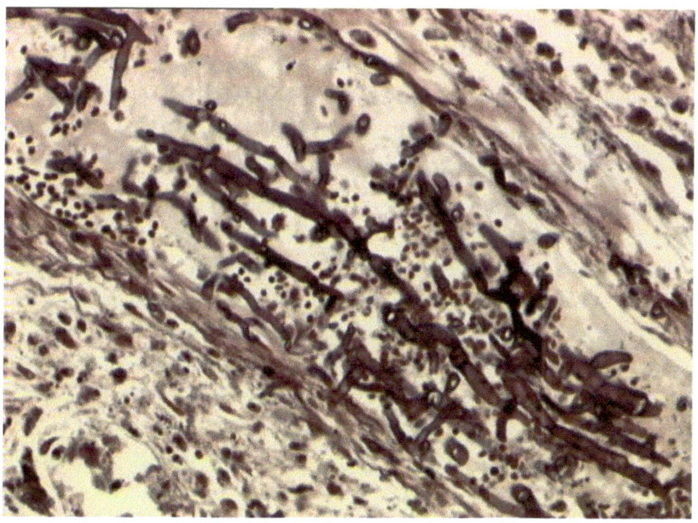

Fig. 2.57 *Penicillium* hyphae present in blood vessels of African grey parrot (Grocott stain) (*Courtesy* Giovanni Lanteri, University of Messina, Italy)

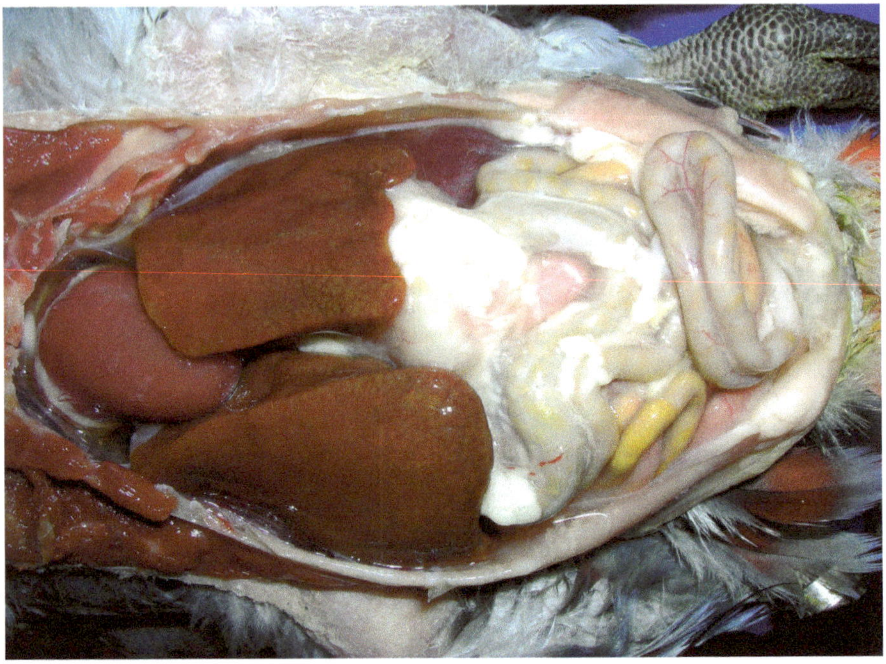

Fig. 2.58 Numerous yellowish white nodules in the liver of African grey parrot infected with *Penicillium* spp. (*Courtesy* Giovanni Lanteri, University of Messina, Italy)

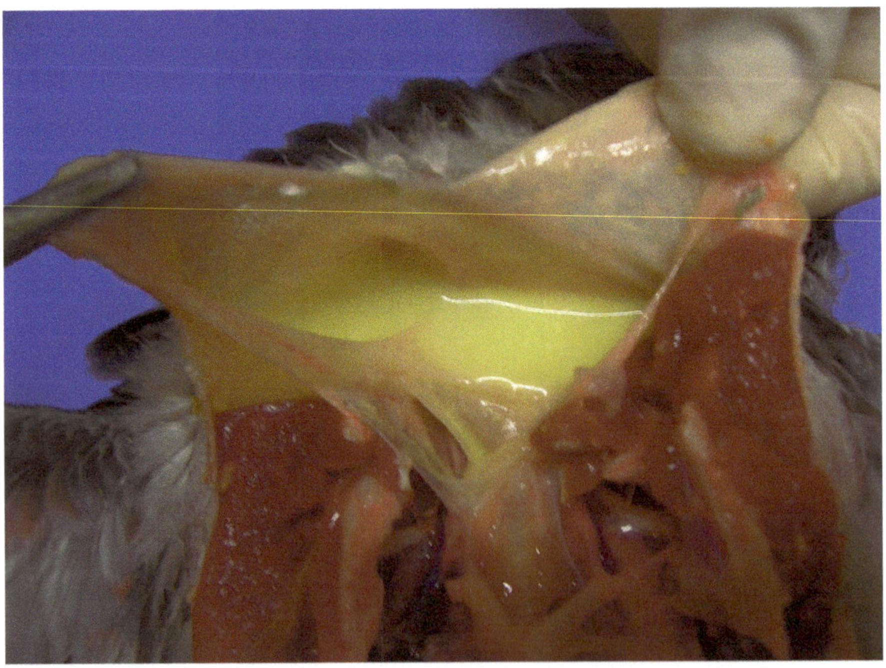

Fig. 2.59 Dense yellowish fluid in the crop of African grey parrot infected with *Penicillium* spp. (*Courtesy* Giovanni Lanteri, University of Messina, Italy)

Table 2.19 Other important fungal infection of pet birds

Fungi	Susceptible hosts	Clinical signs/Gross lesions	Treatment
Penicillium chrysogenum	African gray parrot (*Psittacus erithacus*)	Cheesy yellowish deposit in oral cavity, yellowish fluid in crop, pyogranulomatous lesions and whitish nodules in lung, liver, kidney	Enilconazole (6 mg/kg body weight, 12 h interval, oral)
Histoplasma capsulatum	Eclectus parrot (*Eclectus roratus*)	Oral and peri-ocular masses, chronic lameness	–
Absidia corymbifera	African gray parrot (*Psittacus erithacus*)	Lesions in air sac and kidney	–
Rhizomucor pusillus	African gray parrot (*Psittacus erithacus*)	Diarrhoea, unsteady gait, twisted neck	–
Rhizopus microsporus var. *chinensis*	Eclectus parrot (*Eclectus roratus*)	Concomitant infection with *Candida krusei* and produces acute necrotising ventriculitis. It is characterized by vomition and bright green faeces	–

Bibliography

Abe N, Makino I, Kojima A (2012) Molecular characterization of *Giardia psittaci* by multilocus sequence analysis. Infect Gen Evol 12:1710–1716

Abe N, Matsuo K, Makino I (2015) *Ascaridia nymphii* n. sp. (Nematoda: Ascaridida) from the alimentary tract of a severely emaciated dead cockatiel—*Nymphicus hollandicus*. Parasitol Res 114:4281–4288

Abegg MA, Cella FL, Faganello J et al (2006) *Cryptococcus neoformans* and *Cryptococcus gattii* isolated from the excreta of Psittaciformes in a southern Brazilian Zoological Garden. Mycopathol 161:83–91

Abulreesh HH, Paget T, Goulder R (2006) *Campylobacter* in waterfowl and aquatic environments: incidence and methods of detection. Environ Sci Technol 40:7122–7131

Ackerman LJ, Benbrook SC, Walton BC (1974) *Mycobacterium tuberculosis* infection in a parrot (*Amazona farinosa*). Am Rev Respir Dis 109:388–390

Adler HE (1957) Isolation of a pleuropneumonia-like organism from the air sac of a parakeet. J Am Vet Med Assoc 130:408–409

Ainsworth GC, Rewell RE (1949) The incidence of aspergillosis in captive wild birds. J Comp Pathol Ther 59:213–217

Alexander DJ (2000a) A review of avian influenza in different bird species. Vet Microbiol 74:3–13

Alexander DJ (2000b) Newcastle disease and other avian Paramyxoviruses. Rev Sci Tech Off Int Epiz 19:443–462

Alvarez-Perez S, Mateos A, Dominguez L et al (2010) Polyclonal Aspergillus fumigatus infection in captive penguins. Vet Microbiol 144:444–449

Andersen AA (1991) Serotyping of Chlamydia psittaci isolates using serovar specific monoclonal antibodies with the micro-immunofluorescence test. J Clin Microbiol 29:707–711

Anderson JF, Johnson RC, Magnarelli LA et al (1986) Involvement of birds in the epidemiology of the Lyme disease agent *Borrelia burgdorferi*. Infect Immun 51:394–396

Ashby E (1907) Parakeets moulting. The Emu 6:193–194

Authenrieth CF (1823) Ueber die hitzige Kopf-Krankheit der Pferde. Auf Verlangen des Münsinger Vereins zur Beförderung der Pferdezucht auf der Alp, und zunächst für diese Gegend, bey Heinrich Laupp. Tübingen

Bakonyi T, Erdelyi K, Ursu K et al (2007) Emergence of Usutu Virus in Hungary. J Clin Microbiol 45:3870–3874

Bakonyi T, Gajdon GK, Schwing R et al (2016) Chronic West Nile virus infection in kea (*Nestor notabilis*). Vet Microbiol 183:135–139

Barin A, Arabkhazaeli F, Rahbari S et al (2010) The housefly, *Musca domestica*, as a possible mechanical vector of Newcastle disease virus in the laboratory and field. Med Vet Entomol 24:88–90

Becker WB (1966) The isolation and classification of tern virus: influenza virus A/tern/South Africa/1961. J Hyg 64:309–320

Beernaert LA, Pasmans F, Haesebrouck F et al (2008) Modelling *Aspergillus fumigatus* infections in racing pigeons (*Columba livia domestica*). Avian Pathol 37:545–549

Bessell PR, Robinson RA, Golding N et al (2014) Quantifying the risk of introduction of West Nile virus into Great Britain by migrating passerine birds. Transbound Emerg Dis. doi:10.1111/tbed.12310

Blanco G (2015) Multiresistant *Salmonella* Serovar Typhimurium monophasic in Wintering Red Kites (*Milvus Milvus*) in Segovia, Central Spain. J Raptor Res 49:337–341

Blanco K, Niehaus C, Salinas A et al (2013) Apparent lack of evidence on selected infectious agents in wild Yellow-naped Amazon parrots: implications for releasing attempts. Rev Ciencias Vet 31:7–17

Bollinger O (1878) Ueber eine neue Wild-und Rinderseuche, welche im Sommer 1878 in der Umgebung von München beobachtet wurde

Bonar CJ, Lewandowski AH (2004) Use of a liposomal formulation of amphotericin B for treating wound aspergillosis in a Goliath heron (Ardea goliath). J Avian Med Sur 18:162–166

Boosinger TR, Winterfield RW, Feldman DS et al (1982) Psittacine pox virus: virus isolation and identification, transmission, and cross-challenge studies in parrots and chickens. Avian Dis 26:437–444

Boseret G, Losson B, Mainil JG et al (2013) Zoonoses in pet birds: review and perspectives. Vet Res 44:36–54

Bounous DI, Schaeffer DO, Roy A (1989) Coagulase-negative *Staphylococcus* sp. septicemia in a lovebird. J Am Vet Med Assoc 195:1120–1122

Bozeman LH, Kleven SH, Davis RB (1984) *Mycoplasma* challenge studies in budgerigars (*Melopsittacus undulatus*) and chickens. Avian Dis 28:426–434

Breithaupt A, Kalthoff D, Dale J et al (2011) Neurotropism in Blackcaps (*Sylvia atricapilla*) and Red-Billed Queleas (*Quelea quelea*) after highly pathogenic avian influenza virus H5N1 infection. Vet Pathol 48:924–932

Brilhante RSN, Castelo-Branco DSCM, Soares GDP et al (2010) Characterization of the gastrointestinal yeast microbiota of cockatiels (*Nymphicus hollandicus*): a potential hazard to human health. J Med Microbiol 59(6):718–723

Brilhante RSN, Rodrigues TDJS, Castelo DDSCM et al (2014) Antifungal susceptibility and virulence attributes of animal-derived isolates of *Candida parapsilosis* complex. J Med Microbiol 63:1568–1572

Broggi J, Garcı´a O, Miranda F et al (2013) Immune response to Newcastle Disease virus vaccination in a wild passerine. J Wildl Dis 49:1004–1008

Broman T, Waldenstrom J, Dahlgren D et al (2004) Diversities and similarities in PFGE profiles of *Campylobacter jejuni* isolated from migrating birds and humans. J Appl Microbiol 96:834–843

Brown MB, Butcher GD (1991) *Mycoplasma gallisepticum* as a model to assess efficacy of inhalant therapy in budgerigars (*Melopsittacus undulatus*). Avian Dis 35:834–839

Burgess EC (1989) Experimental Inoculation of Mallard Ducks (*Anas Platyrhynchos Platyrhynchos*) with *Borrelia Burgdorferi*. J Wildl Dis 25:99–102
Burr EW (1981) Intranasal caseous fibrinous plug causing upper respiratory distress in two African grey parrots (*Psittacus erithacus*). Avian Dis 25:542–544
Busquets N, Alba A, Allepuz A et al (2008) Usutu Virus Sequences in Culex pipiens (Diptera: Culicidae), Spain. Emer Infect Dis 14:861–863
Busse O (1894) Uber parasitare Zelleinschlusse und ihre Zuchtung. Centralbl Bakt Parasit 16
Butron O, Brightsmith DJ (2010) Testing for *Salmonella* spp. in released parrots, wild parrots, and domestic fowl in lowland Peru. J Wildlife Dis 46:718–723
Buur J, Saggese MD (2012) Taking a rational approach in the treatment of avian Mycobacteriosis. Vet Clin Exot Anim 15:57–70
Carboni DA, Nevarez JG, Tully TN Jr et al (2008) West Nile virus infection in a sun conure (*Aratinga solstitialis*). J Avian Med Surg 22:240–245
Carini A (1911) Infection spontanée du pigeon et du chien due au *Toxoplasma cuniculi*. Bull Soc Pathol Exot 4:518–519
Carrasco L, Bautista MJ, De Las Mulas JM et al (1993) Application of enzyme-immunohistochemistry for the diagnosis of aspergillosis, candidiasis, and zygomycosis in three lovebirds. Avian Dis 37:923–927
Catár G (1974) Toxoplazmóza v ekologickych podmienkach na Slovensku (in Slovakian). Biologické Práce (Bratislava) 20:1–138
Centanni E and Savonuzzi E (1901) La peste aviaria I & II. Communicazione fatta all'accademia delle scienze mediche e naturali de Ferrara
Chahota R, Katoch RC, Batta MK (1997) Prevalence of *Chlamydia psittaci* among feral birds in Himachal Pradesh, India. J Appl Anim Res 12:89–94
Chaves A, Sotomayor-Bonilla J, Monge O et al (2016) West Nile virus in resident birds from Yucatan, Mexico. J Wildl Dis 52:159–163
Chen H, Li Y, Li Z et al (2006) Properties and dissemination of H5N1 viruses isolated during an influenza outbreak in migratory waterfowl in western China. J Virol 80:5976–5983
Circella E, Pugliese N, Todisco G et al (2011) Chlamydia psittaci infection in canaries heavily infested by Dermanyssus gallinae. Exp Appl Acarol 55:329–338
Clarke IN (2011) Evolution of *Chlamydia trachomatis*. Ann N Y Acad Sci 1230:E11–E18
Clavijo A, Robinson Y, Booth T et al (2000) Velogenic Newcastle disease in imported caged birds. Can Vet J 41:404–406
Clipsham RC, Britt JO Jr (1983) Disseminated cryptococcosis in a macaw. J Am Vet Med Assoc 183:1303–1305
Cobbett L (1917) The causes of tuberculosis: together with some account of the prevalence and distribution of the disease. CUP Archive
Comstedt P, Bergström S, Olsen B et al (2006) Migratory passerine birds as reservoirs of Lyme Borreliosis in Europe. Emerg Infect Dis 12:1087–1095
Cong W, Huang SY, Zhang XX et al (2014) *Chlamydia psittaci* exposure in pet birds. J Med Microbiol 63:578–581
Copetti MV, Segabinazi SD, Flores ML et al (2004) Pulmonary aspergillosis outbreak in the stitchbird in New Zealand. J Wildl Dis 35:481–486
Cork SC, Alley MR, Johnstone AC et al (1999) Aspergillosis and other causes of mortality in *Rhea americana* in southern Brazil. Mycopathol 157:269–271
Coyle MB, Carlson LC, Wallis CK et al (1992) Laboratory aspects of "Mycobacterium genavense" a proposed species isolated from AIDS patients. J Clin Microbiol 30:3206–3212
Current WL, Upton SJ, Haynes TB (1986) The life cycle of *Cryptosporidium baileyi* n. s. (Apicomplexa, Cryptosporidiidae) infecting chickens. J Protozool 33:289–296
Curtis F (1896) Contribution à l'étude de la saccharomycose humaine. Ann Inst Pasteur 10: 449–468

Damborg P, Broens EM, Chomel BB et al (2015) Bacterial zoonoses transmitted by household pets: state-of-the-art and future perspectives for targeted research and policy actions. J Comp Pathol. doi:10.1016/j.jcpa.2015.03.004
Daut EF, Lahodny G Jr, Peterson MJ et al (2016) Interacting effects of Newcastle Disease transmission and illegal trade on a wild Population of white-winged parakeets in Peru: a modeling approach. PLoS ONE 11:e0147517
Dawson CO, Wheeldon EB, McNeil PE (1976) Air sac and renal mucormycosis in an African gray parrot (*Psittacus erithacus*). Avian Dis 20:593–600
Deem SL, Noss AJ, Cue´llar RL et al (2005) Health evaluation of free-ranging and captive blue-fronted amazon parrots (*Amazona aestiva*) in the Gran Chaco, Bolivia. J Zoo Wildlife Med 36:598–605
Devriese LA, Chiers K, De Herdt P et al (1995) *Enterococcus hirae* infections in psittacine birds: epidemiological, pathological and bacteriological observations. Avian Pathol 24:523–531
Dhama K, Mahendran M, Tiwari R et al (2011) Tuberculosis in birds: insights into the *Mycobacterium avium* infections. Vet Med Int 2011:712369
Dhondt AA, Dhondt KV, McCleery BV (2008) Comparative infectiousness of three passerine bird species after experimental inoculation with *Mycoplasma gallisepticum*. Avian Pathol 37: 635–640
Dimitrov KM, Ramey AM, Qiu X et al (2016) Temporal, geographic, and host distribution of avian paramyxovirus 1 (Newcastle disease virus). Infect Gen Evol 39:22–34
Dobell C (1932) Antony Van Leeuwenhoek and His "little Animals", Swets & Zeitlinger Amsterdam
Doneley RJT (2009a) Bacterial and parasitic diseases of parrots. Vet Clin Exot Anim 12:417–432
Doneley RJT (2009b) Bacterial and parasitic diseases of parrots. Vet Clin Exot Anim 12:417–432
Dorrestein GM (2009) Bacterial and parasitic diseases of Passerines. Vet Clin Exot Anim 12: 433–451
Downie JC, Laver WG (1973) Isolation of a type A influenza virus from an Australian pelagic bird. Virol 51:259–269
Dubey JP (2002) A review of toxoplasmosis in wild birds. Vet Parasitol 106:121–153
Dubey JP, Hamir AN (2002) Experimental toxoplasmosis in budgerigars (*Melopsittacus undulatus*). J Parasitol 88:514–519
Dubska L, Literak I, Kocianova E et al (2009) Differential role of Passerine birds in distribution of *Borrelia* spirochetes, based on data from ticks collected from birds during the Post breeding migration period in Central Europe. Appl Environ Microbiol 75:596–602
Duneau D, Boulinier T, Go´ mez-Dı´az E et al (2008) Prevalence and diversity of Lyme borreliosis bacteria in marine birds. Infect Gen Evol 8: 352–359
Evans EE (1949) An immunologic comparison of twelve strains of *Cryptococcus neoformans* (*Torula histolytica*). Exp Biol Med 71:644–646
Evans EE (2011) Zoonotic diseases of common pet birds: psittacine, passerine, and columbiform species. Vet Clin North Am Exot Anim Pract 14:457–476
Fenwick B, Takeshita K, Wong A (1985) A moluccan cockatoo with disseminated cryptococcosis. J Am Vet Med Assoc 187:1218
Ferreira Junior FC, Donatti RV, Marques MVR et al (2012) Fatal Toxoplasmosis in a Vinaceous Amazon Parrot (*Amazona vinacea*). Avian Dis 56:774–777
Filippich LJ, McDonnell PA, Munoz E et al (1998) *Giardia* infection in budgerigars. Aust Vet J 76:246–249
Fischer JR, Stallknecht DE, Luttrell P et al (1997) Mycoplasmal conjunctivitis in wild songbirds: the spread of a new contagious disease in a mobile host population. Emerg Infect Dis 3:69–72
Flach EJ, Stevenson MF, Henderson GM (1990) Aspergillosis in gentoo penguins (Pygoscelis papua) at Edinburgh Zoo, 1964 to 1988. Vet Rec 126:81–85
Flammer K, Drewes LA (1988) Species-related differences in the incidence of gram negative bacteria isolated from the cloaca of clinically normal psittacine birds. Avian Dis 32:79–83

Fogell DJ, Martin RO, Groombridge JJ (2016) Beak and feather disease virus in wild and captive parrots: an analysis of geographic and taxonomic distribution and methodological trends. Arch Virol. doi:10.1007/s00705-016-2871-2

Forshaw JM (2010) Parrots of the world. CSIRO Publishing, Collingwood

Frank B (1889) Ueber die Pilzsymbiose der Leguminosen. Ber Deut Bot Ges 7:332–346

Frasca S Jr, Hinckley L, Forsyth MH et al (1997) Mycoplasmal conjunctivitis in a European starling. J Wildl Dis 33:336–339

Fuller TL, Saatchi SS, Curd EE et al (2010) Mapping the risk of avian influenza in wild birds in the US. BMC Infect Dis 10:187–199

Gamino V, Höfle U (2013) Pathology and tissue tropism of natural West Nile virus infection in birds: a review. Vet Res 44:39–55

Gelis S, Gill JH, Oldfield T et al (2006) Mycobacteriosis in gang gang cockatoos (*Callocephalon fimbriatum*). Vet Clin North Am Exot Anim Pract 9:87–94

Giles N, Carter MJ (1980) *Yersinia enterocolitica* in budgerigars. Vet Rec 107:362–363

Glavis J, Larsen RS, Lamberski N et al (2011) Evaluation of antibody response to vaccination against West Nile virus in thick billed parrots (*Rhynchopsitta pachyrhyncha*). J Zoo Wildl Med 42:495–498

Gomez G, Saggese MD, Weeks BR et al (2011) Granulomatous encephalomyelitis and intestinal ganglionitis in a spectacled Amazon parrot (*Amazona albifrons*) infected with *Mycobacterium genavense*. J Comp Pathol 144:219–222

Gomez-Villamandos JC, Mozos E, Sierra MA et al (1991) Mortality in psittacine birds resembling Pacheco's disease in Spain. Avian Pathol 20:541–547

González-Acuña D, Silva GF, Moreno SL et al (2007) Detection of some zoonotic agents in the domestic pigeon (*Columba livia*) in the city of Chillán, Chile. Revista Chilena de Infectología 24:199–203

Goodpasture EW, Anderson K (1962) Isolation of a wild avian pox virus inducing both cytoplasmic and nuclear inclusions. Am J Pathol 40:437–453

Gough RE, Alexander DJ (1993) Pacheco's disease in psittacine birds in Great Britain 1987 to 1991. Vet Rec 132:113–115

Gough RE, Drury SE, Culver F et al (2006) Isolation of a coronavirus from a green-cheeked Amazon parrot (Amazon viridigenalis Cassin). Avian Pathol 35:122–126

Graczyk TK, Cockrem JF (1995) Aspergillus spp. seropositivity in New Zealand penguins. Mycopathol 131:179–184

Graham CL (1978) Poxvirus infection in a spectacled Amazon parrot (*Amazona albifrons*). Avian Dis 22:340–343

Graham CL, Graham DL (1978) Occurrence of *Escherichia coli* in feces of psittacine birds. Avian Dis 22:717–720

Graham DL, Calnek BW (1987) Papovavirus infection in hand-fed parrots: virus isolation and pathology. Avian Dis 31:398–410

Grespan A, Camera O, Knöbl T et al (2012) Virulence and molecular aspects of *Bordetella avium* isolated from cockatiel chicks (*Nymphicus hollandicus*) in Brazil. Vet Microbiol 160:530–534

Griner LA, Walch HA (1978) Cryptococcosis in columbiformes at the San Diego Zoo. J Wildl Dis 14:389–394

Grodio JL, Buckles EL, Schat KA (2009) Production of house finch (*Carpodacus mexicanus*) IgA specific anti-sera and its application in immunohistochemistry and in ELISA for detection of *Mycoplasma gallisepticum*-specific IgA. Vet Immunol Immunopathol 132:288–294

Grund CH, Werner O, Gelderblom HR et al (2002) Avian paramyxovirus serotype 1 isolates from the spinal cord of parrots display a very low virulence. J Vet Med B Infect Dis Vet Public Health 49:445–451

Gryczyńska A, Zgódka A, Płoski R et al (2004) *Borrelia burgdorferi* sensu lato infection in passerine birds from the Mazurian Lake region (Northeastern Poland). Avian Pathol 33:67–73

Halberstaedter L and von Prowazek S (1907) Ueber Zelleinschlüsse parasitärer Natur beim trachom. Arbeiten aus dem Kaiserlichen Gesundheitsamte 26:44–47

Hamer SA, Hickling GJ, Keith R et al (2012) Associations of passerine birds, rabbits, and ticks with *Borrelia miyamotoi* and *Borrelia andersonii* in Michigan, U.S.A. Parasit Vectors 5: 231–242

Han GZ, He CQ, Ding NZ et al (2008) Identification of a natural multi-recombinant of Newcastle disease virus. Virol 371:54–60

Han Y, Hou G, Jiang W et al (2012) A survey of avian influenza in tree sparrows in China in 2011. PLoS ONE 7:e33092

Hannon DE, Bemis DA, Garner MM (2012) *Mycobacterium marinum* infection in a blue-fronted Amazon parrot (*Amazona aestiva*). J Avian Med Surg 26:239–247

Harkinezhad T, Schautteet K, Vanrompay D (2009) Protection of budgerigars (*Melopsittacus undulatus*) against *Chlamydophila psittaci* challenge by DNA vaccination. Vet Res 40:61–72

Harkins GW, Martin DP, Christoffels A et al (2014) Towards inferring the global movement of beak and feather disease virus. Virol 450–451:24–33

Hartup BK, Mohammed HO, Kollias GV et al (1998) Risk factors associated with mycoplasmal conjunctivitis in house finches. J Wildl Dis 34:281–288

Hasle G (2010) Dispersal of ticks and tick-borne pathogens by birds: Dynamics of birds' transport of ticks to Norway. PhD thesis, University of Oslo, Norway

Hatch TP (1975) Utilization of L-cell nucleoside triphosphates by *Chlamydia psittaci* for ribonucleic acid synthesis. J Bacteriol 122:393–400

Hawkins MG, Crossley BM, Osofsky A et al (2006) H5N2 avian influenza A in a red-lored Amazon parrot (*Amazona autumnalis autumnalis*). J Am Vet Med Assoc 228:236–241

Heffels-Redmann U, Enderlein D, Herzog S et al (2011) Occurrence of Avian Bornavirus infection in captive psittacines in various European countries and its association to Proventricular Dilatation Disease. Avian Pathology 40:419–426

Hegler C (1930) Psittakose (Klinische Erfahrungen beim Menschen). Dt Tiera¨ rztl Wschr 38: 677–681

Heller EL, Wright CL, Nadolny RM et al (2015) New records of *Ixodes affinis* (Acari: Ixodidae) parasitizing avian hosts in Southeastern Virginia. J Med Entomol 53:441–445

Hernandez J, Lindberg P, Waldenstrom J et al (2012) A novel *Salmonella* serovar isolated from Peregrine Falcon (*Falco peregrinus*) nestlings in Sweden: *Salmonella enterica enterica* serovar Pajala (*Salmonella* Pajala). Infect Ecol Epidemiol 2:7373

Hidasi HW, Hidasi NJ, Moraes DM et al (2013) Enterobacterial detection and *Escherichia coli* antimicrobial resistance in parrots seized from the illegal wildlife trade. J Zoo Wildl Med 44:1–7

Hill FI, Woodgyer AJ, Lintott MA (1995) Cryptococcosis in a North Island brown kiwi (*Apteryx australis mantelli*) in New Zealand. J Med Vet Mycol 33:305–309

Hinshaw WR (1933) Tuberculosis of human origin in the Amazon parrot. Am Rev Tuberc Pulm Dis 28:273–278

Hirschel B, Chang HR, Mach N et al (1990) Fatal infection with a novel unidentified mycobacterium in a man with the acquired immunodeficiency syndrome. N Engl J Med 323:109–113

Honkavuori KS, Shivaprasad HL, Williams BL et al (2008) Novel Borna virus in Psittacine birds with Proventricular Dilatation Disease. Emerg Infect Dis 14:1883–1886

Hoop RK (2002) *Mycobacterium tuberculosis* infection in a canary (*Serinus canaria* L.) and a blue-fronted Amazon parrot (*Amazona amazona aestiva*). Avian Dis 46:502–504

Hoop RK, Bo¨ttger EC, Pfyffer GE (1996) Etiological agents of Mycobacterioses in pet birds between 1986 and 1995. J Clin Microbiol 34:991–992

Hoop RK, Bottger EC, Ossent P et al (1993) Mycobacteriosis due to *Mycobacterium genavense* in six pet birds. J Clin Microbiol 31:990–993

Horner RF, Parker ME, Abrey AN et al (1992) Isolation and identification of psittacid herpesvirus 1 from imported psittacines in South Africa. J South Afr Vet Assoc 63:59–62

Howe L, Hunter S, Burrows E et al (2014) Four cases of fatal Toxoplasmosis in three species of endemic New Zealand birds. Avian Dis 58:171–175

Hulin V, Oger S, Vorimore F et al (2015) Host preference and zoonotic potential of *Chlamydia psittaci* and *C. gallinacea* in poultry. FEMS Pathogen Dis 73:1–11

Huynh M, Carnaccini S, Driggers T et al (2014) Ulcerative dermatitis and valvular endocarditis associated with Staphylococcus aureus in a hyacinth macaw (*Anadorhynchus hyacinthinus*). Avian Dis 58:223–227

Imada T, Yamaguchi S, Kawamura H et al (1980) Isolation of an Influenza A virus from the budgerigar, Melopsittacus undulatus. Natl Inst Anim Health Q (Tokyo) 20:30–31

Irokanulo EOA, Makinde AA, Akuesgi CO et al (1997) *Cryptococcus neoformans* var. *neoformans* isolated from droppings of captive birds in Nigeria. J Wildl Dis 33:343–345

Jackson B, Heath A, Holyoake, C et al (2013) A mite-y problem: what is causing feather loss in Red crowned parakeets in Tiritiri Matangi Island, New Zealand. In: Wildlife Disease Association Australasian Section annual conference, Grampians, Australia

Jackson MK, Phillips SN (1996) Necrotizing hepatitis in pet birds associated with Pseudomonas fluorescens. Avian Dis 40:473–476

Jansson DS, Bröjer C, Mattsson R et al (2008) Mycotic proventriculitis in gray partridges (*Perdix perdix*) on two game bird farms. J Zoo Wildl Med 39:428–437

Jarmin S, Manvell R, Gough RE et al (2006) Avipoxvirus phylogenetics: identification of a PCR length polymorphism that discriminates between the two major clades. J Gen Virol 87: 2191–2201

Jones DM, Carroll CMM (1977) Debilitating syndrome in budgerigars (*Melopsinacus undulatus*). Vet Rec 101:188

Jones JC, Sonnberg S, Koçer ZA et al (2014) Possible role of songbirds and parakeets in transmission of influenza A (H7N9) virus to humans. Emerg Infect Dis 20:380–385

Jones JC, Sonnberg S, Webby RJ et al (2015) Influenza A(H7N9) Virus Transmission between Finches and Poultry. Emerg Infect Dis 21:619–628

Julian L, Piasecki T, Chrzastek K et al (2013) Extensive recombination detected among beak and feather disease virus isolates from breeding facilities in Poland. J Gen Virol 94:1086–1095

Kaltenboeck B, Kousoulas KG, Storz J (1991) Detection and strain differentiation of *Chlamydia psittaci* mediated by a two-step polymerase chain reaction. J Clin Microbiol 29:1969–1975

Kano R, Sakamoto Y, Hanahachi A et al (2001) Molecular identification of *Candida parapsilosis* from crop mucosa in a cockatiel. J Vet Diag Invest 13:437–439

Katoh H, Ohya K, Ise K et al (2010) Genetic analysis of beak and feather disease virus derived from a cockatiel (*Nymphicus hollandicus*) in Japan. J Vet Med Sci 72:631–634

Kaye D, Shinefield HR, Hook EW (1961) The parakeet as a source of Salmonellosis in man. Report of a case. N Engl J Med 264:868–869

Keymer IF (1958) A survey and review of the causes of mortality in British birds and the significance of wild birds as disseminators if disease. Vet Rec 70:713–720

Kistler AL, Gancz A, Clubb S, Skewes-Cox P, Fischer K, Sorber K, Chiu CY, Lublin A, Mechani S, Farnoushi Y and Greninger A (2008) Recovery of divergent avian bornaviruses from cases of proventricular dilatation disease: identification of a candidate etiologic agent. Virol J 5:88

Knittler MR, Sachse K (2015) *Chlamydia psittaci*: update on an underestimated zoonotic agent. FEMS Pathogen Dis 73:1–15

Kozdruń W, Czekaj H, Styś N (2015) Avian zoonoses—a review. Bull Vet Inst Pulawy 59:171–178

Kraft V, Teufel P (1971) Detection of a pox-virus in lovebirds (Agapornis personata and Agapornis roseicollis). Berl Munch Tierarztl Wochenschr 84:83–87

Kuhn JH, Dürrwald R, Bào Y et al (2015) Taxonomic reorganization of the family *Bornaviridae*. Arch Virol 160:621–632

Lagrou K, Van eldere J, Keuleers S et al (2005) Zoonotic transmission of *Cryptococcus neoformans* from a magpie to an immunocompetent patient. J Int Med 257:385–388

Lane RS, Kucera TF, Barrett RH et al (2006) Wild turkey (*Meleagris Gallopavo*) as a host of Ixodid ticks, lice, and Lyme Disease spirochetes (*Borrelia Burgdorferi* sensu lato) in California State Parks. J Wildl Dis 42:759–771

Lanteri G, Marino F, Reale S et al (2011a) *Mycobacterium tuberculosis* in a red-crowned parakeet (*Cyanoramphus novaezelandiae*). J Avian Med Sur 25:40–43

Lanteri G, Sfacteria A, Macrì D et al (2011b) Penicilliosis in an African grey parrot (*Psittacus erithacus*). J Zoo Wildl Med 42:309–312

Lebarbenchon C, Chang CM, van der Werf S et al (2007) Influenza A virus in birds during spring migration in the Camargue, France. J Wildl Dis 43:789–793

Ledwoń A, Dolka I, Dolka B et al (2015) Multidrug therapy of *Mycobacterium avium* subsp. *avium* infection in experimentally inoculated budgerigars (*Melopsittacus undulatus*). Avian Pathol 12:1–18

Leibovitz (l.) (1962) Unusual bird-parasite cases and overall parasite incidence found in a diagnostic laboratory during a five year period. Avian Dis 6:141–144

Lencˇa´kova´ D, Hizo-Teufela C, Pet'kob B et al (2006) Prevalence of *Borrelia burgdorferi* s.l. OspA types in *Ixodes ricinus* ticks from selected localities in Slovakia and Poland. Int J Med Microbiol 296:108–118

Lennox AM (2007) Mycobacteriosis in companion psittacine birds: a review. J Avian Med Sur 21:181–187

Lenny BJ, Shanmuganatham K, Sonnberg S et al (2015) Replication capacity of avian influenza A (H9N2) virus in pet birds and mammals, Bangladesh. Emerg Infect Dis 21:2174–2177

Ley DH, Moresco A, Frasca S (2012a) Conjunctivitis, rhinitis, and sinusitis in cliff swallows (*Petrochelidon pyrrhonota*) found in association with *Mycoplasma sturni* infection and cryptosporidiosis. Avian Pathol 41:395–401

Ley DH, Moresco A, Frasca S Jr (2012b) Conjunctivitis, rhinitis, and sinusitis in cliff swallows (*Petrochelidon pyrrhonota*) found in association with *Mycoplasma sturni* infection and cryptosporidiosis. Avian Pathol 41:395–401

Lignières J (1900) Contribution a l'étude et la classification des septicemies hemorragiques. Impimerie Coni Frères

Literaka I, Nortec AC, Núnciod MS et al (2015) Ticks on passerines from the Archipelago of the Azores as hosts of Borreliae and Rickettsiae. Ticks Tick Borne Dis 6:607–610

Lugarini C, Goebel CS, Condas LAZ et al (2008) *Cryptococcus neoformans* isolated from passerine and psittacine bird excreta in the state of Paraná, Brazil. Mycopathol 166:61–69

Luttrell MP, Stallknecht DE, Kleven SH et al (2001) *Mycoplasma gallisepticum* in house finches (*Carpodacus mexicanus*) and other wild birds associated with poultry production facilities. Avian Dis 45:321–329

Madadgar O, Salehi TZ, Ghafari MM et al (2009) Study of an unusual paratyphoid epornitic in canaries (*Serinus canaria*). Avian Pathol 38:437–441

Madewell BR, McChesney AE (1975) Salmonellosis in a human infant, a cat, and two parakeets in the same household. J Am Vet Med Assoc 167:1089–1090

Malbrant R (1942) Game, Tsetse Files and Trypanosomiasis. Rev Sci Med Pharm et Vet de V Afrique Francaise Libre 1:73–87

Manarolla G, Bakonyi T, Gallazzi D et al (2010) Usutu virus in wild birds in northern Italy. Vet Microbiol 141:159–163

Manarolla G, Liandris E, Pisoni G et al (2007) *Mycobacterium genavense* and avian polyomavirus co-infection in a European Goldfinch (*Carduelis carduelis*). Avian Pathol 36:423–426

Mannelli A, Nebbia P, Tramuta C et al (2005) *Borrelia burgdorferi* sensu lato Infection in Larval *Ixodes ricinus* (Acari: Ixodidae) Feeding on Blackbirds in Northwestern Italy. J Med Entomol 42:168–175

Marietto-Goncalves GA, Almeida SMD, Lima ETD et al (2010) Isolation of *Salmonella enterica* Serovar Enteritidis in Blue-Fronted Amazon Parrot (*Amazona aestiva*). Avian Dis 54:151–155

Marietto-Gonçalves GA, Martins TF, Andreatti Filho RL (2012) Chewing lice (Insecta, Phthiraptera) parasitizing birds in Botucatu, SP, Brazil. R Bras Ci Vet 19:206–212

Mase M, Imada T, Sanada Y et al (2001) Imported parakeets harbor H9N2 influenza A viruses that are genetically closely related to those transmitted to humans in Hong Kong. J Virol 75: 3490–3494

Matsumoto A, Manire GP (1970) Electron microscopic observations on the effects of penicillin on the morphology of *Chlamydia psittaci*. J Bacteriol 101:278–285
Mattsson R, Haemig PD, Olsen B (1999) Feral pigeons as carriers of Cryptococcus laurentii, Cryptococcus uniguttulatus and Debaryomyces hansenii. Med Mycol 37:367–369
McDonald SE, Lowenstine LJ, Ardans AA (1981) Avian pox in blue-fronted Amazon parrots. J Am Vet Med Assoc 179:1218–1222
McHugh JM, de Kloet SR (2015) Discrepancy in the diagnosis of avian *Borna disease virus* infection of *Psittaciformes* by protein analysis of feather calami and enzyme-linked immunosorbent assay of plasma antibodies. J Vet Diag Invest 27:150–158
Meyer KF, Eddie B (1935) Avian psittacosis. J Bacteriol 29:67
Mihaylov G, Petrov V, Marutsov P et al (2008) A case of aspergillosis in a bearded vulture (Gypaetos barbatus). Trakia J Sci 6:144–146
Miller TD, Millar DL, Naqi SA (1979) Isolation of Pacheco's disease Herpesvirus in Texas. Avian Dis 23:753–756
Miringa EN (1975) Pasteurellosis in African grey parrots (*Psittacus erithacus* L.). Avian Dis 19:812–813
Mitura A, Szymańska-Czerwińska M, Niemczuk K, Anara J (2014) *Chlamydia* in birds-occurrence, new species and zoonotic potential—a review. Bull Vet Inst Pulawy 58:503–506
Moroney JF, Guevara R, Iverson C et al (1998) Detection of chlamydiosis in a shipment of pet birds, leading to recognition of an outbreak of clinically mild psittacosis in humans. Clin Infect Dis 26:1425–1429
Moulder JW (1962) Some basic properties of the psittacosislymphogranuloma venereum group of agents. Structure and chemical composition of isolated particles. Ann NY Acad Sci 98:92–99
Muir M, Raidal SR (2012) Necrotising ventriculitis due to combined infection with *Rhizopus microsporus* var. *chinensis* and *Candida krusei* in an eclectus parrot (*Eclectus roratus*). Aust Vet J 90:277–280
Nakamura AA, Homem CG, da Silva AMJ et al (2014) Diagnosis of gastric cryptosporidiosis in birds using a duplex real-time PCR assay. Vet Parasitol 205:7–13
Napier JE, Hinrichs SH, Lampen F et al (2009) An outbreak of avian mycobacteriosis caused by *Mycobacterium intracellulare* in little blue penguins (*Eudyptula minor*). J Zoo Wildl Med 40:680–686
Nardoni S, Ceccherelli R, Rossi G et al (2006) Aspergillosis in Larus cachinnans micaellis: survey of eight cases. Mycopathol 161:317–321
Nemeth NM, Oesterle PT, Bowen RA (2008) Passive Immunity to West Nile virus provides limited protection in a common Passerine species. Am J Trop Med Hyg 79:283–290
Nemeth NM, Oesterle PT, Poulson RL et al (2013) Experimental infection of European starlings (*Sturnus vulgaris*) and house sparrows (*Passer domesticus*) with pandemic 2009 H1N1 and swine H1N1 and H3N2 triple reassortant influenza viruses. J Wildl Dis 49:437–440
Nicholls EB (1914) Bird life in the National Park, N.S.W. The Emu 14:84–88
Nikolay B (2015) A review of West Nile and Usutu virus co-circulation in Europe: how much do transmission cycles overlap? Trans R Soc Trop Med Hyg 109:609–618
Njabo KY, Zanontian L, Sheta BN et al (2016) Living with avian FLU-Persistence of the H5N1 highly pathogenic avian influenza virus in Egypt. Vet Microbiol 187:82–92
Norte AC, Ramos JA, Gern L et al (2013) Birds as reservoirs for *Borrelia burgdorferi* s.l. in Western Europe: circulation of *B. turdi* and other genospecies in bird–tick cycles in Portugal. Environ Microbiol 15:386–397
Nosanchuk JD, Shoham S, Fries BC et al (2000) Evidence of zoonotic transmission of *Cryptococcus neoformans* from a pet cockatoo to an immunocompromised patient. Ann Intern Med 132:205–208
Okuda H, Ohya K, Shiota Y et al (2011) Detection of *Chlamydophila psittaci* by using SYBR Green Real-Time PCR. J Vet Med Sci 73:249–254
Olias P, Wegelin M, Zenker W et al (2011) Avian malaria deaths in parrots, Europe. Emerg Infect Dis 17:950–952

Olsen B, Gylfe A, Bergström S (1996) Canary finches (*Serinus canaria*) as an avian infection model for Lyme borreliosis. Microb Pathog 20:319–324

Ortiz-Catedral L, Ismar SMH, Baird K et al (2009) No evidence of *Campylobacter*, *Salmonella* and *Yersinia* in free-living populations of the red-crowned parakeet (*Cyanoramphus novaezelandiae*). New Zeal J Zool 36:379–383

Pacheco G and Bier O (1930) Epizootie chex les perroquets du Brésil. Relations avec le psittacose. C R Soc Biol 105:109–111

Pal M (1992) Disseminated Aspergillus terreus infection in a caged pigeon. Mycopathol 119: 137–139

Palmieri C, Franca M, Uzal F et al (2011) Pathology and immunohistochemical findings of West Nile Virus infection in Psittaciformes. Vet Pathol 48:975–984

Palmieri C, Roy P, Dhillon AS et al (2013) Avian Mycobacteriosis in Psittacines: a retrospective study of 123 cases. J Comp Path 148:126–138

Palomar AM, Santibáñez P, Mazuelas D et al (2012) Role of birds in dispersal of etiologic agents of tick-borne zoonoses, Spain, 2009. Emerg Infect Dis 18:1188–1191

Panigrahy B, Elissalde G, Grumbles LC et al (1978) *Giardia* infection in parakeets. Avian Dis 22:815–818

Panigrahy B, Harmon BG (1985) Bacterial septicemias in two psittacine birds. J Am Vet Med Assoc 186:983–984

Pass DA, Perry RA (1984) The pathology of psittacine beak and feather disease. Aust Vet J 61:69–74

Pass DA, Prus SE, Riddell C (1987) A papova-like virus infection of splendid parakeets (*Neophema splendida*). Avian Dis 31:680–684

Pasteur L (1880) *Sur les maladies virulentes et en particulier sur la maladie appele´e vulgairement chole´ra des poules*. CR Acad Sci (Paris) 90:239–248

Payne S, Covaleda L, Jianhua G et al (2011) Detection and characterization of a distinct Bornavirus lineage from healthy Canada Geese (*Branta canadensis*). J Virol 85:12053–12056

Perkins LEL, Swayne DE (2003) Varied pathogenicity of a Hong Kong-origin H5N1 avian influenza virus in four passerine species and budgerigars. Vet Pathol 40:14–24

Perpiñán D, Garner MM, Wellehan JFX et al (2010) Mixed infection with Reovirus and *Chlamydophila* in a flock of Budgerigars (*Melopsittacus undulatus*). J Avian Med Sur 24: 316–321

Peters M, Prodinger WM, Gu¨mmer H et al (2007) *Mycobacterium tuberculosis* infection in a blue fronted amazon parrot (*Amazona aestiva aestiva*). Vet Microbiol 122: 381–383

Piasecki T, Chrząstek K, Wieliczko A (2012) Detection and identification of *Chlamydophila psittaci* in asymptomatic parrots in Poland. BMC Vet Res 8:233–238

Pillai SPS, Suarez DL, Pantin-Jackwood M et al (2008) Pathogenicity and transmission studies of H5N2 parrot avian influenza virus of Mexican lineage in different poultry species. Vet Microbiol 129:48–57

Pilny AA, Quesenberry KE, Bartick-Sedrish TE et al (2012) Evaluation of *Chlamydophila psittaci* infection and other risk factors for atherosclerosis in pet psittacine birds. J Am Vet Med Assoc 240:1474–1480

Platonov AE, Karan LS, Kolyasnikova NM et al (2011) Humans infected with relapsing fever spirochete *Borrelia miyamotoi*, Russia. Emerg Infect Dis 17:1816–1823

Portaels F, Realini L, Bauwens L et al (1996) Mycobacteriosis caused by *Mycobacterium genavense* in birds kept in a zoo: 11-year survey. J Clin Microbiol 34:319–323

Qi M, Wang R, Ning C et al (2011) *Cryptosporidium* spp. in pet birds: genetic diversity and potential public health significance. Exp Parasitol 128:336–340

Quist EM, Belcher C, Levine G et al (2011) Disseminated histoplasmosis with concurrent oral candidiasis in an Eclectus parrot (*Eclectus roratus*). Avian Pathol 40:207–211

Raidal SR, Butler R (2001) Chronic Rhinosinusitis and Rhamphothecal Destruction in a Major Mitchell's Cockatoo (Cacatua leadbeateri) due to Cryptococcus neoformans var gattii. J Avian Med Sur 15:121–125

Raphael BL, Iverson WO (1980) Coligranuloma and psittacosis in an Amazon parrot. J Am Vet Med Assoc 177:927–929

Raso TF, Teixeira RHF, Carrasco AOT et al (2013) *Chlamydophila psittaci* infections in Hyacinth Macaws (*Anodorhynchus Hyacinthinus*) confiscated in Brazil. J Zoo Wildlife Med 44:169–172

Raso TF, Werther K, Miranda ET et al (2004) Cryptococcosis outbreak in psittacine birds in Brazil. Med Mycol 42:355–362

Raue R, Gerlach H, Muller H (2005) Phylogenetic analysis of the hexon loop 1 region of an Adenovirus from psittacine birds supports the existence of a new psittacine Adenovirus (PsAdV). Arch Virol 150:1933–1943

Rauter C, Hartung T (2005) Prevalence of *Borrelia burgdorferi* sensu lato genospecies in *Ixodes ricinus* ticks in Europe: a metaanalysis. Appl Environ Microbiol 71:7203–7216

Ravich ML, Reavill DR, Hess L et al (2014) Gastrointestinal Cryptosporidiosis in captive psittacine birds in the United States: a case review. J Avian Med Sur 28:297–303

Rinder M, Ackermann A, Kempf H et al (2009) Broad tissue and cell tropism of Avian Bornavirus in parrots with Proventricular Dilatation Disease. J Virol 83:5401–5407

Ritchie PA, Anderson IL, Lambert DM (2003) Evidence for specificity of psittacine beak and feather disease viruses among avian hosts. Virol 306:109–115

Ritter J (1879) Uber Pneumotyphus, eine Hausepidemie in Uster. Dtsch Arch klin Med 25:53

Rosell J, Máinez M, Didier ES et al (2016) *Encephalitozoon hellem* infection in aviary passerine and psittacine birds in Spain. Vet Parasitol 219:57–60

Rosenbusch CT and Merchant IA (1939) A study of the hemorrhagic septicemia Pasteurellae. J Bacteriol 37(1):69

Rossi G, Piccinini A, Heikema AP et al (2015) *Campylobacter jejuni* and Proventricular Dilatation Disease in parrots: there may be a correlation? In: Proceedings of ICARE-2015, Paris

Rosskopf WJ, Woerpel RW (1984) Cryptococcosis in a thick billed parrot. In: Proc Assoc Avian Vet, pp 281–288

Roy P, Venugopalan AT, Selvarangam R et al (1998) Velogenic Newcastle disease virus in captive wild birds. Trop Anim Health Prod 30:299–303

Ryser Degiorgis MP, Hirschheydt JV, Origgi FC (2012) Epidemic of Salmonellosis in passerine birds in Switzerland with spillover to domestic cats. Vet Pathol 50:597–606

Saggese MD, Tizard I, Gray P et al (2014) Evaluation of multidrug therapy with azithromycin, rifampin, and ethambutol for the treatment of Mycobacterium avium subsp. avium in ring-neck doves (Streptopelia risoria): an uncontrolled clinical study. J Avian Med Sur 28:280–289

Samanta I (2013) *Salmonella*. In: Samanta I (ed) Veterinary bacteriology, 1st edn. New India Publishing Agency, New Delhi, India, pp p138–p150

Samanta I (2015) Veterinary mycology, 1st edn. Springer

Samanta I, Joardar SN, Das PK et al (2014) Prevalence and antibiotic resistance profiles of Salmonella serotypes isolated from backyard poultry flocks in West Bengal, India. J Appl Poult Res 23:536–545

Sammann B, Diaz L, Cristina M et al (1994) *Cryptococcus neoformans* (San Felipe) Vuillemin en aves confinadas y recintos hospitalarios de la región Metropolitana (Chile). Bol Micol 9:65–72

Salmon DE (1885) The Virus of Hog Cholera. Public Health Pap Rep 11:73.

Sanfelice F (1894) Contributo alla morfologia e biologia dei blastomiceti che si sviluppano nei succhi de alcuni frutti. Ann Ist Ig R Univ Roma 4:463–469

Sargeant K, Sheridan A, O'Kelly J et al (1961) Toxicity associated with certain samples of groundnuts. Nat 192:1096–1097

Sarker S, Ghorashi SA, Forwood JK et al (2014) Phylogeny of beak and feather disease virus in cockatoos demonstrates host generalism and multiple-variant infections with in Psittaciformes. Virol 460–461:72–82

Sarker S, Moylan KG, Ghorashi SA et al (2015) Evidence of a deep viral host switch event with *beak and feather disease virus* infection in rainbow bee-eaters (*Merops ornatus*). Sci Rep 5:14511

Sassa Y, Horie M, Fujino K et al (2013) Molecular epidemiology of avian bornavirus from pet birds in Japan. Virus Genes 47:173–177
Schafer JR, Kawaoka Y, Bean WJ et al (1993) Origin of the pandemic 1957 H2 influenza-A virus and the persistence of its possible progenitors in the avian reservoir. Virol 194:781–788
Schafer W (1955) Vergleichende sero-immunologische untersuchungen uber die viren der influenza unf klassichen geflugelpest. Zeitschrift fur Naturforschung 10B:81–91
Schmittdiel E (1966) Studies of parrots, parakeets and their excrement to demonstrate the presence of the causative agent of psittacosis following treatment of the birds with chlortetracycline. Berl Munch Tierarztl Wochenschr 79:398–399
Schremmer C, Lohr JE, Wastlhuber U et al (1999) Enteropathogenic *Escherichia coli* in Psittaciformes. Avian Pathol 28:349–354
Scott GR, Gibson MA, Danskin D (1956) The reappearance of Newcastle disease in Kenya. Bull Epiz Dis Afr 4:65–68
Scott GR, Winmill AJ (1960) Newcastle Disease in the Grey Parrot (*Psittacus Erithacus* L.). J Comp Pathol Ther 70:115–119
Scott JD, Lee MK, Fernando K et al (2010) Detection of Lyme disease spirochete, *Borrelia burgdorferi* sensu lato, including three novel genotypes in ticks (Acari: Ixodidae) collected from songbirds (Passeriformes) across Canada. J Vector Ecol 35:124–139
Seeley KE, Baitchman E, Bartlett S et al (2014) Investigation and control of an attaching and effacing *Escherichia coli* outbreak in a colony of captive budgerigars (*Melopsittacus undulatus*). J Zoo Wildl Med 45:875–882
Shihmanter E, Weisman Y, Lublin A et al (1998) Isolation of avian serotype 3 paramyxoviruses from imported caged birds in Israel. Avian Dis 42:829–831
Shitaye JE, Halouzka R, Svobodova J et al (2010) First isolation of *Mycobacterium genavense* in a blue headed parrot (*Pionus menstruus*) imported from Surinam (South America) to the Czech Republic: a case report. Vet Med 55:339–347
Shojaei TR, Tabatabaei M, Shawky S et al (2015) A review on emerging diagnostic assay for viral detection: the case of avian influenza virus. Mol Biol Rep 42:187–199
Simpson CF, Hanley JE, Gaskin JM (1975) Psittacine herpesvirus infection resembling pacheco's parrot disease. J Infect Dis 131:390–396
Simpson VR (1991) *Leucocytozoon*-like infection in parakeets, budgerigars and a common buzzard. Vet Rec 129:30–32
Slavin D (1955) *Cryptosporidium meleagridis* (sp. nov.). J Comp Pathol 65:262–266
Slemons RD, Johnson DC, Osborn JS et al (1974) Type-A influenza viruses isolated from wild free-flying ducks in California. Avian Dis 18:119–124
Slemons RD, Johnson DC, Malone TG (1973) Influenza type A isolated from imported exotic birds. Avian Dis 17(2):485–489
Smith KA, Bradley KK, Stobierski MG et al (2005a) Compendium of measures to control *Chlamydophila psittaci* (formerly *Chlamydia psittaci*) infection among humans (psittacosis) and pet birds, 2005. J Am Vet Med Assoc 226:532–539
Smith KE, Anderson F, Medus C et al (2005b) Outbreaks of Salmonellosis at elementary schools associated with dissection of owl pellets. Vector Zoon Dis 5:133–136
Smithburn KC, Hughes TP, Burke AW et al (1940) A neurotropic virus isolated from the blood of a native of Uganda. Am J Trop Med Hyg 20:471–492
Socolovschia C, Reynaud P, Kernif T et al (2012) Rickettsiae of spotted fever group, *Borrelia valaisiana*, and *Coxiella burnetii* in ticks on passerine birds and mammals from the Camargue in the south of France. Ticks Tick Borne Dis 3:354–359
Splendore A (1913) Bull Soc Pathol Exot 6:318–323
Sréter T, Varga I (2000) Cryptosporidiosis in birds-a review. Vet Parasitol 87:261–279
Staeheli P, Rinder M, Kaspers B (2010) Avian Bornavirus associated with fatal disease in psittacine bird. J Virol 84:6269–6275
Steinmetz HW, Bakonyi T, Weissenbock H et al (2011) Emergence and establishment of Usutu virus infection in wild and captive avian species in and around Zurich, Switzerland-genomic and pathologic comparison to other central European outbreaks. Vet Microbiol 148:207–212

Steinmetz HW, Rutz C, Hoop RK et al (2006) Possible human-avian transmission of *Mycobacterium tuberculosis* in a green-winged macaw (*Ara chloroptera*). Avian Dis 50:641–645

Stockman J, Hawkins MG, Burns RE et al (2010) West Nile virus infection in a green-winged macaw (*Ara chloropterus*). Avian Dis 54:164–169

Strauch D, Rott R (1958) Comparative studies on isolation of psittacosis virus from the parakeet with aid of the mouse experiment and cultivation in chick embryo. Z Hyg Infektionskr 145:5–13

Sychra O (2006) *Neopsittaconirmus vendulae*, a new species of louse (Phthiraptera: Philopteridae) from the Cockatiel *Nymphicus hollandicus* (Psittaciformes: Cacatuidae). Zootaxa 1270:57–68

Takahashi T, Masuda M, Tsuruno T et al (1997) Phylogenetic analyses of *Chlamydia psittaci* strains from birds based on 16S rRNA gene sequence. J Clin Microbiol 35:2908–2914

Tel OY, Bozkaya F, Keskin O (2013) *Salmonella*, *Campylobacter* and *Chlamydophila* in bald ibis (*Geronticus eremita*) faeces in Turkey. J Zoo Wildl Med 44:21–26

Tresierra-Ayala A, Bendayan ME (1998) Thermotolerant *Campylobacter* species isolated from Psittaciformes in the Peruvian Amazon region. Rev Inst Med Trop S Paulo 40:263–264

Trevisan V (1887) Sul micrococco della rabbia e sulla possibilità di riconoscere durante il periode d'incubazione, dall'esame del sangue della persona moricata, se ha contratta l'infezione rabbica. Rend Ist Lombardo Sez (ser. 2), 20:88–105

Tsai SS, Hirai K, Itakura C (1992) Histopathological survey of protozoa, helminths and acarids of imported and passerine local psittacine and birds in Japan. Jpn J Vet Res 40:161–174

Tunca R, Toplu N, Kırkan Ş et al (2012) Pathomorphological, immunohistochemical and bacteriological findings in budgerigars (*Melopsittacus undulatus*) naturally infected with *S.* Gallinarum. Avian Pathol 41:203–209

Tyzzer EE (1907) A protozoan found in the peptic glands of the common mouse. Proc Soc Exp Biol Med 5:12–13

Tyzzer EE (1910) An extracellular coccidium *Cryptosporidium muris* (gen. et sp. nov.) of the gastric glands of the common mouse. J Med Res 23:487–509

Tyzzer EE (1929) Coccidiosis in gallinaceous birds. Am J Hyg 10:269–383

van der Meulen KM, Pensaert MB, Nauwynck HJ (2005) West Nile virus in the vertebrate world. Arch Virol 150:637–657

Vandervort WJ (1954) *Salmonella* gastroenteritis; report of two cases ascribed to a pet duck. Del Med J 26:159–160

Vazquez A, Jimenez-Clavero MA, Franco L et al (2011) Usutu virus-potential risk of human disease in Europe. Euro Surveill 16:pii19935

Vergara-Alert J, Ferna´ndez-Bellon H, Busquets N et al (2011) Comprehensive serological analysis of two successive heterologous vaccines against H5N1 avian influenza virus in exotic birds in zoos. Clin Vaccine Immunol 18:697–706

Vigo GB, Origlia J, Gornatti D et al (2009) Isolation of *Salmonella* Typhimurium from Dead Blue and Gold Macaws (*Ara ararauna*). Avian Dis 53:135–138

Vuillemin P (1901) Les blastomycètes pathogènes. Rev Gen Sci Pures Appl 12:732–751

Ward MP, Raim A, Yaremych-Hamer S et al (2006) Does the roosting behavior of birds affect transmission dynamics of West Nile virus? Am J Trop Med Hyg 75:350–355

Ward MP, Ramer JC, Proudfoot J et al (2003) Outbreak of Salmonellosis in a Zoologic Collection of Lorikeets and Lories (*Trichoglossus*, *Lorius*, and *Eos* spp.). Avian Dis 47:493–498

Washko RM, Hoefer H, Kiehn TE et al (1998) *Mycobacterium tuberculosis* infection in a green-winged macaw (*Ara chloroptera*): report with public health implications. J Clin Microbiol 36:1101–1102

Weisman J, Singer N (1976) Isolation of *Yersinia* (*Pasteurella*) *pseudotuberculosis* from the Palm dove (*Streptopelia senegalensis*). Avian Dis 20:202–204

Weissenbock H, Kolodziejek J, Fragner K et al (2003) Usutu virus activity in Austria, 2001–2002. Microbes Infect 5:1132–1136

Weissenbock H, Kolodziejek J, Url A et al (2002) Emergence of Usutu virus, an African mosquito borne Flavivirus of the Japanese encephalitis virus group, central Europe. Emerging Infect Dis 8:652–656

White DJ, Hall RJ, Jakob-Hoff R et al (2015) Exudative cloacitis in the kakapo (*Strigops habroptilus*) potentially linked to *Escherichia coli* infection. N Z Vet J 63:167–170

Williams MC, Simpson DI, Haddow AJ et al (1964) The isolation of West Nile Virus from man and of Usutu virus from the bird biting mosquito *Mansonia aurites* (Theobald) in the Entebbe area of Uganda. Ann Trop Med Parasitol 58:367–374

Wörz JJ (1858) Die halb-acute Gehirn-Entzündung oder Kopf-Krankheit der Pferde. Ebner und Seubert; Stuttgart, German Confederation

Wünschmann A, Rejmanek D, Cruz-Martinez L et al (2009) *Sarcocystis falcatula*-associated encephalitis in a free-ranging great horned owl (*Bubo virginianus*). J Vet Diagn Invest 21: 283–287

Xenoulis PG, Gray PL, Brightsmith D et al (2010) Molecular characterization of the cloacal microbiota of wild and captive parrots. Vet Microbiol 146:320–325

Yogasundram K, Shane SM, Harrington KS (1989) Prevalence of *Campylobacter jejuni* in selected domestic and wild birds in Louisiana. Avian Dis 33:664–667

Zuydam DM (1952) Isolation of Newcastle disease virus from the osprey and the parakeet. J Am Vet Med Assoc 120:88–89

3 Systemic Clinical and Metabolic Diseases

Samiran Bandyopadhyay

3.1 Overview of Systemic Illness

Like other animals pet and companion birds are also prone to systemic illness. This is presented in the form of certain clinical signs and symptoms which is known as "sick-bird syndrome." The pet birds have their unique physiology and anatomy and their metabolic turnover is quite high. The feeding practice adopted by owner may have an impact over their health status. Birds are kept confined with limited exercise and they are deliberately kept isolated from their foraging behaviour. All these may have an impact over their health status. Diagnosis of these diseases is necessary to adopt a correct treatment or management protocol. However, this is rather complex in caged birds as many of the birds may not exhibit any clinical manifestation as such. By the time these symptoms are presented birds become really sick. Many of systemic diseases are presented with same cluster of symptoms. Moreover, they have a tendency to hide their illness as common tendency of all prey animals. Therefore, the observable and detectable clinical signs indicate a grave concern. Even a little manipulation may lead to serious consequences.

There are certain symptoms which must be borne into mind indicating towards general illness

1. Huddling of the birds together
2. Consistent ruffled feather or any change in plumage
3. Change of wings position (dropped/elevated)
4. Sitting low on perch or only at one side or bottom of the cage
5. Any observable change of colour of the beak
6. Convulsion/ataxia/trembling
7. Ambulatory disturbance/walking in an imbalanced manner
8. Walking in circle

I. Samanta and S. Bandyopadhyay, *Pet Bird Diseases and Care*,
DOI 10.1007/978-981-10-3674-3_3

9. Any detectable lump or swelling on any part of its body
10. Any kind of weakness
11. Anorexia
12. Self-mutilation
13. Discharge from eyes or nares
14. Any change or abnormality in color, volume, consistency, and number of droppings
15. Respiratory difficulty
16. Lameness or pain.

The symptoms which require emergency interventions are

1. Dyspnoea
2. Ascites
3. Convulsion
4. Trauma and bleeding
5. Collapse
6. Burn injury.

Behavioural changes are also good indicators of systemic illness. The changes which must be looked after are as follows—

1. The affected birds do not response to any kind of stimuli
2. They will have abnormal changes in vocalizations
3. Lethargy and inactivity
4. Mental stupor or over aggression
5. Reluctant to play
6. Any change in eating or drinking behaviour.

There are certain points which must be taken seriously when the dietary formulation of the companion birds are followed as many of systemic diseases originate from there itself.

1. Many of caged or companion birds are highly sensitive to dyes preservatives used in many feed formulation. This can cause anaphylactic reaction with serious health problem.
2. High calorie diet with all seed or nuts often cause the serious problem like cardiomyopathy, atherosclerosis, fatty liver syndrome etc.
3. Mycotoxin in diet is a great source of problem like pneumonia and toxic hepatopathy leading to micro hepatica, hepatic cirrhosis and toxic nephropathy
4. Ample potable drinking water is required to maintain electrolyte balance and good hydration and keep the renal system in good health.

3.2 Disorders of Endocrinological Origin

3.2.1 Avian Goiter/Thyroid Enlargement or Hyperplasia

Thyroid gland enlargement or thyroid hyperplasia is detected in various pet birds like pigeons, macaw and other birds. It is generally caused due to dietary deficiency of iodine. The birds which are kept on seed based diet are more prone to develop this disease particularly when the seeds are grown on soil deficient on iodine. Therefore, pellet based diet is preferable when thyroid hyperplasia is detected in a flock. This is very common in various species of pet birds like pigeons, canaries, cockatiels, macaw, and budgerigars (budgies). Goiter has been detected in penguins also. Other than neoplasia, this is the most common cause of death of budgerigars. Certain breeds of pigeons are also more prone like White Carneau.

3.2.1.1 Etiology

Various factors were detected to cause thyroid hyperplasia or goiter in pet birds. The most common cause is iodine deficiency. The birds which are solely placed under iodine deficient seed or grain diet are more likely to develop thyroid hyperplasia. Again more consumption of goitrogenic agents (kale, cabbage, broccoli, soybean, flax, rapeseed, turnips etc.) or their incorporation in the diet may lead to develop goiter. Again chronic infection of thyroid gland or toxicity due to ingestion of organophosphates and chlorinated biphenols were reported to cause thyroid hyperplasia.

3.2.1.2 Pathophysiology

Due to iodine deficiency (Fig. 3.1) the production of thyroxine hormone is diminished and decreased blood concentration of thyroxine give a signal to brain which sensitizes the thyroid gland for optimum production of thyroxine through

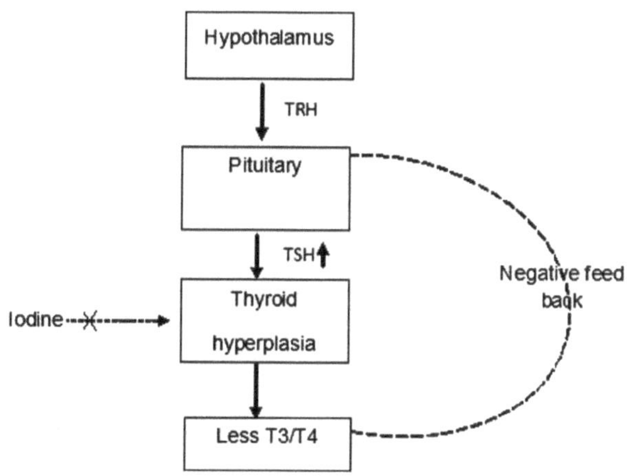

Fig. 3.1 Insufficient dietary iodine leads to goiter

thyroid stimulating hormone (TSH). Thus the thyroxin producing cells are increased in number leading to development of thyroid hyperplasia.

3.2.1.3 Clinical Findings

- The thyroid enlargement may be large enough to be visible by clinicians. The gland is located in thyroid cartilage near upper respiratory tract.
- The enlargement may sometimes occlude the passage of trachea with development of dyspnoea or respiratory distress.
- The pressure on oesophagus and crop may lead to difficulty in swallowing which may lead to apathy to take food resulting in weight loss. Recurrent vomiting or regurgitation may be developed due to pressure on proventriculus.
- Likewise pressure on heart, major vessels, lung or air-sac may lead to convulsion and sudden death.
- Change of voice is often noticed in the affected birds.
- Due to diminished concentration of thyroxine, the birds may be lethargic, depressed with reduced metabolic rate and develop dermatological symptoms like loss of feathers and chronic itching.
- Reproductive problems like retardation in sexual maturity, embryonic death and reduced hatchability may be noticed in birds with goiter.
- The clinical presentation of goiter is different in pigeon to some extent. The affected birds may be lethargic and obese with palpable enlarged mass at thoracic inlet (Fig. 3.2). The birds become immunocompromised and presented with ruffled or distorted feathers and plumage. There is accumulation of fluid under skin with puffy appearance (myxedema). Affected pigeons exhibit lower hatchability and reduced fertility.

Fig. 3.2 A pigeon suffering from avian goiter showing enlargement of thyroid gland. The bird was presented with complications like dyspnoea, painful swallowing and progressive emaciation (*Courtesy* Suman Biswas, ARD Department, Government of West Bengal, India)

3.2.1.4 Pathology

Gross pathological changes include bilateral enlargement of thyroid gland with more number of translucent and distended follicles in the capsule with dark red or black haemorrhages. Histologically, the hyperplastic glands are detected with enlarged columnar or cuboidal epithelial cells with no apparent colloid or follicular lumen. Thyroid hyperplasia may be confused with thyroid adenoma or adenocarcinoma. In carcinoma, there is severe distortion and destruction of the whole gland. The poorly differentiated cells may be observed to infiltrate the capsule and surrounding tissue.

3.2.1.5 Diagnosis

- Thyroid enlargement with change of voice may be the cardinal sign.
- X-ray of the neck region of the affected birds will reveal thyroid enlargement with dorsal or ventral displacement of trachea or oesophagous.
- Blood examination for estimation of serum thyroxine concentration.
- By this method T4 level is estimated. However, no separate kit is available for avian species for T4 estimation and as the concentration of serum T4 is too low in birds, it is not always possible to precisely determine the concentration using feline or canine testing kits, especially when it drops down below 0.2 μg/dl.
- Many workers suggested TSH stimulation test as better alternative for diagnosis of hypothyroidism. In this case 1.0–2.0 U of TSH is injected intramuscularly and two samples are collected at 4 h before or 6 h after the administration of TSH. In healthy birds, T4 level should at least double in concentration following administration of TSH. In pigeons, the second blood collection is usually done after 24 or 32 h of 0.1 or 1 U of TSH injection. In healthy birds, serum T4 level should increase by 2.5 folds above the basal level.

3.2.1.6 Treatment

Iodine supplementation is the mainstay of therapy. In severe cases, sodium/ potassium iodide may be injected. For a 30–35 g of budgerigar the dietary requirement of iodine is about 20 μg per week. This is usually done by injecting 0.01 ml/budgie once, IM in the form of 20% sodium iodine in saline. For oral supplementation a stock solution of 2 ml Lugol's iodine in 30 ml water should be prepared and one drop of stock solution is to be mixed in 250 ml drinking water. This is to be used daily for treatment, 2–3 times per week for prevention.

L-thyroxine is often advocated in psittacine birds. It is generally given at the dosage of 0.02–0.04 mg/kg body weight at every 24 h. However, its administration needs special care as in healthy or euthyroid birds; it may create the problems like cardiomyopathy and congestive heart failure. Goiter may be prevented by supplementation of trace amount of iodine in the diet.

3.2.2 Diabetes Mellitus

Diabetes mellitus (DM) is frequently reported in granivorous birds including the domestic pigeons and the disease is characterized by polyphagia, polydipsia, polyuria and chronic weight loss. Persistent hyperglycemia and glycosuria are mainly demonstrated in the affected birds. It is more common in budgerigars, cockatiels and galahs. Among others, larger psittacine, toucans, mynahs are also affected.

3.2.2.1 Pathophysiology

The underlying mechanism of diabetes is poorly understood in birds. Unlike mammals where insulin has a predominant role in DM, in birds blood glucose level seems to be controlled by a complex hormonal milieu. In comparison to mammalian pancreas, avian pancreas has a low proportion of insulin secreting cells and 5–6 times higher number of glucagon producing cells. Circulating glucagon concentration in avian blood is 10–50 times higher than the mammalian blood. Glucagon as a catabolic hormone plays a pivotal role in gluconeogenesis, lipolysis and glycogenolysis to augment the blood glucose level, while insulin controls the entry of glucose in the cells and its utilization. In birds particularly in granivorous species, Glucagon is considered to play a more relevant role for development of DM. However, other factors like somatostatin, growth hormone, epinephrine, thyroxine, prolactin, pancreatic polypeptides and corticosteroid may have a modulatory role in development of persistent hyperglycemia. Hyper production of any of these hormones either due to tumor of the hormone producing cells or due to paraneoplastic syndrome may lead to such condition. Islet cell carcinoma with DM has been described in a parakeet.

In general, DM may occur in three forms in birds—

1. Type I DM: It is purely of pancreatic origin due to selective destruction of pancreatic cells. This form of the DM is more common in toucans and parrots.
2. Type II DM: The type II DM is associated with some other diseases or conditions like obesity and iron storage diseases.
3. Type III DM: It is linked with pancreatic diseases like pancreatic neoplasia, pancreatic insufficiency and pancreatitis. Some insulin inhibitory chemicals or drugs like megestrol acetate, medroxyprogesterone acetate or corticosteroids.

3.2.2.2 Clinical Findings

Clinical manifestation of the birds is straightforward—

1. Polyuria
2. Polyphagia
3. Polydipsia
4. Chronic weight loss.

Sometimes, this disease is associated with other non-specific signs like obesity, vomiting and lethargy. However, the affected birds generally maintain a good appetite.

3.2.2.3 Diagnosis

The main diagnosis is based on the detection of persistent hyperglycemia and glycosuria. However, detection of blood glucose level is tricky in birds as the avian blood glucose level is higher than in mammalian. Therefore, consistently higher level of blood glucose 38–44 mmol/l is indicative of DM in birds. Other than glucose, presence of ketone bodies in urine is also indicative of DM in birds. Clinical symptoms in some other conditions often resemble with the classical symptoms of DM in birds. Polyuria/polydipsia (PU/PD) syndrome is very common in pigeons which fed the squabs crop milk possibly due to decreased concentration of circulating prolactin. Likewise, infection of paramyxovirus serotype-1 may cause lasting PU/PD syndrome in pigeons. Again psychogenic PU/PD is also not uncommon in caged birds. Further, glycosuria has been detected in birds like African Grey Parrot without hyperglycemia. Thus glycosuria, even though an important indicator, cannot be considered alone for diagnosis of DM in avian. Moreover, other conditions like diabetes insipidus, medication with corticosteroids, diuretics, progesterone, renal or hepatic insufficiency, other hormonal irregularities should be considered before definite diagnosis of DM in birds.

3.2.2.4 Treatment

Management of DM is always a challenge. In birds it is more difficult as it is not easy to monitor the blood glucose level and thereby to evaluate the effect of the hypoglycemic drugs. The main objective of the treatment is to lower down the blood glucose level and maintain it. The treatment constitutes of insulin and other hypoglycemic drugs like sulfonylurea. The acute rise of blood glucose level can be treated with short acting insulin @ 0.1–0.2 U/kg. However, to maintain the blood glucose level persistently long acting insulin is prescribed with a varying dose (0.067–3.3 U/kg OD/BID) depending upon the clinical improvement and the blood glucose or urinary glucose level. Insulin is recorded to cause hypoglycemic shock in human. However, such instances were not recorded in birds. But possibility of such fallout cannot be denied. To avoid such happenings repeated sampling is required to monitor the blood glucose level making this quite difficult in birds. Oral hypoglycemic agents like sulfonylurea is good alternative. Sulfonylurea is known to stimulate the pancreatic beta cell to secrete endogenous insulin and to increase tissue sensitivity towards insulin. However, its efficacy in birds with insulin resistance is not properly known. Although insulin has been reported for treatment of DM in birds, its real efficacy is controversial in granivorous birds which are known to exhibit insulin insensitivity. Carnivorous birds may be more amicable to insulin therapy.

Dietary management of the affected birds is good option. In many cases dietary management itself is enough to control DM in birds without any antidiabetic drug intervention. It is important to avoid high calorie diet or diet with high fat like sunflower oil. This can control obesity. Increasing fiber supple mention is also an important part for clinical resolution of DM in birds.

3.3 Cardiovascular Diseases

Due to some predisposing factors cardiac diseases are not uncommon in pet birds. Depending upon the data of retrospective pathology, almost 10–40% pet birds are presumed to be affected by cardiac diseases. Nevertheless, these abnormalities are rarely detected ante-mortem. Restricted movement with lack of adequate exercise, nutritional deficiency, stresses associated with acclimatization and persisting hypertension are few of the factors that are responsible for cardiac diseases in pet birds.

Among various congenital heart diseases, intraventricular septal defect, duplicitas cordis, multiplicatis cordis and ectopis cordis were reported in pet birds. Again most of these conditions were recorded during post-mortem examination. As the avian heart is always under substantial pressure, most of the congenital disorders lead to early embryonic death.

3.3.1 Pericardial Diseases

Two kinds of pericardial diseases are frequently noted in pet birds—pericarditis and pericardial effusion either in the form of hemopericardium or hydropericardium. Pericarditis is the inflammation of pericardium and its associated tissue occurs mostly due to infectious organism. Mycotic infection originated from respiratory tract, *Mycobacterium* infection or trichomoniasis in pigeons may lead to pericarditis. Visceral gout with deposition of uric acid crystals may also cause severe pericarditis in caged birds. Hemopericardium develops mostly from trauma whereas hydropericardium (Fig. 3.3) results from dietary protein deficiency and congestive heart failure.

3.3.1.1 Clinical Findings

Most of the conditions are difficult to diagnose as clinical symptoms are nonspecific and clinicians have to rely on radiography showing increased cardiac silhouette. On the other hand, other test like echocardiography and endoscopy may be used as reliable alternative.

Birds with pericardial effusion may show few characteristic symptoms like—

Decreased exercise intolerance,
Coughing,
Respiratory trouble or dyspnoea
Abdominal distension.

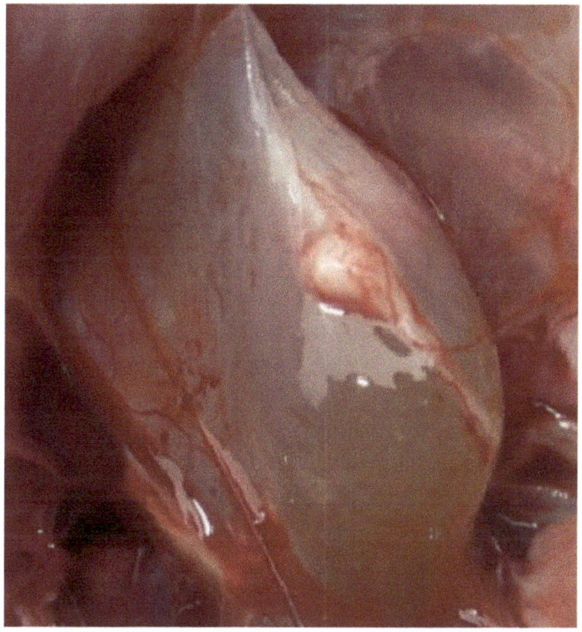

Fig. 3.3 Hydropericardium in a bird (*Courtesy* Amrit Dhara)

Congestion of liver and ascites may be detected additionally in case of hydropericardium.

3.3.1.2 Treatment

Pericardial effusion especially hydropericardium is usually treated with furosemide @ 0.15–2 mg/kg IM, SC or PO q12–24 h or some other diuretics along with acetylcholine esterase inhibitors. Overdose of furosemide can lead to dehydration and electrolyte imbalance. Lorikeets are very much sensitive to this drug, therefore, extra precaution is required.

3.3.2 Myocardial Diseases

Myocarditis or myocardial diseases are recorded in pet birds mainly resulting from systemic viral, bacterial, mycotic or protozoal infection. Infection with polyomavirus, chlamydia, sarcocystis and *E. coli* is not uncommon in birds. Moreover, chronic toxicity such as consumption of furazolidone often leads to myocarditis in birds. Increased preload during contraction as happened due to atherosclerosis or pulmonary hypertension may lead to aberrations in myocardial contractility. Haemosiderosis along with vulvar insufficiency was detected to cause myocardial failure in myna birds. Besides, vitamin E or selenium deficiency, right ventricular hypertrophy/dilatation was found to be associated with myocardial diseases. Chronic systemic mycosis often results in right sided enlargement of atrium and ventricle resulting in failure of myocardial contractility. Prolonged stress during transportation may also lead to myocardial failure.

3.3.3 Arrhythmias

3.3.3.1 Cardiac Arrhythmia

Cardiac arrhythmia is known as irregular heart beat where the heartbeat is too irregular, fast or slow. Many of the arrhythmia are asymptomatic and may not require treatment where as others is life threatening. Most of the arrhythmia noted in the pet birds are normal in nature and has been traced due to various predisposing factors. Sinus bradycardia has been observed in birds due to hypothermia. Long term anaesthesia often leads to hypothermia and bradycardia. Several organophosphorus or organochlorine compounds are also known to cause bradycardia via vagal stimulation. Again blood potassium concentration is an important parameter to control the cardiac rhythm. Hyperkalaemia or hypokalaemia, thiamine deficiency and deficiency of vitamin E may cause sinus bradycardia. Similarly sinus tachycardia with elevated heart rate is often asymptomatic and do not cause a potential damage unless the heart rate is too high to reduce the cardiac output. Most of the sinus tachycardia cases are associated with pain, nervousness and stress factors.

3.3.3.2 Atrial Tachycardia

It occurs when the electrical impulses are generated in some abnormal places (ectopic pacemakers) other than SA node and characterized by consistent rapid heart rate. This condition may be characterized by P on T wave in ECG where the P and T waves are super imposed. Atrial tachycardia may be seen with avian influenza virus infection. Atrial fibrillation is characterized by rapid and irregular heart beat and characterized by no P wave and an irregular ventricular rate in ECG. The QRS complexes are wide and are with higher amplitude indicating ventricular hypertrophy with inconsistent SS interval. The electrical impulses are generated from atrium rapidly and in an irregular manner and reaches to the AV node. There may be other type of supraventricular tachycardia where the rapid and irregular impulses are generated near the AV node junction.

3.3.3.3 AV Node Arrhythmia or Heart Block

AV node arrhythmia or heart block is not very infrequent in birds. It is especially due to incoordination of the atrial and ventricular depolarisation due to the incoordination in the electrical conduction system within heart. The first-degree atrioventricular block or PR prolongation the electrical conduction from atria to ventricle through AV node is delayed and reflected in ECG in the form of increased PR interval. Several diseases are known to cause such conditions like medication or application of anaesthetic agents like halothane, cardiotoxicants, excessive exertion, electrolyte imbalance, myocardial infarction, myocarditis and AV nodal disease. In ECG, PR interval may be increased almost 3–4 times. Many of the cases lead to severe bradycardia which can be reversed by atropine. Second degree block occurs when one or more atrial impulses fail to be conducted to the ventricle and Mobitz I/Wenckebach phenomenon which is characterized by gradual prolongation of PR interval followed by dropped "P wave" (or absence of QRS complex) has been recorded in racing pigeons, raptors and parrots. Third degree heart block or complete heart block where the impulses generated in SA node of the atrium does not propagate to ventricle and both the atrium and ventricle behave independently. In ECG the PR interval will vary abruptly and there will be no relationship between P wave and QRS complex. The birds may be under haemodynamic instability and the condition was noted in electrolyte imbalance like hypokalemia in birds.

3.3.3.4 Ventricular Premature Complex

Ventricular premature complex or premature ventricular contraction is noticed in a number of conditions where ventricular contraction occurs before the atrial contraction takes place leading to insufficient cardiac output and body is haemodynamically compromised. This is noted as skipped beat and the unlike normal heart beat where the heartbeat is initiated at SA node, in VPC, the heartbeat starts from Purkinji fibre. The VPC appears in large and bizarre shape and can be differentiated from the normal ECG finding very easily. VPC normally is recorded when the blood circulation to the cardiac muscle is severely compromised like—myocardial ischemia. There are other several conditions where such VPC can be recorded like hypoxia, smoking, thyroid problem, myocarditis, pericardial effusion, hypokalaemia, vitamin E or thiamine deficiency, lead poisoning etc. Ventricular tachycardia is a condition where the heart rate becomes too rapid due to improper electrical activity in ventricle. Such condition is observed in birds mostly due to myocardial hypoxia and myocardial infarction with abnormal ECG findings. It may lead to ventricular fibrillation where the heart loosed its pumping activity leading to cardiac arrest and death. The ECG records abnormal QRS complexes which are not properly formed and shaped.

3.3.3.5 Treatment

1. Cardiac glycosides like digoxin @ 0.02–0.05 mg/kg bodyweight at every 12 h followed by 0.01 mg/kg bodyweight as maintenance dose are effective to slow down the heart rate, improve the coronary circulation and reduce the oxygen demand in cardiac muscle with positive ionotropic effect.
2. Angiotensin converting enzyme (AGE) inhibitors like enalapril @ 5 mg/kg/day followed by a maintenance dose @ 1 mg/kg/day is known to induce diuresis, reduce blood pressure and cardiac load.
3. Beta-blocker (propranolol) @ 0.2 mg/kg/day is effective to overcome arrhythmia in birds.

3.3.4 Congestive Heart Failure

Congestive heart failure (CHF) occurs due to the persistent failure of the heart to pump blood effectively and that can lead to congestion in the systemic circulation resulting in fluid accumulation. Generally pulmonary venous congestion occurs due to left sided heart failure whereas systemic venous congestion starts with right sided heart failure. CHF is associated with incapacitated heart to maintain the cardiac output.

In general the pathophysiology of CHF is very complex. Owing to venous congestion, the blood circulation is reduced to kidney stimulating the renin-angiotensin mechanism and activation of the angiotensin I and II and aldosterone leading to retention of water and sodium. Ultimately in LHF, pulmonary oedema occurs whereas hepatomegaly and splenomegaly occurs with ascites and anasarca in RHF.

There are various causes for CHF in birds like

1. Any damage of the heart like disorders of pericardium, myocardium and endocardium which may interfere the blood circulation resulting in CHF.
2. Myocardial weakness, ischemia and infraction.
3. The right sided AV node thickening in birds may lead to valvular insufficiency and CHF.

3.3.4.1 Clinical Findings

The birds may be presented with prolonged weakness, ascites, coughing, complains of exercise intolerance, dyspnoea, syncope and cyanosis. Hepatomegaly may be observed in few cases.

3.3.4.2 Diagnosis

1. LHF may be accompanied by muffled lung sound due to pulmonary fluid accumulation where as in RHF, heart sound may be muffled due to hepatomegaly.
2. Chest radiograph may give the indication of cardiomegaly, pulmonary oedema and pleural or pericardial effusion.
3. Electrocardiography may be used to detect arrhythmia, atrial or ventricular defect.

3.3.4.3 Treatment

The birds should not be put under any stress or exertion. Strenuous exercise or activities should be avoided.

1. Oedema should be treated with diuretics like furosemide @ 0.15–2 mg/kg q12–24 h. To avoid sudden drop of blood pressure due to increased dose of furosemide it may be given on alternate day.
2. Angiotensin converting enzyme inhibitor enalapril may be given to inhibit renin-angiotensin-aldosterone activity @ 0.5 mg/kg two times a day.
3. Further to support myocardial function digoxin may be given @ 0.02–0.05 mg/kg once a day.

3.3.5 Arteriosclerosis and Atherosclerosis

Arteriosclerosis means thickening and hardening of the arteries and there are various kinds of arteriosclerosis and of them atherosclerosis is very common among captive birds. The word atherosclerosis has come from the Greek word "athero" (gruel or porridge) and "sclerosis" which means hardening. The incidence is exuberantly high among few species of birds particularly among Psittaciformes (parrots, parakeets etc.), Anseriformes (swan, geese, duck etc.), Columbiformes (pigeons, doves, etc.) and Galliformes (fowl, pheasants, etc.). In pet birds particularly those which are affected by senility between 8–15 years of age and those belong to psittacine birds like African grey parrots, amazons and cockatoos are frequently affected by atherosclerosis. This is a chronic inflammatory reaction particularly affecting the layer between tunica intima and elastic lamina of the major vessels of heart, arteries and peripheral vessels leading to deposition of lipid followed by formation of fibrous plaques. These plaques are constituted of lipids, cholesterol, proteoglycans, collagen, cellular waste substances, calcium, foam cells, macrophages, and other leukocytes. Although, the process starts at a very young age, clinical signs may take time to develop and in most of the cases this remains undetected.

3.3.5.1 Etiology

Although it is assumed that development of atherosclerosis may be spontaneous, several factors are implicated to facilitate its development like sedentary life style of the captive birds, lack of exercise, diet rich in fats or cholesterol and hyperlipidaemia. Excessive fats in diet are thought to be responsible for inducing chronic inflammation leading to atherosclerosis. Endothelial inflammation and formation of immune complexes are responsible for such plaque formation. Fatty liver disease was also recorded to be associated with atherosclerosis.

3.3.5.2 Clinical Symptoms

The symptoms of atherosclerosis are often misleading. In many cases it remains undetected throughout the life span. In certain cases it may be exhibited as fainting, falling and sudden death. The affected birds may show the symptoms of dyspnoea, exercise intolerance, coughing, nervous symptoms, paresis, paralysis, seizures, limb weakness etc. The gradual plaque formation and deposition causes the thickening of intima and intraluminal protrusion which causes narrowing of the blood vessels with reduction of blood supply and oxygen to the major organs and peripherals. This subsequently leads to development of clinical signs. Gradual thinning of arterial membrane may lead to formation of aneurism, sudden rupture and death due to severe blood loss.

3.3.5.3 Diagnosis

Blood cholesterol level has a significant correlation with development of atherosclerosis and therefore, periodic measuring the blood cholesterol level is required to predict the possible occurrence of the disease. Many workers have suggested that elevated level of blood cholesterol, VLDL (very-low-density lipoprotein), LDL (low-density lipoprotein) and low level of HDL (high-density lipoprotein) may give indication of such occurrence. Radiography is useful for detection of formation of increased atheriosclerotic plaques as reflected by increased density of the major arteries (Fig. 3.4).

3.3.5.4 Treatment and Management

The affected birds should be kept on low fat and more fibre diet. At attempt must me make to make them exercise and play so as to reduce obesity and risk of further damage. Application of bet-blockers (like propranolol) or calcium channel blockers (diltiazem, verapamil and enalapril @ 0.25–0.5 mg/kg/day, PO) may be effective. Isoxpurine was also reported to be effective in a separate study.

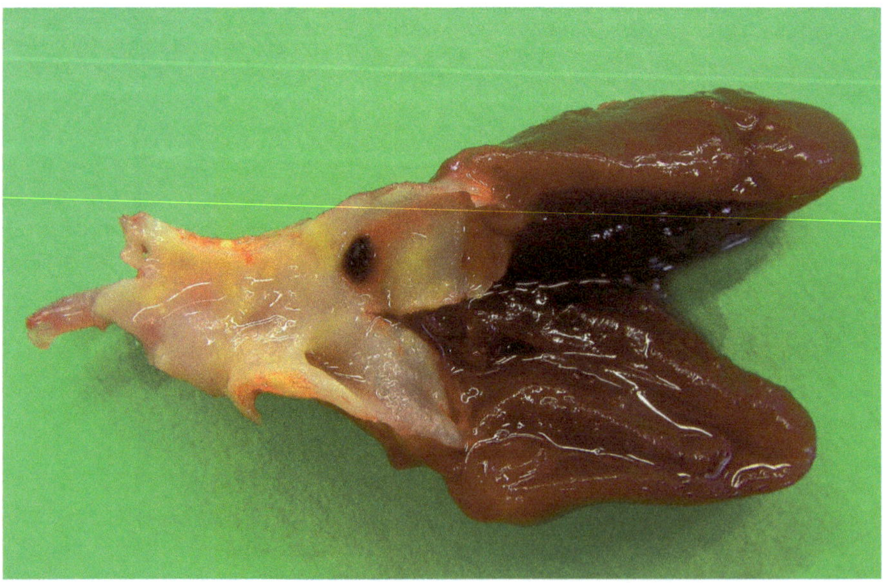

Fig. 3.4 Presence of plaques at the blood vessel of a bird with atherosclerosis (*Courtesy* Prof. Richard Hoop, University of Zurich, Switzerland)

Dietary PUFAs, especially omega-3 fatty acids (alpha-linolenic acid (ALA), eicosapentaenoic acid (EPA), and docosahexaenoic acid (DHA)), play an important role in the prevention of atherosclerosis by inhibiting inflammation within the blood vessels, reducing plaque formation, and by other means that protect the vessels. EPA is known to reduce the risk of atherosclerosis by shifting the plasma and platelet fatty acid profile. Fishes, flaxseed, camelina, rapeseed, chia seed, and walnuts are good source of omega-3 fatty acids.

3.4 Ophthalmic Problems

3.4.1 Conjunctivitis

Conjunctivitis is one of the most common problems encountered in the pet birds. In general, it may be of primary origin only affecting the conjunctiva or may have been originated from the problems of the eye lids or other periorbital parts. Both the palpebral or bulbar conjunctiva along with the nictating membrane is affected. Most of the case of the conjunctivitis are originated from bacterial, viral, mycoplasma or chlamydia infection which may be primary or may be secondary to respiratory infection specially sinusitis. Many times systemic infection like septicaemia may lead to development of conjunctivitis. Mycoplasma infection often exhibits the symptoms of respiratory infection with conjunctivitis. Likewise adeno viral infection also causes systemic illness along with conjunctivitis. Pox virus infection often produces characteristic lesions and can be easily diagnosed with detection of intracytoplasmic inclusion bodies whereas cytomegalovirus infection, a common nuisance in Gouldian finches, is responsible for swollen and congested conjunctiva and accumulation of huge amount of serous exudate. Chlamydial infection also causes injected conjunctiva with watery or serous discharge in cockatiels. Similarly, mycoplasma is also known to cause swollen conjunctiva in cockatiels, budgerigars and house finches. Fungal infection, (*Aspergillus* sp., *Candida* sp., and *Cryptococcus neoformans*) is associated as secondary to bacterial infection in conjunctivitis. However, they are more associated with immune-suppressed birds with systemic illness.

Foreign bodies may also lodge in the conjunctiva and cause such abnormality. Irritation from chemicals such as smoke aerosols or chemical fumes may cause such irritable and chronic conjunctivitis. Metaplasia with hyperkeratosis of the conjunctival epithelium was observed in birds associated with vitamin A deficiency.

3.4.2 Keratitis

Inflammation of cornea is often seen in birds along with conjunctivitis and termed as keratoconjunctivitis. Keratitis usually exhibits similar kind of symptoms like conjunctivitis with reddening of the bulbar conjunctiva with increased vascularization;

however, mostly keratitis appears in the form of corneal opacity with accumulation of muco-purulent discharge or pus in the cornea. Keratitis occurs frequently due to bacterial or fungal infection apart from injuries from the foreign objects or conditions like stromal degeneration as recorded in pet birds. Vitamin A deficiency in birds especially in caged birds is not uncommon due to a particular type of diet and this may lead to drying up of the cornea leading to a condition known as Xeropthalmia. If the condition persists for a long period it may be developed in Keratoconjunctivitis sicca.

Persisting mycotic infection often leads to development of corneal ulceration with accumulation of white–yellowish proliferative exudate (Fig. 3.5). Microsporidial keratitis is not uncommon in psittacine birds and frequently appears in chronic form with conjunctival reddening and corneal opacity. Infective microsporidia can be recovered from infected conjunctival or corneal swab following staining with Gram stains and trichrome stains. Birds with mycosporidial keratoconjunctivitis have a

Fig. 3.5 Severe keratoconjunctivitis in a bird (*Courtesy* Suman Biswas, ARD Department, Government of West Bengal, India)

record of long history of suffering from non-healing conjunctival inflammation or ulceration despite rigorous antibiotic or antifungal therapy. Topical flubiprofen or systemic application of flunixin (2–4 mg/kg BID) or meloxicam (0.2–0.5 mg/kg BID) along with antibiotic therapy like amoxicillin-clavulinic acid @ 125 mg/kg BID for 5 days or cefotaxime @ 75–100 mg/kg IM BID for 5 days is effective. Tarsorrhaphy using the third eye lid is often practised to narrow the opening of the eyelid; however, its success is still to be determined as expertise is required.

3.4.3 Cataracts

This is often common in birds; however, it is quite difficult to determine the aetiologies' leading to development of cataracts. Generally, in many species of the birds, cataracts are mostly due to hereditary origin linked to an autosomal recessive gene like Falconiformes and Cuculiformes (Yorkshire and Norwich canaries). On the other hand in a wide range of species it is acquired following some metabolic diseases like diabetes, toxicity or local and severe infection of the eye. Physical trauma, UV or microwave radiation and lightening stroke may lead to such condition. Viral infection (like avian encephalomyelitis infection), vitamin E and selenium deficiency may also be responsible for such cataract development. The condition can be detected with corneal opacity with cloudiness of the eyes, vision impairment and reluctance of the birds to detect or spot the feeds and to move. The affected birds are less responsive to normal stimuli. Treatment is not rewarding in most of the cases, however, phacoemulsification with intraocular lens implantation is practised in many cases with success. In general, clinicians should go for aggressive application of ophthalmic preparation of antibiotics with NSAID in the initial stages if suspected to be associated with local infection and trauma.

3.4.4 Uveitis

This is a condition associated with inflammation of the middle part of the eye—that the junction area between sclera and conjunctiva. In addition to the trauma and infection, uveitis may result from neoplasia and immune mediated inflammation. Rupture of the lens, reo-virus infection and infestation by *Toxoplasma* are the leading causes for uveitis. Systemic infection with *Mycoplasma, Pasteurella*, *Salmonella* and other pathogens may also lead to septicaemia and uveitis. The affected birds exhibit the signs like blepharospasms, photophobia, presence of blood clots (hyphema) or inflammatory exudate or pus (hypopyon) in anterior chamber, miosis (excessive constriction of pupil), dyscoria (abnormal shaped pupil) and corneal edema.

Treatment includes topical application of antibiotics and anti-inflammatory drugs.

3.4.5 Exophthalmos

It is relatively rare in pet birds. However secondary disease may lead to such conditions like periorbital traumatic injuries and haemorrhages, orbital fractures, orbital abscesses which often spread from peri-nasal sinuses as recorded in Amazon and African grey parrots. Retro-bulbar sinusitis is another cause of exophthalmos. In psittacines, infection of the Harderian gland may develop into exophthalmos. Neoplastic diseases of the adjoining areas like-glioma of the optic nerve, osteosarcoma of the orbital or nasal bones or adenocarcinoma may also lead to exophthalmos.

3.4.6 Panopthalmitis

It is the inflammation of the whole eye including, eyelids, uvea, conjunctiva and sclera and periocular diffuse swelling is noticed in and around the eyeball including infra-orbital sinus and nasal glands.

This is mostly due to traumatic injury or infection and extension of the injuries from adjoining areas of the eyes. Several infective agents like *Staphylococcus* spp., *Escherichia coli*, *Streptococcus* spp. *Pasteurella multocida Actinobacillus* spp. *Pseudomonas* spp *Plasmodium* and *parvovirus* are known to cause the condition. Apart from these chemical or physical irritants like smokes, chemical fumes, environmental toxins are such irritants to precipitate the condition. Many times panopthalmitis is associated with infraorbital sinusitis particularly, in psittacines and develops puffiness over the orbital cavity.

Clinical symptoms are associated with the similar pictures as noticed in other cases described above. Local and systemic antibiotic and anti-inflammatory therapy are warranted with debridement of the scar tissues formed commonly in viral infection.

3.4.7 Optical Neuropathy

Optical nerve degeneration and neuritis is commonly due to traumatic injury or infection leading to partial blindness or reduced light reflexes in birds. Moreover, adenoma or neoplastic changes in the pituitary gland may also cause compression over the optic nerve with degeneration in case of parakeets. Persisting viral infection and immune-reaction may cause optical nerve injury.

3.4.8 Examination of Eyes

Examination of the ocular reflex is one of the major criteria to examine the vision of the birds. Menace reflex is tested by quick approach of any object or head towards eye and the positive response can be recorded with eye blink, rapid withdrawal of the head or aggressive behaviour with beaks. Palpebral reflex may be evaluated by touching the skin at the lateral or medial edge of the eyelid. A quick covering with nictating membrane is seen in healthy cases. Corneal reflex can be tested with touching of the cornea with moist cotton symmetrically in both the eyes.

3.4.8.1 Schirmer Tear Test

It is an effective test to evaluate the tear production in animals using paper strips. However, its efficacy in birds is not well established due to wide variation in results. Generally the value ranges between 3–12 mm in healthy birds. An effective alternative of this test is thread test using pH indicator phenol red. The variation in results is much reduced in this test because of the less irritating reaction of the test materials.

3.4.8.2 Other Methods

Tonometry using a commercially available pressure indicator may be used to measure the intraocular pressure. In healthy birds it ranges from 9–25 mmHg. Apart from these modern techniques like ultrasonography, electroretinography and fluorescein angiography are used to evaluate the eye. Electroretinography is normally used to detect the retinal disease and electrodes are placed on cornea and skin at the sides of the eye. The eyes are subjected to various types of stimuli and their reflexes are noted electrically to evaluate the functions of various retinal cells like rods and cones. Fluorescein angiography is conducted to study the blood flow and blood vessels in the eyes following intravenous injection of a special dye—fluorescein. This can detect minute haemorrhages, atrophy of vessels and retinal diseases.

3.5 Pulmonary and Airway Diseases

The respiratory system consists of air passage the tubes that supply air to the lung and respiratory system for passage of oxygenated blood through the capillary bed lying on lung parenchyma and alveoli. The very purpose of the respiratory tract is the exchange of fresh and foul air. Due to constant exposure of the environmental air the capillary bed of alveoli and air sac are very vulnerable to gas, vapors, toxic ingredients, various microbes and toxic particles. At the same time the system is also prone to injury from the haematogenous insults. To protect from the environmental hazard the bronchioles are layered with ciliated epithelial cells and there is also a water protein layer enriched with lysozyme to protect from the pathogens. The non-ciliated granular clara cells lining the bronchioles are important for detoxification of the foreign proteins and drugs. However, alveoli contain a very delicate structure consisting of three layers—endothelium, alveolar interstitium and alveolar epithelium consisting of type I and II pneumocytes. Pulmonary surfactant with phospholipid is instrumental to maintain the surface tension and to stabilize the alveoli preventing its collapse during expiration. Besides, the pulmonary defense mechanism is coordinated by mucociliary defense mechanism and mononuclear phagocyte system of the respiratory system (alveolar and interstitial macrophages). Mucus is produced by goblet and submucosal cell and contains water, glycoprotein, lipid and salt. Mucociliary defense mechanism can protect the initial insult by dissolving the foreign agents like bacteria, toxic gas and foreign injurious particles and propelling them out.

Alveolar macrophages are derived from blood monocytes and are highly active phagocytic cells which can readily ingest bacteria and other particles reaching to the alveoli. They usually kill the ingested bacteria by releasing lysozyme and incorporating them within phagosome.

3.5.1 Sinusitis

Because of the complex and tortuous nature of infra-orbital sinus, many of the pet birds usually suffer from sinusitis.

3.5.1.1 Etiology

Besides, bacterial (*Staphylococcus, Pseudomonas, Pasteurella, E. coli, Klebsiella, Haemophilus, Mycobacterium* in few occasions), viral (reovirus, poxvirus, herpes virus, ILT, avian influenza virus) and fungal (*Aspergillous*, *Candida* and *Mycoplasma*) infection, irritation of the sinus mostly due to ammonia intoxication is seen in birds when they are kept poorly ventilated areas/farms or in congested manner in the cages. Besides, excessive air pollution, intensive use of smoking for mosquito repletion may lead to sinusitis. Insertion of foreign body in the sinus, neoplastic or benign growth may result in sinusitis. Different pet birds may be detected with sinusitis with cockatiels, macaws and pigeons having higher incidences.

3.5.1.2 Clinical Findings

Sinusitis is usually represented with acute or chronic oculo-nasal discharge, nasal blockage, open mouth breathing, distension of the nasal orifices/nares, hyperinflation of the cervico-cephalic air sac and exophthalmos with dyspnea with occulo-nasal discharge mating the feathers around the beaks or head. Constant irritation of the nares may be noticed with incessant scratching of the head or cervical areas and shaking of the head. In few cases, due to excessive accumulation of exudates, the birds may be unable to open their beak—commonly referred as “locked jaw syndrome” as detected in animals with tetanus. If aggressive therapy is not undertaken, animals may die. This is most common in cockatiels.

3.5.1.3 Diagnosis

In general, clinical symptoms are indicative of sinusitis. However, diagnosis is rather tricky if the clinicians need to identify the foreign body or any growth in the sinus. The clinicians may take the help of contrast radiograph, CT scan or MRI and endoscopy to find out the cause of infection.

3.5.1.4 Treatment and Management

Successful recovery of sinusitis requires aggressive therapy as failure of timely treatment may lead to life-threatening situation for the affected birds. Removal of the nidus of infection like foreign body is the first step. It substantially reduces the symptoms and suffering of the affected bird. Correction of nutritional deficiency like deficiency of zinc or vitamin A leading to hyperplasia of the epithelial lining needs to be corrected with proper supplementation.

First the blocked sinus should be flushed with lukewarm saline to clear the blockage. Acetylcysteine dilution may be another option which can effectively remove the thickened mucus for their mucolytic property. If blockage is severe enough, surgical debridement and drainage may be required. Although it is effective, caution must be taken to avoid injury to nasal or ocular nerve. Nasal drop containing sorbitol and beta adrenergic blocker like xylometazoline/oxymetazoline hydrochloride and nasal spray with fluticasone propionate are good alternative to offer relief to the affected birds—as we experienced clinically. If inflammation is severe enough local application of NSAID and dry hot fomentation are useful. Oral or parenteral antibiotic like potentiated amoxicillin and fluroquinones are good choice. However, prior confirmation with antibiotic sensitivity test is better before initiation of therapy.

3.5.2 Rhinitis

3.5.2.1 Etiology

Like any other animals and birds, rhinitis is very common in pet birds. Mostly rhinitis or rhinotracheitis may occur due to allergic origin following irritations caused by infectious pathogens, dust, and gas. Hypertrophy of ceres, development of rhinolith and *Knemidocoptes* infection may lead to the development of rhinitis. In general, rhinitis is mostly detected among parrots, cockatoo and budgerigars.

Choanal atresia—failure of the choana membrane to close properly may be important reason for persistent rhinitis. During such cases, the nasal secretion fails to come in the oral cavity with swelling of infraorbital sinus (Fig. 3.6). This is more common in cockatoos and parrots. Rhinolith is another anatomical obstruction which may lead to constant nasal irritation and rhinitis.

3.5.2.2 Clinical Findings

Common symptoms are severe nasal and ocular discharge of watery consistency in the initial stage. Further, the discharge turns into muco-purulent nature mostly due to secondary bacterial infection. There may be asymmetry of the size of the nares.

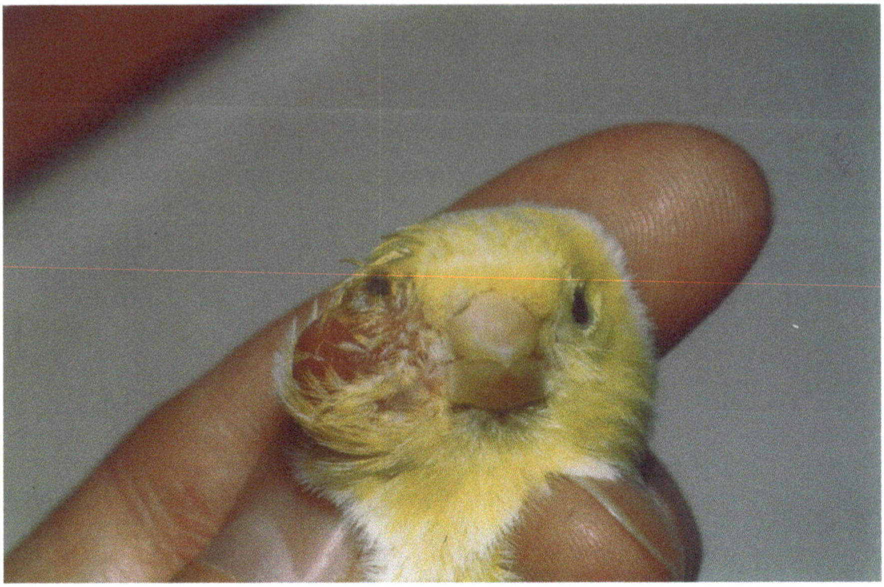

Fig. 3.6 Infraorbital sinusitis in a canary (*Courtesy* Petra Maria Burgmann, Canada)

The birds may manifest symptoms like constant sneezing, rubbing of the head and nares due to constant tickling and irritation.

3.5.2.3 Diagnosis

Diagnosis is mostly dependent on clinical observation. However, rhinolith is not always detected at the external nares. The choanal atresia needs to be detected with flushing of the saline water at the nares and the saline fails to go into the oral cavity which ascertains the presence of persistent membrane or bony plate at the palate of the nasal cavity. Rhinography using contrast media may also be used for the same purpose for definite diagnosis.

3.5.2.4 Treatment and Management

Surgical correction for removal of the boney plate or membrane and manual removal of the rhinolith may improve the clinical condition. Simple rhinitis often improves with oral anti-allergic drugs like chlorpheniramine maleate and cetirizine hydrochloride. Suspect of excess mucus accumulation in URT can be managed with mucolytic or expectorant of ambroxal hydrochloride. Although, clinical improvement was reported by many clinicians with these drugs, it should be monitored with strict clinical supervision. Vitamin C is good adjunct therapy used by many veterinarians to reduce stress and enhance the immunity of the caged birds.

3.5.3 Tracheitis

3.5.3.1 Etiology

Tracheitis is seen in pigeons, parrots, cockatoos and cockatiels. Bacterial infection like *Pseudomonas aeruginosa, E. coli, K. pneumonia* are very common. Other than bacterial infection various viral (herpesvirus, ILT, paramyxovirus, adenovirus and cytomegalovirus) parasitic (*Sternastoma tracheacolum*, *Trichomonas gallinae* or *Syngamus trachea*) and fungal (*Mucor* and *Aspergillous*) infection were reported previously. Foreign body lodgment in the trachea as occurred in cockatiels with millet, or pigeons and parrots with rice bran or gram seed may cause severe tracheitis or tracheal obstruction with secondary infection. Extramural masses with goiter, neoplastic growth and boney growth may cause tracheal compression or tortuousness leading to severe dyspnea.

3.5.3.2 Clinical Findings

The disease starts with acute onset of dyspnea. The extreme breathing trouble often initiates with stretching of the neck and open mouth breathing. Further, the bird may exhibits the symptoms like wing drooping and bobbing of the tail. Extramural growth may be seen or felt with palpation of neck region. Breathing or respiration may accompany with wheezing sound with rales which are audible from considerable distance. Infection or further involvement of larynx may be reflected with a change in voice.

3.5.3.3 Diagnosis and Treatment

The tracheal compression can be best judged with contrast radiography whereas the tracheal obstruction may be visualized with tracheal endoscopy.

Airway catheterization is the only option to relieve the birds with tracheal obstruction/tracheal compression/stenosis. Surgical removal or debridement may be helpful for correction of formation of caseous or diphtheritic membrane, lodgment of foreign bodies or formation of granulomas. Tracheotomy is another helpful technique to relieve the birds. Oral or parenteral medication with antibiotics (Cephalosporins) in case of suspected bacterial infection and antifungal (amphotericin B, itraconazole) in fungal infection is advocated along with anti-allergic medication. To clear out the mucus or exudate or to facilitate the expectoration ambroxal hydrochloride and acetyl cysteine are useful. Nebulization with amphotericin B is very effective technique.

3.5.4 Diseases of the Lung

There are many factors responsible for diseases of lung and its associated structure in the pet birds. However, pneumonia and bronchitis (Fig. 3.6) develops mostly from bacterial (*K. pneumonia, E. coli* and *P. multocida*, *Mycobacterium* and *Mycoplasma*) viral (paramyxovirus, herpesvirus, avian influenza), fungal (Mucor,

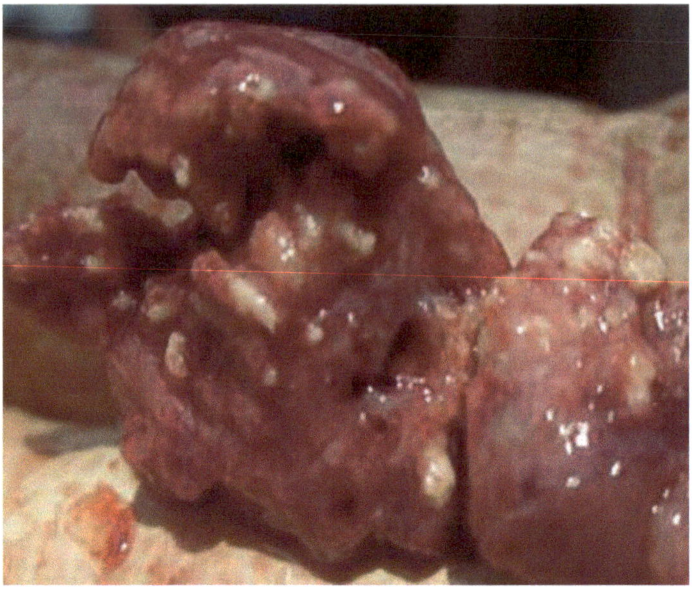

Fig. 3.7 Severe pneumonia is a bird (*Courtesy* Amrit Dhara)

Aspergillous, Cryptococcus) or parasitic (*Cryptosporidium* spp., *Toxoplasma* spp., *Sternastoma* spp., *Sarcocystis* spp., *Atoxoplasma* spp.) infection (Fig. 3.7).

Air sac mite is one such parasite which may affect any part or entire respiratory tract. Such parasite may be present in the nasal orifice of the infested parasite to the tiny air sacs. The air sac mites are more common among the birds like canaries, budgies, cockatiels and finches. Among them canaries and finches commonly suffer from air sac mites. The air sac mite is otherwise known as *Stemostoma tracheacolum*. This mite usually spends their whole life span within the respiratory tract of the birds. From hatching of eggs to complete their life cycle it normally took not more than seven days.

Air sac mites are commonly transmitted by infested birds. The coughed up materials, moist droplets and discharges may carry such parasites and enter in the air sac of healthy birds. Contaminated drinking water is the primary source of such pathogens.

The clinical manifestation depends upon the severity of infestation. The mite—*Stemostoma tracheacolum* has a peculiar habit to travel all along the respiratory tract. This is very irritating and may be responsible for repeated coughing in the affected birds. Mild infestation may not exhibit any noticeable symptom. In severe infection variety of symptoms may be developed and that may be confusing with common bacterial infection of the respiratory tract. Many birds develop sudden change in their voice and that can be a notification for air mite infection. Breathing trouble is a very common symptom and the affected birds may be seen with open mouth breathing. Excessive salivation and shaking of the tail may be seen in some occasions. Generally the birds are seen to make whistling and clicking sounds while they take breathing. Such sounds are usually audible from distance. Symptoms are worse when the birds are in stress. Exercise, sudden fear, even excessive handling of the birds may produce such results. Severe air sac mite infestation may lead to the death of the birds also. Common respiratory symptoms like coughing, sneezing, mucopurulent discharge from nasal or ocular areas, labored breathing are seen among these birds.

Ivermectin and moxidection are recommended for treatment. Intradermal application of ivermectin is usually given for three times in a weekly interval.

3.5.5 Pulmonary Hypersensitivity in Birds

One of the major discussed hypersensitivity affecting the lung of human being is known as farmer's lung and this is due to type III hypersensitivity reaction to the inhaled fungal spore of actinomycete. Similar kind of pulmonary hypersensitivity has been noticed in macaw birds due to feather dander of other birds housed in same farm like cockatoos. Such hypersensitivity may also occur due to ingestion of foreign protein leading to congestion, edema, hemorrhage and emphysema although; they are less common among pet birds. Inhalation of toxic gases like nitric oxide, hydrogen sulphide, ammonia may be hazardous causing severe bronchiolitis, edema and interstitial pneumonia with fibrosis. Further these toxic gases may destruction or necrosis of bronchial cells and pneumocyte type I cells. Affected birds display chronic sinusitis, watery or mucopurulent nasal discharge, chronic coughing and poor exercise tolerance. Treatment option includes antihistaminic drugs like pheniramine malleate, cetrizine and corticosteroids depending upon the severity of symptoms. Underlying cause of allergic reactions must be removed to provide a permanent solution.

3.6 Diseases of Bone and Muscles

3.6.1 Rickets and Osteomalacia

Both rickets and osteomalacia occur due to defective mineralization of the bone owing to deficiency of vitamin D, calcium and phosphorous. Rickets is a condition of the young or juvenile birds when both the bones and cartilages undergoing endochondral ossification are affected while osteomalacia or osteoporosis is a disease of mature skeleton in adult birds.

3.6.1.1 Clinical Findings and Pathophysiology

Active form of vitamin D is responsible for intestinal absorption of calcium, maintenance of normal calcium level in blood and its proper deposition for mineralization of bones. Inadequate vitamin D3 level due to lack of exposure of UV radiation/sunlight, lack of Vitamin D3 receptors, hypoparathyroidism, defective vitamin D3 activation, renal failure and defective gastrointestinal absorption may lead to such inadequate vitamin D3 level. Dietary deficiency of calcium may cause such condition due to improper and inadequate mineralization of the osteoid and cartilaginous matrix. Similar condition may occur in phosphorous deficiency and it is an essential ingredient of calcium phosphate required for such mineralization.

The most prominent changes in both the condition are enlargement of the ends of the long bones and costochondral junctions. The long bones may bend under body weight. Bones and beaks are extremely soft like rubber. The thickening of the costochondral junctions may appear like string of beads—"rachitic rosary". Abnormality and stiffness of gait and soft pliable appearance of bones, beaks and claws are noticed with severe retardation in growth. Affected lay birds show reduced hatchability and thin and soft shelled eggs.

3.6.1.2 Diagnosis

Radiographic examination shows bending and bowing of long bone, distortion and widening of the growth plates.

3.6.1.3 Treatment

External application or parenteral vitamin D3 therapy is required; however, it must be borne in the mind that exuberant vitamin D3 therapy may promote dystrophic mineralization leading to soft tissue calcification. After initial vitamin D3 injection, the affected birds may be subjected to UV irradiation by sunray exposure. Chronic vitamin D3 deficiency may be corrected by dietary changes especially with specialized diet. However, corrective therapy may require long period for clinical recovery. Surgical intervention may be required for correction of excessive bone deformity taking a due consideration of its feasibility.

3.6.2 Osteopetrosis

The word "osteopetrosis" came from a Latin word "petra" which means stone. This name itself denotes that this condition is related with increased bone formation and density. This is also known as "osteoscerosis" from the Greek word "sclerosis" means hardening.

3.6.2.1 Clinical Findings and Pathophysiology

This is a condition mostly seen in case of laying hen. This is characterized by formation of increased bone formation in the medullary cavity of the long bone like radius, ulna, femur and petroid bones. Other than birds, some other domestic animals like cattle, horse, sheep and dogs are also affected. In many domestic animals it is generally linked to a hereditary defect leading to decreased bone resorption owing to osteoclast abnormality and defective bone remodelling. The bone density and bone mass both are increased and the affected bone becomes cartilaginous with absolutely no medullary cavity. However, the affected bones are brittle and are susceptible to fracture. In birds it is mostly occurred due to stimulation caused by estrogen. In reproductively active hens the condition is very common as such hyper estrogenic response is common to meet the higher demand of calcium for laying eggs. However, condition like hyper estrogenic diets, cystic ovarian diseases or ovarian hyperplasia and ovarian carcinoma may promote such condition.

3.6.2.2 Diagnosis

Radiological evidence of increased bone density and history of frequent fractures may give the clinician the clue of such pathological condition.

3.6.2.3 Treatment

In case of laying birds where the condition is purely physiological, no treatment is required. However, for pathological conditions, hormonal therapy like ovarian neoplasia or cystic ovarian diseases, partial or complete ovariectomy may be done. As it is quite difficult to execute and may not be possible, hormonal therapy like leuprolide acetate, HCG may be given as non-invasive approach.

3.6.3 Osteodystrophy

Osteodystrophy is a general term referring to disorder of bone arising from faulty or defective nutrition and such condition in birds is characterized by increased osteoclastic resorption of bone and further replacement by fibrous tissue.

This condition basically appears due to demineralization of bone mostly due to feeding of the birds either a calcium deficient diet or feeding a diet containing excessive phosphorous. The birds are often fed with seed only diet which have high phosphorous: calcium ratio. Similarly, diet like nuts, fruits and vegetables also lack optimum calcium. Thus feeding of such kind of diet excessively leads to stimulation of the parathyroid gland to release parathyroid hormone which causes resorption of calcium from bone to maintain calcium homeostasis in blood. Excessive phosphorous in the diet interfere with gastrointestinal calcium absorption.

The lesions are characterized by increased osteoclastic resorption of cancellous and cortical bone and replacement by proliferated fibrous tissue. The articular surfaces of the long and weight bearing bones collapse and tend to fracture. The bones may bend and bow with deformity in the vertebrae and ribs.

The condition is mostly taken care of by nutritional rectification. Surgical intervention is required when there is urgent need for repairing the fractures.

3.6.4 Osteitis and Osteomyelitis

Inflammation of bone is known as osteitis and that of the medullary cavity is known as osteomyelitis. It is a chronic condition characterized by necrosis and destruction of bone, their removal and subsequent replacement by new bone.

3.6.4.1 Clinical Findings and Pathophysiology

The condition is associated with compound fracture or neoplasia caused by numerous kinds of pathogens including bacteria (both aerobic or anaerobic), virus and fungus (like *Candida* and *Aspergillous*). Although it is not uncommon that infection has started following entry of the pathogen via open fracture, infection may also be initiated via hematogenous route. Bacteria are generally localized at metaphyseal area where the medullary veins join with capillaries and there is establishment of infection due to slow turbulence and flow of blood. Due to low phagocytic activity in this area, it is easy for the bacterial pathogens to kick start infection from this area. Further the infection spreads to medullary cavity resulting in osteomyelitis.

Clinically the osteomyelitis is characterized by painful gait and posture. The affected bird is unable to move or fly with painful lameness. In radiography, there will be definite lysis of the affected bone and soft tissue inflammation around the affected bone may be observed. Formation of abscess often leads enlargement of the affected area with expansion of the bone outline.

3.6.4.2 Diagnosis

Diagnosis is often based on clinical and radiological interpretation. Hematological observations include heterophilia, leukocytosis and basophilia.

3.6.4.3 Treatment

Before initiation of therapy a culture sensitivity test must be performed to determine optimum antimicrobial therapy. A number of antibiotics are effective depending on the type of infective organism like gentamicin, neomycin, ceftiofur, lincomycin, tobramycin and cefalothin. Antibiotic coated biomaterials like polymethylmethacrylate beads should be placed to obtain effective and high concentration of antibiotic following surgical correction and debridement of the affected bone and tissues. Systemic antibiotic therapy should be continued for at least 2–3 months.

3.6.5 Nutritional Myopathy

Nutritional myopathy occurs in the birds which are kept on the meal rich in unsaturated fatty acids resulting in vitamin E and selenium deficiency. Both vitamin E and selenium play important role in protecting the cell membrane of the muscle from free radicals. In the time of deficiency, the cell membrane becomes physiologically defective with rapid influx of calcium in the cytosol and then in mitochondria. The mitochondria fail to work further with rapid fall in cellular energy supply. Thus there will be wide spread destruction of the myositis with muscle fibre necrosis. Such condition results in muscle weakness and cardiomyopathy. Nutritional rectification may recover the affected birds rapidly.

3.7 Diseases of Skin, Feather, Beak and Cere

3.7.1 Feather Cyst

Development of feather cyst is a common problem in pet birds. It is commonly detected in macaw, canaries and parrots.

3.7.1.1 Cause

Feather cyst usually develops in the course of new feather generation. The newly grown feathers sometimes fail to grow and instead curl within and transforms in the cyst. The cysts are usually painful and impede in developing new feather. Generally injury or infection at the point of newly grown feathers is also associated with feather cyst. However, canaries are thought to be genetically predisposed for development of feather cyst. In canaries feather cyst is believed to be benign neoplasm of the feather follicle. Several dermal papillae forms within a single follicle result in entangling of the feathers.

3.7.1.2 Clinical Findings

Feather cysts are usually painful. Once developed and if not removed by surgical intervention, the cyst usually continues to grow with the growing feather and large swelling develops. It may develop at any region over the body. In parrots it is more common at the feathers of wing. Canaries usually develop more than one cyst at a time and surgical intervention is not pragmatic in those birds where many feather cysts are detected at a time.

3.7.1.3 Treatment

Topical application of antibiotic and periodic washing with the antiseptic is advocated to reduce the pain, infection and pruritus. Frequent application of 5% tincture iodine along with neomycin and tolnaftate solution helps to check the infection.

3.7.2 Feather Duster Diseases

This is the disease of budgerigar. It is also known as Chrysanthemum syndrome. This is thought to be caused by a lethal recessive genetic disorders or budgerigar herpes virus. The affected bird is lethargic and prostrated, unable to fly with excessive growth of the flight or contour feathers and tail. Most of the affected birds die at a very young age.

3.7.3 Straw Feather Diseases

Likewise canaries also suffer from a congenital disorder known as straw feather disease where the feather fails to grow from the sheath and the feathers appear like straw. A similar condition is also noted in homer and fantail pigeons which are known as "porcupine feather".

3.7.4 Alopecia and Baldness

3.7.4.1 Occurrence and Presentation

In canaries, cockatiels and zebra finches, baldness is noticed with loss of feathers at the back of the head. In each of these birds, it is very difficult to detect the exact causes for the loss of feathers in head. Cockatiels are noted to suffer from baldness on head due to loss of crest feathers. This is also known to be hereditary disorder. Certain color patterns of the feathers are also known to cause such loss of feathers. In canaries, three factors are known to play for such baldness like—hereditary, endocrinological or hormonal deficiency and sex of the birds. In canaries, baldness is more common among male than the female birds. In contrast, finches of either sex suffer from such bald ness in head and it is believed to be mostly caused by feather picking by a dominant group of birds.

3.7.4.2 Treatment

As baldness in budgerigars and cockatiels is not properly understood, treatment of the affected bird is very difficult. Supplementation of thyroid and testosterone is practiced by many. In canaries, the dominant group should be separated to avoid such feather picking. L-thyroxine is to be supplemented orally at the dose of at 0.02–0.04 mg/kg q24 h if hypothyroidism is suspected.

3.7.5 Self-mutilation

This is probably the most complicated syndrome affecting the pet birds causing a severe skin problem which is difficult to treat.

3.7.5.1 Etiology

To date, the etiological factors leading to such self-mutilation is unexplored and may vary depending upon the management practices and the types of the birds.

In budgerigars and love birds formation of polyfolliculosis with thickening of the pulp cap and feather sheath may lead to extreme pruritus and such self-mutilation.

Again *Agapornis pox* or love bird pox virus infection may lead to such pruritus and mutilation in lovebirds.

In cockatiels, such self-mutilation is very common as a result of intestinal protozoal infection like giardiasis. Such infection often leads to vitamin E and selenium deficiency and protein loosing enteropathy. This leads to dry flakes and poor feathering often tempting the birds for self-mutilation.

Herpes virus infection is another factor to incite for such temptation.

3.7.5.2 Clinical Findings

Birds are extremely pruritic, irritating and damage their own body parts with repeated biting, picking or chewing. Skin of the chest, tail, wing and other region of the trunks may be affected. Sometimes, the cage-mates may mutilate each other resulting in severe feather or skin damage.

3.7.5.3 Diagnosis

Diagnosis of self-mutilation, feather chewing, biting or picking is difficult as it is poorly understood. However, complete blood profile, serum chemistry and presence of other diseases should be investigated to eliminate the predisposing factors, if any. Stress factors including extreme or adverse climate conditions like extreme cold climate of winter or hot temperature in summer, social stress may instigate such condition. Low humidity in winter may induce extreme pruritus leading to self-mutilation. Lack of fresh air, sunlight, light/dark cycle is the important factors to be considered. Nutritional deprivation is another cause for such condition. Basic diets are often deficient in essential skin nutraceuticals like zinc, vitamin A, E or omega-3—fatty acids that may cause such skin or feather abnormality. Again, preservatives, pesticides, dye present in diet may have such deleterious consequences. The general systemic disorders like neoplasm, hepatic dysfunction, coelomic cavity granuloma, hypothyroidism, allergy and zinc toxicities should be looked for as these may induce such intense pruritus. Other psychological factors must be considered when such mutilating behavior is noticed like prolonged captivity, lack of interaction with the flock or with the owner, lack in the scope of reproductive behavior.

3.7.5.4 Treatment

The stress factors must be minimized by giving the scope for sound sleep, good physical exercise.

3.7.6 Psittacine Beak and Feather Disease

Psittacine Beak and Feather Disease (PBFD), also known as neonatal feather dystrophy is an infectious disease caused by Circo virus.

3.7.6.1 Clinical Findings

This disease is characterized by lethargy, loss of appetite, regurgitation, diarrhoea, deformity of the feathers and loss of feathers. Sometimes, infection may be accompanied by other bacterial infection and Polyoma virus. A large number of caged birds like—cockatoos, macaws, African grey parrots, parakeets, and Love-birds are affected. The infection mostly affects the young birds with a profound damage in several organs like liver, brain and immune system. Due to severe immune-suppression, the birds may die of secondary bacterial and fungal infection. Other than the dystrophic feather, beaks may also show some abnormality like overgrowth of beaks and symmetrical lesions on beaks.

3.7.6.2 Diagnosis and Treatment

Diagnosis is based on haemagglutination inhibition test. Currently no therapy is available for the diseases. Only efforts may be made to overcome secondary infection.

3.7.7 Feather Destructive and Plucking Behavior

It is associated with self-mutilation behavior discussed earlier. This condition has been noticed in a number of birds including parrots (such as Amazons or African Greys), Cockatoos, Macaws and Cockatiels. Its successful treatment is very difficult as it is not possible in most of the time to identify the underlying cause of such syndrome.

3.7.7.1 Etiopathology

It is quite difficult to determine the exact etiology of such condition as it may be of multifactorial origin. In many cases physical, psychological and other factors may contribute for such development. The main factors are discussed here—

1. Endocrinological complication: Exclusively hypothyroidism often contributes to such feather plucking behavior. Such low level of thyroid usually leads to loss of feather with a thickened greasy appearance of skin.
2. Malnutrition: Deficiency of some key minerals like zinc, hypovitaminosis A, E and lack of essential fatty acid may lead to loss of skin and feather texture in birds. Dry skin with abnormal keratinization may cause such syndrome.
3. Boredom and stress: Psittacine birds are intelligent which need a constant physical and mental stimulation for their wellbeing. If the birds are left alone and the birds are not in interaction with other birds of the flock or they are not getting intimate touch of human beings such destructive behavior may be developed.
4. Improper wing trimming often causes as irritation which instigate the birds to develop such habit.
5. Heavy metal toxicity, aerosol contamination, smoke, excessive humid environment are other triggering factors.
6. It may completely psychological when the birds are trying to capture the attention of the owner or some kind of anxiety or stress may stimulate the birds

to develop such kind of syndrome. This ultimately leads to obsessive compulsive or stereotyped behavior pattern of feather destruction.

7. During vitellogenesis with enlargement of oviduct and ovary, the affected birds may feel discomfort and may exhibit such feather plucking in the areas of thigh and ventral abdomen.
8. Parasitic infection, particularly giardiasis is associated with such problem.
9. Systemic illness like hepatopathy, pancreatitis, renal diseases, neoplasia, and osteomyelitis may also triggers such behaviour due to chronic stress and discomfort exerted on the birds (Fig. 3.8).

3.7.7.2 Treatment and Management

Once the problem is identified, efforts must be made to relieve the stress of the birds either by keeping it busy with some destructible toys and varieties of feed. The environment of the birds must be changed. It will extenuate its stress and boredom. The birds may be let to spend some time outside its cage. Owner may spent time with it playing and talking.

Counselling or behavioral modification of the birds may be rewarding. However, it may require the identification of the proper cause. The birds may be sensitized for a brief exposure to such cause and then adjusted with extending the degree or time of such exposure. This ultimately changes the behavioral pattern of the birds. The birds may be left with a favorite toy to play or it may be allowed to be distracted with TV or radio.

Few clinicians recommend Elizabethan collar or some kind of such physical barrier. However, it may further increase the anxiety and increase the problem.

Fig. 3.8 Feather damage in a ring necked parrot (*Courtesy* Mousam Das, Animal Resources Development Department, Government of West Bengal, India)

Behavior or mood modifying psychoactive drugs

1. Benzodiazepines are good mood elevator. It works by inhibition of dopamine and potentiation of GABA.
   Diazepam @ 0.5 mg/kg PO q8–12 h or 0.25–0.5 mg/kg IM or IV or lorazepam @ 0.1 mg/kg PO q12 h may be given.
2. Tricyclic antidepressants potentiate serotonin and are sedative and anxiolytic with anticholinergic activity. Following drugs may be tried—
   Amitriptyline @ 1–5 mg/kg PO q12–24 h for at least 30 days or
   Clomipramine 0.5–2.0 mg/kg PO q12–24 h for 2–3 weeks or
   Doxepin @ 0.5–1.0 mg/kg PO q12 h at least for 2 weeks
3. Butyrophenones like haloperidol @ 0.1–0.4 mg/kg PO q12–24 h or 1–2 mg/kg IM every 2–3 weeks may be advocated to control the symptoms.
4. Medroxyprogesterone acetate @ 5–25 mg/kg IM every 4–6 weeks have a good anti-inflammatory effect over the integumentary system.
5. Antihistaminics like diphenhydramine @ 2–4 mg/kg PO q12 h and hydroxyzine @ 2 mg/kg PO q8 h may be used when the pruritus is extreme or some allergic disorders are suspected.

3.7.8 Skin and Feather Disease Associated with Endocrinological Disorders

There are few instances where skin and feather diseases are associated with hypothyroidism or thyroiditis. In such cases, birds usually exhibit symptoms like change in the colour or complex of feather or plumage, non-pruritic feather loss and anemia. Many birds may be obese which will hardly respond to dietary restriction and exercise. Blood profile may reflect the changes like non-regenerative anemia, hypercholesterolemia, mild leukocytosis and heterophilia. The birds failed to moult for more than one year. In cockatiels, hypothyroidism is associated with loss and darkening of feathers with a greasy appearance. The diagnosis of hypothyroidism is difficult in birds as the basal T4 level is very low and even beyond detection. TSH stimulation test may be confirmative; however, commercial unavailability of avian TSH stands in the way of such diagnostic test. The suspected cases may be treated with thyroxine therapy.

3.7.9 Delayed Moulting

Delayed moult is a frequent problem encountered in many pet or caged birds.

3.7.9.1 Etiology

Several factors are implicated for such delayed moult in birds like lack of essential nutrients, endocriological imbalance like hypothyroidism, chronic egg laying syndrome or other concurrent illness including hypothyroidism. Lack of diurnal rhythm is an important factor responsible for delayed moult in caged birds.

3.7.9.2 Clinical Findings

The affected birds display symptoms like loss of feathers, abnormality of plumage or feathers, bald spot, untidy plumage, brittle, frayed and discolored feathers. The male birds may be associated with complete or partial loss of vocalization. There may be drastic loss or excessive egg production in female birds with depression, chronic weight loss and lethargy.

3.7.9.3 Treatment and Management

Management of such delayed moult is difficult as it requires many tricky manipulations. Nutritional supplementation is essential to manage the preexisting nutrient deficiency. Care must be taken to tackle the endocrinal imbalance like hypothyroidism or chronic egg laying syndrome. The diurnal rhythm for the affected birds must be established with 12–14 h of sleep time.

3.7.10 Abnormality of Beaks

Abnormal presentation of the beak has been reported in many caged as well as companion birds including Amazon parrots, macaw, cockatoos, rosella and quaker. This condition causes great disturbance in picking the feeds for the birds. Moreover, it reduces the beauty of the pet birds kept in houses as show purpose.

3.7.10.1 Etiology

Till date the cause of abnormal beaks is poorly understood. However, it may be due to acquired or congenital. Congenital abnormality is presumed to be associated with improper incubation temperature, aeration, ventilation, humidity and incorrect turning of eggs. This is mostly seen in case of rosella and Amazon parrot. The acquired abnormality is detected due to mechanical damage or compression as recorded frequently due to exuberant pressure applied when the birds are hand fed. Malnutrition or poor calcification may also cause softening and bending of the beaks. Other causes like mechanical trauma, PFBD virus or *Knemidocoptes* infestation may lead to such condition.

3.7.10.2 Clinical Findings

The maxilla of the affected birds grows abnormally on the side (Fig. 3.9). This lateral deviation usually starts from the cere or tip of the beak. The affected birds usually fail to pick the feed particles and feed dropping from the mouth parts is very common. This situation is very commonly observed in macaws. Manibular prognathism is common in young cockatoos.

Fig. 3.9 Overgrown beak (*Courtesy* Petra Maria Burgmann, Canada)

3.7.10.3 Treatment and Management

Clinical management of beak abnormality is easy when it is started at a very young age and the beak is rather soft. Gentle figure pressure applied over the abnormal beak may facilitate the beak to be reshaped in normal position. Such light digital pressure must be applied at least 10 times for 3–4 occasions in a day in 4 h interval. The birds may be allowed or encouraged to pick hard feed pellets. If the affected birds are very old or the beak has been hard with too much calcification, it is rather difficult to manage the abnormality with digital or manual pressure. In such cases, beak may be trimmed off with grinding the overgrown maxilla to allow it back in the normal position. Acrylic ramps or prostheses may be applied to correct the maxillary deviation. Orthodontic manipulation like trans-sinus pinning may be applied to rectify the lateral deviation. Similarly, such manipulation may be applied to correct the mandibular prognathism with extension of the craniofacial hinge joint into normal positioning.

3.7.11 Hyperextension of Maxilla and Mandible

Hyperextension of maxilla is a common nuisance detected among the macaws possibly due their constant habits of biting or picking the hard surface or solid object. This causes a forceful extension of the maxilla leading to subluxation of palatine bone. Traumatic injury may also lead to such condition. Hyperextension of the mandible may be associated with mechanical injury, damage to germinal epithelial layers, malnutrition, PFBD or *Knemidocoptes* infection and neoplasia.

In many cases, this is associated with chronic liver disease. The maxillary hyperextension is managed under general anesthesia where a blunt artery forceps is introduced through infraorbital sinus to lift the palatine bone and disengage the infraorbital plate for normal positioning of maxilla. Mandibular extension may be managed with proper nutritional supplementation, corrective measures to manage the liver diseases and trimming of the part when required.

3.7.12 Problem of Cere

Cere is a naked skin structure situated at the base of the maxilla and the external nares is situated within cere in case of parrots, pigeons and owls. In many birds cere may be hypertrophied due a number of factors especially due to hyperoestrogenism in female budgerigars and in many birds due to hypovitaminosis A and *Knemidocoptes* infection. Such problem may also cause discoloration of cere. Chronic illness or sartoli cell tumors in male budgerigars may cause such discoloration.

Cere abscess is a common condition in cage birds during the time of winter due to chronic respiratory infection. The condition can be easily detected with abnormal swelling, painful hyperemia, chemosis and accumulation of pus materials in the cere. Asymmetry of the nares may be detected in the birds suffering from such chronic respiratory infection or chronic rhinitis. There may be accumulation of creasy exudates and hardened materials in the form of calculi in the nares. These are known as rhinoliths. In general, treatment protocol involves extensive flushing of the area or surgical debridement to remove all the tissue exudates with prolonged antibiotic therapy. Care must be taken to remove the rhinoliths.

3.8 Diseases of the Urinary Tract

3.8.1 Kidney Diseases

Kidney diseases are not uncommon in birds either in the form of nephritis or nephrolithiasis. Avian kidneys are bilaterally symmetrical and lie within renal fossa. The kidney is moderately vascularized with supply of vasculature from aorta, ischiadic and external iliac arteries in the form of cranial, caudal and middle renal arteries. Interestingly, the avian kidneys have both reptilian and mammalian type of nephrons. The reptilian nephrons are more in cortex region where as the mammalian types are redundant in the cortico-medullary junction.

3.8.2 Renal Hypoplasia or Aplasia

Unilateral renal aplasia or hypoplasia is not uncommon in birds, however, it is mostly asymptomatic and detected during necropsy. Divisional aplasia particularly the cranial kidney region has been noted in birds. Rarely, aplasia of middle or caudal kidney is observed.

3.8.3 Renal Cyst

It is another congenital disorder detected in birds. It may be solitary or multiple. Generally solitary cyst revealed no clinical significance. However, severe lesions with multiple cysts may lead to renal failure.

3.8.4 Nephritis with Renal Failure

It is the inflammation of kidney associated with various systemic symptoms and is mostly occurred due to infectious and non-infectious causes.

3.8.4.1 Infectious Causes

A number of viral infections like Adenovirus, Circovirus, Coronavirus, Herpesvirus, Orthomyxovirus, Polyomavirus, Paramyxovirus, Poxvirus, and Retrovirus were noted to cause nephritis in birds. Many psittacine birds were known to be affected with adeno virus and the disease is characterized by non-specific and general enlargement of kidney. Although in many of the birds adeno viral infection is incidentally observed in necropsy (in love birds), polyomavirus infection may cause severe glomerulonephritis in non-budgerigar psittacine birds due to large scale deposition of IgG and viral antigen complexes and type III hypersensitivity reactions. In general, polyomaviral infection causes slight enlargement of kidneys. Non-budgerigar parrots are detected with anasarca and ascites with polyoma viral infection probably due to protein losing nephropathy or virus induced hepatic necrosis leading to hepatic failure to produce albumin in sufficient amount to maintain the colloidal hydrostatic pressure. Among other viral pathogens, important is the Paramyxovirus 1 which causes interstitial lymphoplasmacytic nephritis in pigeons.

Several bacterial pathogens are known to cause nephritis and bacteria mostly spread from ureter or via haematogenous route. Infection via blood mostly localize in the glomerular region. However, birds died of persisting and overwhelming bacteraemia or septicaemia may have bacterial burden mostly in the cortical regions with extensive multifocal lesions and interstitial nephritis. Wide varieties of gram positive or negative bacteria like *Staphylococcus, Streptococcus, Enterobacteriaceae, Listeria* sp., *Erysipelothrix rhusiophathiae, Pasteurella* sp., *Chlamydia* and *Mycobacterium* spp. are known to cause kidney infection in finches, canaries and other psittacine birds.

Fungal infection is very severe generally arising from the air sac or mycotic pneumonitis. Sometimes, infection in the form of fungal thrombosis may elicit severe necrotizing reaction in kidney tissue through haematogenous route.

Parasitic infestation like *Cryptosporidium*, *Isospora* and *Eimeria* are known to cause severe monocytic and plasmocytic interstitial nephritis in goose and ducks, however, these infections are rarely severe in pet birds. *Encephalitozoon hellem* is a potential cause of mononuclear interstitial nephritis in love birds and budgerigars with several focal lesions.

3.8.4.2 Non-infectious Causes

Apart from the infectious causes, dehydration may lead to deposition of uric acid crystals in the renal tubules with severe inflammatory reactions and nephritis. Gross necropsy reveals renal swelling with white coloured chalky deposits. Transient dehydration is reversible and kidneys are able to cope up with, however, prolonged dehydration may be potentially damaging for kidneys. Similarly excessive salt ingestion may also lead to such condition.

Metastatic or dystrophic calcification with nephritis or nephrosis is commonly detected in nestling budgerigars, cockatiels and macaws particularly which are on excess calcium or vitamin D3 in the diet. In general diets having more that 0.7% of calcium usually trigger the development of such calcification. Vitamin D3 or their analog facilitates increased intestinal absorption of calcium and hypercalcemia resulting in calcium deposits in soft tissues including kidney. On the other hand deficiency of vitamin A may also cause degeneration of ureter epithelial cells with deposition of crystals and renal damage.

Besides, renal amyloidosis, renal lipidosis, exertional or high intensity exercise induced myoglobinuric nephrosis, disseminated intravascular coagulation (DIC), hemochromatosis, haemoglobin deposits may cause renal failure. Renal lipidosis is mostly associated with the birds on high fat/cholesterol diet or the birds like parrots and cockatiels with chronic active hepatitis. Toxic nephropathies are not uncommon in pet birds. Nephrotoxicity was detected among the birds particularly which are exposed to high concentration of aminoglycosides as a part of therapeutic or prophylactic strategies. Heavy metals like lead, zinc, several mycotoxins, including oosporein, citrinin, and ochratoxin may lead to such renal failure and nephritis. Sudden fluid loss may cause renal hypo-perfusion with decreased blood supply or ischemic damage to kidneys with tubular necrosis, proteinuria, and urate deposition.

Renal tumours and carcinoma is not uncommon in pet birds particularly in budgerigars. The most common symptom is unilateral or bilateral lameness/paralysis due to compression over the ischiadic nerve that passes through kidney.

3.8.4.3 Clinical Findings

1. Anorexia
2. Polydipsia and polyuria
3. Regurgitation

4. Articular or visceral gout
5. High degree of prostration
6. Abdominal distension and ineffective straining mostly detected when associated with nephrolithiasis or urolithiasis.

3.8.4.4 Diagnosis

It is based on clinical findings indicating towards renal failure or symptoms of nephritis or nephrosis. Urine samples must be examined meticulously for any change. Urine samples must be checked for traces of protein, sugar, ketone bodies, blood or haemoglobin. Persistent low specific gravity of urine may indicate towards chronic renal failure. Change in pH (normally 6.0–7.5) also gives indication towards acidosis or alkalinity due to bacterial metabolism. Urine sediments should be examined after staining with methylene blue for presence of mononuclear cells like lymphocytes or neutrophils (>2–3/HPF indicates infection), bacteriuria, cellular, granular or other casts and desquamated epithelial cells.

Radiography enables the clinicians to assess normal shape and size of kidney and dystrophic or metastatic calcification can be detected with radio-dense deposits in the renal parenchyma.

Water deprivation test may be conducted confirm diabetes insipidus. The birds should be kept on gradual water restriction @ 10% for 3–5 days and then on complete withdrawal of water. The failure to concentrate urine indicates towards DI. Plasma protein, body weight and PCV should be closely monitored during the test. Similarly, to distinguish between neurogenic or psychogenic and nephrogenic vasopressin response test is done where the bird is given an oral dose of desmopressin acetate @ 0.02–0.2 mg/kg and reduction in polyuria or polydipsia within 30 min may confirm the neurogenic DI.

3.8.4.5 Treatment

In case of hyperuricaemia the birds should be treated with the following drugs—

1. Allopurinol: 100 mg tablet should be crushed in 10 ml of water and 1 ml of the same should be diluted in 30 ml of drinking water. This should be given several times a day and the birds should be kept under close supervision during treatment as it may induce skin allergies and hepatitis.

Alternative drugs are colchicine @ 0.04–0.2 mg/kg at every 24 h interval and probenecid. However, their use, dosage and safety are still unexplored.

In case of glomerulonephritis which is more common in older birds treatment with aspirin @ 1 mg/kg every day with dietary supplementation of omega-3 and omega-6 fatty acid (1:6) @ 1 ml/kg may give good response.

The affected birds must be kept under good hydration with fluid therapy in nephrolithiasis or acute renal toxicity to boost up the blood supply and eliminate the toxins @ 100 ml/kg/day for first 3 days followed by 50–75 ml/kg/day till the blood uric acid level returns to normal level.

3.8.5 Avian Gout

Gout is one the most important and common problem encountered in birds. It is also common in human and reptiles. The condition is caused by deposition of uric acid on body organs (visceral gout) in joints (articular gout) or in the ureters (renal constipation). It is also known as avian nephropathy or avian urolithiasis.

The disease has been detected in poultry as well as various pet birds like pigeons, cockatiels, and budgerigars (budgies). The birds which are on seed based diet and those which are fed on high protein diet.

3.8.5.1 Etiology

The exact cause of avian gout is often difficult to diagnose and misleading. However, successful management often depends upon determination of the factor responsible for development of gout. In general the causes may be divided in three categories—nutritional, infectious and toxicogenic.

High level of dietary protein is mostly incriminated for development of gout. However, other factors like high calcium in diet, hypervitaminosis D3, deficiency of vitamin A may be accountable for it. High level of calcium in diet may lead to kidney damage. Sometimes marginal phosphorus deficiency may also cause it. Phosphorus is essential to protect the calcium induced kidney damage. On the other hand phosphorous acts as urinary acidifier thus prevent the formation of urate crystals in kidney or ureter. Vitamin A is also required to maintain the epithelial lining of renal tubules. Water deprivation, cold weather and other stress factors may also be responsible for gout in the pet birds. Many pet owners preferred to keep parrots on fruits. High fructose content of the fruits is believed by many to cause gout in human; however, it has not been properly confirmed.

Viral infectious agents like avian nephritis virus and nephropathogenic strains of infectious bronchitis virus may act as important predisposing factor for gout with prior damage to kidney either by virus or by the deposition of antigen-antibody complex on glomerulus with prolonged infection. Polyomavirus infection in parrot is also believed to cause kidney damage and gout. The involvement of avian pathogenic *E. coli* in gout cannot be denied also.

The list of toxins whose role is suspected in gout is too long and difficult to enumerate. However, antibiotics like aminoglycosides, mycotoxins and other toxic agents like ochratoxin A, oosporein, and deoxynivalenol (DON), excess of vitamin D, calcium, sodium all can have a potential role to induce kidney damage.

Further, poor kidney response, water deprivation, cold weather and other impending stress factors interfering the kidney's ability to adequately excrete uric acid may be responsible.

3.8.5.2 Pathophysiology

Primary gout mostly results from abnormal breakdown of the protein which leads to high level of uric acid in the circulation and kidney fails to compensate the overproduction of this metabolite and excretion is no longer possible leading to deposition of uric acid crystals. The secondary gout develops due to malfunction of the kidney which may occur due to chronic kidney diseases, over medications, overeating, consumption of increased dietary proteins, high level of vitamin D and low level of vitamin A, poor blood circulation, poor physical activity which is very common in caged birds, dehydration and stress or other environmental factors which may compromise the ability of kidney to eliminate uric acid.

Uric acid is produced in liver following breakdown of dietary proteins and deamination of the amino acids. Almost 80% of the nitrogenous waste is being excreted in the form of uric acid in birds via tubular secretion. As this process is independent of tubular water reabsorption, hydration status of the birds hardly influence the uric acid excretion. However, impairment of kidney function may interfere the excretion of uric acid leading to hyperuricemia. In human hyperuricemia occurs when the plasma concentration of uric acid is more than 380–400 μmol/l. However, solubility of uric acid is higher in birds due to higher body temperature as well as higher sodium concentration in the avian plasma. However, when renal function is severely compromised at the level of about 70%, plasma uric acid concentration is abruptly elevated with deposition of uric acids in several locations. The high dietary protein level was often recorded to precipitate high uric acid level in the birds. This is common in budgerigars for the protein concentration of the feeding pellets. Surplus protein is catabolized and nitrogenous waste is converted into uric acid. When the total amount of uric acid produced overpower the clearing efficacy of the kidney, hyperuricemia develops and uric acid is deposited in various places. The higher plasma concentration of uric acid often leads to the development of articular gout with formation of deformed limbs and joints. Synovial joints sheaths and joints are preferred predilection sites possibly because of the lower temperature of these areas. Gradual accumulation of uric acid leads to formation of "tophi" and nodules.

Water deprivation, vitamin A deficiency, renal infection and other factors may contribute to development of urate crystals in kidney collecting ducts and tubules. Thus the condition may develop urinary incontinence in the affected birds like oliguria or anuria—a condition called acute obstructive uropathy. During such condition, tubular secretion of uric acid severely compromised leading to elevation of plasma uric acid and precipitation of uric acid in various sites including the articular areas. Several visceral organs are affected with deposition of uric acid. The site of predilection varies like heart, joints, liver and kidney. In several occasions the affected birds may be dead without any significant clinical manifestations which are more common in articular gout. The cause of death is not properly known, however, acute renal tubular failure may result hyperkalaemia leading to cardiac arrest.

3.8.5.3 Clinical Findings

Gout may occur in two forms—visceral and articular form. Articular form mostly affects the joints and visceral form affects the internal organs.

Joints are swollen, enlarged, painful and stiff which is evident by the continuous effort from the birds to shift their weight from one to another foot and exhibit a shuffling gait (Fig. 3.10).

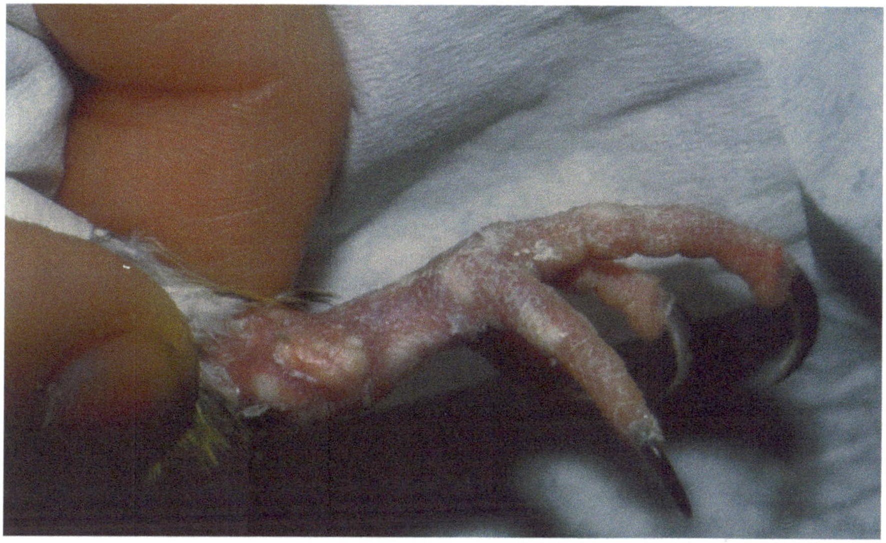

Fig. 3.10 Articular gout in a bird (*Courtesy* Petra Maria Burgmann, Canada)

With their wings affected, birds are unable to fly.

In initial stages, the urate deposition in internal organs is clinically not significant and may not be detected. However, in advanced stages affected birds become offed, lethargic, prostrated. There may be a tendency of chronic weight loss with abnormal droppings.

It is not unusual to see sudden death of the affected birds.

3.8.5.4 Pathology

In case of visceral gout gross pathological changes may be noticed when the dead birds are opened for PM examination. Uric acid precipitate as calcium sodium urate crystals in kidney and on the serous membranes of the liver, heart, air sacs, and joints (Figs. 3.11 and 3.12).

In chronic cases urate deposition may be seen in trachea also. This usually appears as chalky white deposit. There is marked changes of the affected kidney like loss of lobulation and atrophy with white discolouration due to deposits of uric acid. The unaffected kidney is usually enlarged to compensate the loss of function of the affected one. The ureters are enlarged with engorgement by urate crystals. In articular gout the affected portions are swollen. When affected portions are cut open, a white semisolid material comes out.

3.8.5.5 Diagnosis

The diagnosis of avian gout is often straight forward, particularly when it occurs in the form of articular gout. The characteristic changes of the joints and limbs indicate the development of the disease. However, visceral form of the diseases

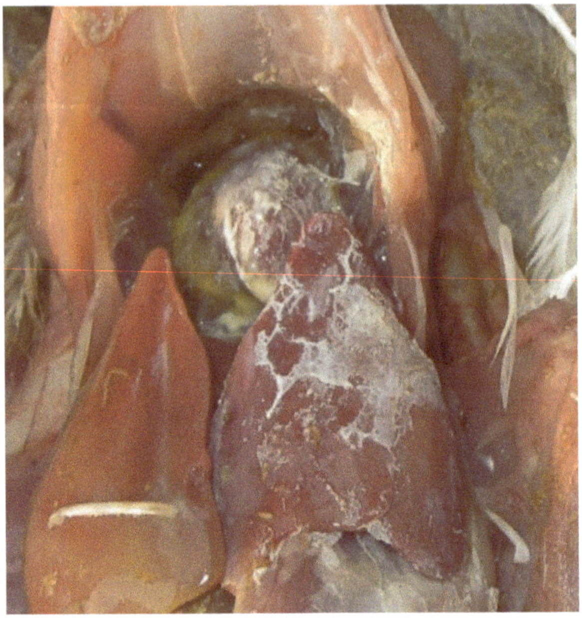

Fig. 3.11 Chalky white deposit of uric acid crystals indicating visceral gout (*Courtesy* Amrit Dhara)

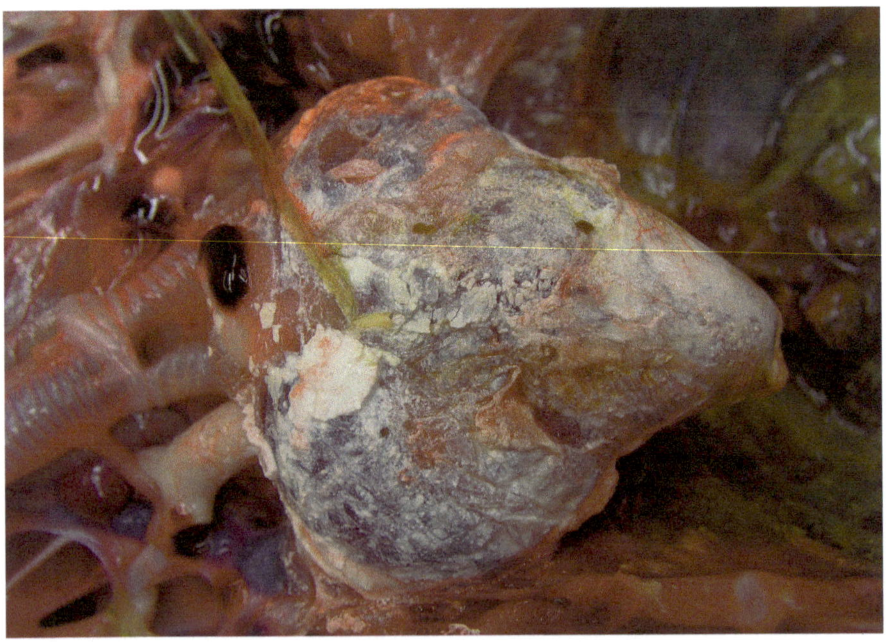

Fig. 3.12 Chalky white deposit of uric acid crystals in pericardium of a bird (*Courtesy* Prof. Richard Hoop, University of Zurich, Switzerland)

appears without any significant clinical changes. The affected birds may die suddenly without any indication either due to cardiac arrest or acute obstructive uropathy. In such cases the bird should be carefully examined with a detailed description of diet, water intake, environmental factors, and other existing health problems. Blood should be checked for elevated uric acid level. PM examination of the dead birds may reveal the presence of chalky white deposition of urate crystals in many places like—joint fluid, heart, kidney and liver.

3.8.5.6 Treatment

The treatment requires a holistic approach for successful treatment of the affected birds. Adequate water supply is the mainstay of therapy. The birds may be given subcutaneous or intravenous fluid like normal saline or supplementation for acute management of the water deprivation. Ringer's lactate is often recommended for restoration of sodium potassium imbalance. In case of articular gout the birds become anorectic and proper dieting is required with high calorie content for restoration of body weight. A low protein diet is preferred to prevent the rise of plasma uric acid concentration. Calcium, phosphorous, magnesium, sodium and vitamin D3 levels should be reduced in affected birds to avoid kidney mineralisation.

Other approaches include supplementation of vitamin A which is essential for repairing the damage of tubular epithelial cells and thus prevent the foci for infection or uric acid deposition. Acidification of urine using ammonium chloride, ammonium sulphate, $_{\mathrm{DL}}$methionine, and methionine hydroxy analog is advocated to dissolve and flush out the crystals.

Medication involves application of allopurinol (10–30 mg/kg, PO, bid) or colchicines (0.04 mg/kg, PO, sid-bid). However, their efficacy is still to be understood in birds. Many studies pointed out that although allopurinol may be an effective therapy of gouts in mammals, same is not true for birds. Allopurinol may cause nephrotoxic damage due to its derivative oxypurinol or due to the crystals of xanthine, hypoxanthine. Furosemide @ 1 mg/kg may be given to counteract the oliguria or anuria. However, its efficacy in acute obstructive uropathy is questionable. Corticosteroids and butorphenol are indicated for management of pain.

The prognosis for a bird with gout is generally poor. Most birds will need to be treated for life or the condition will quickly reappear if therapy is discontinued. If arthritis from gout is severe, it is possible to surgically remove the uric acid crystals from the joint. Often the damage to the joints or organs is irreversible.

3.8.6 Avian Urolithiasis

Avian urolithiasis is the main cause of obstruction in the lower urinary tract. Although it occurs in both males and females, urolithiasis is more common in males in case of mammals due to their long, narrow urethra. In birds the most common site of obstruction is cloaca or ureter. The one or both ureter may be involved. The calculi are mainly composed or uric acid crystals and proteinaceous matrix.

3.8.6.1 Etiology

A common predisposing factor in case of the birds is the period of egg incubation when the birds sit tightly over the egg which causes ineffective purgation of the cloacal content and deposition of the uric acid crystals and formation of urolith. Thus it is disease primarily of the laying flocks. Other factors include—

Deficiency of vitamin A or B vitamins
Bacterial infection leading to cystitis, urethritis, nephritis
Infection with nephropathic strain of infectious bronchitis virus
Excess dietary calcium and phosphorous
Deprivation of water
Alteration of urine pH due to dietary change
Presence of mycotoxin like ochratoxin and oosporein in feed
Nephritis.

3.8.6.2 Clinical Findings

The mortality rate may reach up to 20–50% in the affected population. Sudden death is the prominent symptom noticed in most of the conditions.

Besides the birds may be exhibit
Chronic anorexia
Weight loss
Anaemia with pale comb.

Frequent ineffective or non-productive straining and pasting of white chalking materials around the cloaca is noticed in many birds.

3.8.6.3 Post-mortem Findings

The ureters are generally distended containing hard compacted white mass composed of calcium sodium urate (Fig. 3.13).

The obstruction may lead to swelling or the kidneys—hydronephrosis (Fig. 3.12). It is not unusual to find the atrophied kidneys in long standing cases..

3.8.6.4 Diagnosis

Urethral obstruction can be diagnosed by palpating a distended and painful bladder.

Post-mortem finding may give a definite diagnosis.

Radio-dense calculi can usually be seen under X-ray. However, contrast radiography or intravenous pyelography and ultrasonography are effective to locate radiolucent calculi.

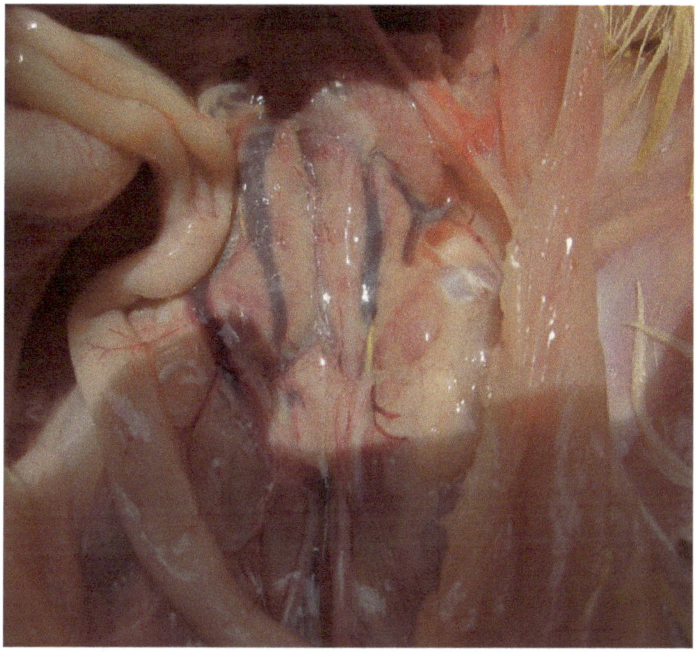

Fig. 3.13 Hydronephrosis due to obstructive urolithiasis (*Courtesy* Amrit Dhara)

3.8.6.5 Treatment

1. Arrange must be taken to ensure optimum uptake of water by the affected birds.
2. Dietary modification to acidify the urine using ammonium chloride, ammonium sulphate and methionine is helpful.
3. In many cases surgical intervention may be helpful but recurrence is common and care must be taken to avoid urethral stricture. Using forceps it is possible to break and remove the cloacal calculi. Post-surgical antibiotic therapy is recommended in such intervention.

3.9 Diseases of the Reproductive Tract

3.9.1 Metritis, Salpingitis and Impacted Oviduct

The disease involves inflammation of the oviduct and shell gland. The metritis and salpingitis along with egg binding or dystocia often leads to impacted oviduct characterized by accumulation of inflammatory exudate, excess mucin, albumen, and soft-shelled or malformed eggs. These materials often become thickened and get adhered to the wall of the oviduct.

3.9.1.1 Etiology

The condition is more commonly seen among the psittacines. Few factors were identified to cause such condition like age, lack of nutrition, calcium deficiency, egg trapping, abnormal egg production, chronic egg laying syndrome or other reproductive diseases. Viral infection like infectious bronchitis, Newcastle disease or hematogenous bacterial infection with *E. coli, Salmonella* spp., *Klebsiella* spp. and *Pseudomonas aeruginosa* may predispose the birds with such condition.

3.9.1.2 Clinical Findings

Depending upon the severity of the lesion and conditions the birds may exhibit the signs like abdominal distension, pain, lethargy, weight loss, dyspnoea and anorexia. There may be history of infertility or embryonic death. The eggs are abnormally shaped, malformed with tinges of blood. Putrid or haemorrhagic discharges may be noticed from the cloaca with flaccid vent.

3.9.1.3 Diagnosis

The clinical symptoms with previous history of excessive egg laying, dystocia, egg binding are suggestive of the disease. Case of metritis or salpingitis without any impaction may be associated with flaccid vent with occasional cloacal discharges. Blood picture reveals the presence of leucocytosis especially with heterophilia and monocytosis indicating systemic inflammatory reaction. Such conditions can also reveal blood biochemical changes like high total protein, cholesterol, and triglyceride concentrations. Radiography, USG may be used to detect abdominal fluid accumulation, egg trapping/dystocia and impacted oviduct with inflammatory exudate.

3.9.1.4 Treatment

1. Broad spectrum antibiotics (Enrofloxacin @ 15 mg/kg IM twice daily; Amoxycillin trihydrate @150–250 mg/kg) along with fluid therapy (isotonic fluid @ 50–100 ml/kg) are advocated to check the infection and rejuvenate body condition.
2. NSAID like meloxicam @ 0.2 mg/kg orally or SC SID may be used to reduce the inflammation and provide relief.
3. PGE2 may be used to relax the uterovaginal sphinter and to increase motility of the oviduct. Salpingohysterectomy is advocated in extreme cases.

3.9.2 Excessive Egg Laying or Chronic Egg Laying Syndrome

This disease or condition is noticed mostly among cockatiels and other small birds like budgerigars and lovebirds. The affected birds lay more than what they usually do or produce more eggs than normal. The birds may exhibit larger clutches of egg laying or repeat clutches. In general, wild cockatiels lay 1–2 clutches of eggs par year. However, pet cockatiels with CES may lay more than 2 clutches of eggs during this period. This condition if left untreated may lead to serious conditions like egg binding, salpingitis, metritis, egg yolk peritonitis and pathological fractures with depletion of body calcium reserve.

3.9.2.1 Etiology

Multiple causes were detected to have a complex relationship with development of such condition. The birds fed on high calorie, carbohydrate or fat rich diets, those which are given sweet food items like fruits or seeds are more prone. Besides, intimate bonding to other bird in the cage or owner, secure and congenial nest or egg laying sites, exposure to extended photoperiod (>10–12 h) and inappropriate diurnal rhythm are known to be important risk factors. Hormonal imbalance may also drive such condition in pet birds.

3.9.2.2 Clinical Findings

The clinical manifestations may not be indicative and no specific symptoms may be observed other than laying of eggs over extended period of time or clutches. The affected birds may show signs of poor health status, depression, lethargy weight loss and frequent pathological fractures. Otherwise poor reproductive tract status may be evident with discharges, dystocia, impacted oviduct, cloacal prolapse, abnormal eggs, and egg yolk peritonitis. The birds may show changes in defecation.

3.9.2.3 Diagnosis

The chronic egg laying behaviour itself is indicative; however, biochemical indication of hypercalcemia, radiographic evidence of hyperostosis and eggs in the oviduct, hyper-triglyceridaemia, and hyper-cholesterolaemia may further reinforce the diagnostic confirmation.

3.9.2.4 Management and Prevention

The management of such syndrome is tricky and may require the expertise of an experienced veterinarian. Multiple factors must be considered before intervention.

Counselling of the affected birds is important when the other companion bird is detected to have an intimate relation with the diseased one. The removal of the companion bird is necessary. The owner must be made aware how his behaviour may bring such changes in the pet. The client education and training is important. He must be taught what kind of his dealing may sexually stimulate the birds. The laid eggs must not be removed from the cage or it may be replaced with an artificial one. This may be helpful to stop such egg laying behaviour at least temporarily.

The change of food habit is very important. A feasible approach can be done by reducing the level of readily available sugar or carbohydrate and fat in the food and increasing the fibre portion. Indulging the foraging behaviour among the pet birds is also an important factor for rapid recovery.

The birds should not be exposed to photoperiod beyond 8 h. The surrounding environment of the birds needs to be changed as the affected birds may have an optimum and congenial atmosphere for laying their eggs. Changing the atmosphere, cage often substantially decrease their attachment and easiness for laying eggs.

As calcium reserve is substantially depleted, a supply of calcium in the form of 10% Calcium gluconate @ 50–100 mg/kg, SC, IM or oral Calcium glubionate @ 25 mg/kg, twice daily should be given.

Hormonal therapy alone is seldom effective unless other interventions are not taken. Leuprolide acetate as GnRH agonist may be given @ 700–800 mcg/kg, IM, every 2–3 wk. Similarly, GnRH agonist deslorelin acetate in the form of implants of 4.7 and 9.5 mg at every 3–6 months is effective. Medroxyprogesterone acetate @ 5–25 mg/kg every 6 weeks is also suggested by many clinicians along with HCG @ 500–1000 IU/kg every 2–4 weeks. As antibodies to HCG develop rapidly concurrent application of corticosteroid is implicated to induce immune suppression.

3.9.3 Egg Binding

Egg binding or dystocia is a common phenomenon among the love birds, cockatiels and budgerigars. This is an emergency condition when the affected birds are dyspnoeic and fainted and corrective measures are immediately required.

3.9.3.1 Etiology

There are various factors which may cause such condition. Very young as well as old birds are more prone. Besides, malnutrition, calcium deficiency, diseases of oviduct—like myositis, metritis and salpingitis, loss of muscular tonicity, malformed egg and abnormally big eggs, vitamin A deficiency, chronic egg laying syndrome, inappropriate environment are commonly associated with such disease. Egg binding in large psitaccine birds is generally attributed to obesity, behavioural abnormalities and environmental effect. Systemic illness, environmental stress or infection involving the urogenital system may also cause such egg binding.

3.9.3.2 Clinical Findings

The birds are generally depressed and show abnormal distension of abdomen. The eggs may not be always palpable. The birds usually lay egg at an interval of 23–26 h, therefore, appropriate time gap must be given and the birds should be kept on watch. In severe cases, birds are dyspnoeic with a "penguin like posture" and may collapse.

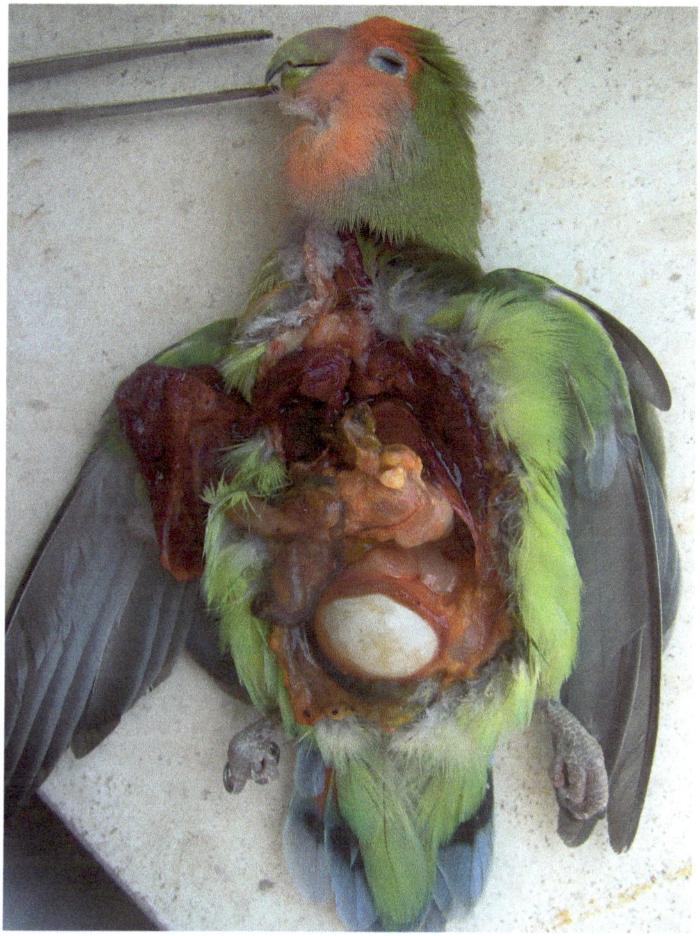

Fig. 3.14 Retention of egg/egg binding in a lovebird (*Courtesy* Prof. Elena Circella, Avian Diseases Unit, Department of Veterinary Medicine, University of Bari, Italy)

3.9.3.3 Diagnosis

Clinical presentation may indicate dystocia (Fig. 3.14). Abdominal manipulation may reveal the presence of eggs. However, the soft shell eggs may not be palpable. Radiography may be helpful to detect the abnormal location of the egg.

3.9.3.4 Management

Birds are kept in a comfortable, stress free warm and humid chamber. To increase the muscle tone of the oviduct, 10% Calcium gluconate @ 50–100 mg/kg, injection should be given at every 6 h. The birds may be fed highly digestible and carbohydrate rich food to provide immediate energy. Intra-cloacal PGE2 gel application may stimulate uterovaginal sphincter dilation and subsequent release of eggs. Manual manipulation may be necessary in few cases. Oxytocin injection @ 5–10 U/kg, IM may be given. If the bird is dyspnoeic, immediate intervention should be done for ovocentosis and egg collapse. The egg content should be extruded out using a large gauge needle and eggs may be collapsed with digital manipulation.

3.9.4 Egg Yolk Peritonitis

This condition is also referred as egg yolk coelomitis or egg peritonitis and is a common cause of abdominal distension in laying hens of all age groups among the pet birds, most commonly in cockatiels.

3.9.4.1 Etiology

It is generally thought to be a common sequela of chronic reproductive disease and detected after salpingohysterectomy as in most of the cases there remains the residual of ovarian tissues and ovulation occurs in the peritoneal or coelomic cavity. There are other possible factors which can trigger such condition like metritis/ salpingitis, cystic ovarian disease, neoplasia, ectopic ovulation, egg impaction, reverse peristalsis, ruptured oviduct and failure of the infundibulum to entrap the egg yolk due to pre-existing diseases like infection, trauma, disease or excessive fat deposition.

3.9.4.2 Pathophysiology

The egg yolk as such is not responsible for more than a mild inflammatory reaction as it is rapidly reabsorbed by the peritoneum. However, as the egg yolk is an

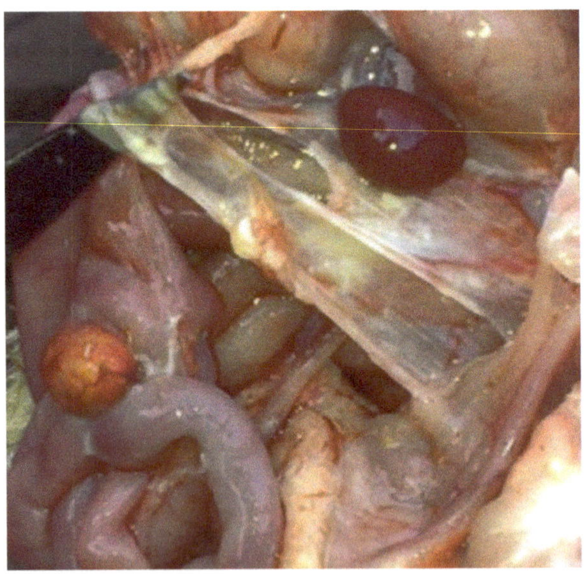

Fig. 3.15 Egg yolk peritonitis in a bird (*Courtesy* Amrit Dhara)

excellent source of nutrient, secondary infection can hardly be avoided causing a severe peritonitis and systemic reaction. Sometimes already contaminated or infected egg yolk spilled in the coelomic cavity may cause the infection. Most common secondary invaders include *E. coli, Staphylococcus* spp., *Salmonella* spp. etc. Ultimately it leads to localized or diffused fibrinous peritonitis (Fig. 3.15) and ascites. The infection may spread and lodge to different organs causing pancreatitis, hepatitis, spleenitis, nephritis, multi-organ failure and death.

3.9.4.3 Clinical Findings

The birds may exhibit severe abdominal distension, fluffed feathers, dyspnoea, lethargy, severe straining, lack of vocalizations, depression, and pain on palpation of the abdominal region, swollen vent soiled with yolk-coloured droppings, ascites, and neurological symptoms due to severe septicaemia. A more common sequel among the broiler birds is abdominal herniation due to increased intra-abdominal pressure with distension of abdomen.

3.9.4.4 Diagnosis

A severe case of peritonitis with septicaemia brings considerable changes in blood profile characterized by leucocytosis and heterophilia. Abdominocentosis may be done when ascites is suspected. The peritoneal fluid will have an albumin like yellow cooked appearance. The exudate may be caseous with offensive smell. The abdominal fluid will contain egg yolk, fat globules, fibrin deposit, bacteria and heterophils, macrophages, lymphocytes or other PMN. Bacteriological investigation with the fluid will be helpful to reveal the infective organisms. Serum amylase level will be high if there is concurrent pancreatitis. Radiography reveals the egg binding, an enlarged oviduct and presence of abdominal fluid. Similarly, ultrasonography is helpful to detect peritoneal fluid accumulation along with inflammation of the internal organs.

3.9.4.5 Treatment

If the affected bird is suffering from respiratory distress, abdominocentosis may be performed to relieve the patient. A long term therapy is required to support the bird with fluid, antibiotics and NSAID. If there is suspicion of oviduct impaction or egg binding or presence of exudate in the oviduct, PGE2 may be used to relax the uterovaginal sphincter and facilitate the clearance of oviductal content with stimulation of contractility. Although most of the birds respond to this management and intervention, salpingohysterectomy is required in extreme cases. In case of broilers hens with many number of ovarian follicles develop a problem known as erratic oviposition and defective egg syndrome (EODES) characterized by abnormal eggs, oviductal herniation, prolapse, internal ovulation and egg peritonitis. This is managed by avoiding light stimulation of the pullets.

3.9.5 Cystic Ovarian Disease

The ovarian cysts may be of congenital origin or acquired following neoplasia or oophoritis. The affected birds are generally not capable of laying eggs for many years. Cysts may be developed either in solitary or multiple forms.

3.9.5.1 Clinical Findings

The birds with small or solitary cyst may not present any symptoms. However, those are with multiple large sized cysts exhibit various symptoms like abdominal distension, swelling, pain, discomfort, ascites and respiratory distress. Fluid accumulation in the abdominal cavity may be confirmed by abdominal palpation. Budgerigars and canaries are commonly detected with such condition.

3.9.5.2 Diagnosis

Radiography and USG may be used to detect accumulated fluid, the cysts and other space occupying lesions. Abdominocentosis may be employed to check the nature of accumulated fluid.

3.9.5.3 Treatment

If the affected birds is dyspnoeic and indicates any emergency, it should be immediately relieved by abdominocentosis.

Antibiotic therapy may be given in case of secondary infection. GnRH agonists, leuprolide acetate and HCG are useful in primary cases. In case of severe infection or when malignancy is suspected ovariectomy may be suggested.

3.9.6 Prolapse of Cloaca

Birds with excessive straining are prone to cloacal prolapse. There are few behavioural factors which have been identified for cloacal prolapse like sexual overwork, masturbation by male cockatoos, intimate dependence or relation with the owner, delayed weaning, tendency to hold the defecation. Besides, oviductal diseases, metritis, salpingitis, diarrhoea, constipation, abdominal straining due to distention of enlargement of visceral organs may lead to such condition.

Cloacal prolapse is easy to correct and treat if detected early. However, if it is left untreated, the prolapsed tissue may develop inflammatory changes, becomes oedematous and necrotic and is very difficult to repose. Hyperosmotic fluid, sugar should be placed to reduce the oedema and the tissue needs to be flushed and cleaned with normal saline. Then the prolapsed portion is to be replaced. It may be lubricated for easy rectification. Silver sulphadiazine, neomycin or gentamicin solution may be given to protect from further infection. Suture may be given to prevent prolapse. Behavioural modification is necessary for permanent cure. Cloacopexy with vent reduction may be required in extreme cases.

3.10 Malignancy and Tumours

Neoplastic diseases are detected among the pet birds in variable frequency. Due to limited works it is difficult to predict on their real time epidemiology, however, compared to the age, the occurrence is quite high. Cutaneous form of neoplasm is quite common. Neoplasms of other organs including the visceral one are also not uncommon. However, no systemic study has been conducted to determine the risk factors associated with such occurrence. Like human and other animals, it is thought that chronic irritation, stress, long term consumption of toxic chemicals, lack of antioxidants in foods and environmental factors are responsible for induction of neoplasia.

3.10.1 Pituitary Neoplasia

Pituitary adenoma is the most frequently reported endocrine neoplasia in birds particularly among the young male budgerigars. The adenoma or adenocarcinoma of pituitary gland involves the chromophobe cells of the anterior lobe.

3.10.1.1 Clinical Findings and Pathophysiology

Due to development of space occupying lesion and increased intracranial pressure with compression of cranial nerve, optic chism and hypothalamus neurological symptoms may be developed like—seizures, convulsion, circling, depression, somnolence, fatigue, loss of consciousness and behavioural changes. Pressure on eyeballs may cause mydriasis, unilateral or bilateral exophthalmos, visual impairment, and blindness with infiltration of neoplastic cells in the optic nerve.

Pituitary adenoma is often associated with polydipsia and polyuria probably mediated by decreased concentration of anti-diuretic hormone and increased concentration of adrenocorticotrophic hormone. Pituitary adenoma with posterior lobe compression may cause decreased transport and storage of ADH. Similarly, increased secretion of ACTH causes cortical hyperplasia, release of corticosteroid and steroid induced diuresis. Changes in pigmentations may be noted in budgerigars and cockatiels as reflected by changes in feather and cere colour.

3.10.1.2 Treatment

In general no treatment option is available. To control seizures or convulsion phenobarbital @ 5–8 mg/kg orally BID may be given. The dose requires to be increased gradually.

3.10.2 Thyroid Adenocarcinoma

The thyroid hyperplasia is very common among the budgerigars reared in iodine deficient soil regions. Similarly, thyroid adenoma adenocarcinoma is not very infrequent among the birds. It is relatively common in human and pet dogs and cats. Thyroid cancers usually arise from cancer of follicular and para-follicular cells of the thyroid gland. The malignant cells may give rise to carcinoma comprising of well differentiated or poorly differentiated anaplastic cells.

Histologically, thyroid cancers may be of various types—

1. Papillary thyroid cancer
2. Follicular thyroid cancer
3. Medullary thyroid cancer
4. Anaplastic and poorly differentiated thyroid cancer
5. Non-invasive follicular thyroid neoplasm.

3.10.2.1 Causes

Among pet birds thyroid cancer has been detected among budgerigars and cockatiels. In general, it is difficult to determine the cause of thyroid neoplasm. However, genetic and environmental factors are often thought to play an important role for such development. Continued exposure to natural or artificial ionizing irradiation may lead to such circumstances.

3.10.2.2 Clinical Findings

The disease starts with formation of unilateral swelling or nodule around the thyroid gland at neck region. Thyroid cancer may occur in euthyroid birds therefore, symptoms of hypo or hyperthyroidism may or may not accompany. Other symptoms include dyspnoea, change or loss of voice due to pressure on recurrent laryngeal nerve or other non-specific symptoms like melena, weight loss and sudden collapse.

3.10.2.3 Diagnosis

The clinical symptoms, plasma thyroid profile along with USG of the thyroid region to detect any mass and status of the thyroid tissue may give some indication. Fine needle aspiration from the mass and its histological examination usually gives the confirmatory diagnosis. Thyroid adenomas are characterized by large follicles which are lined by columnar epithelial or cuboidal cells with minimum or no colloid. The adenocarcinoma reveals poorly differentiated cells which may invade the capsule or surrounding structures. Generally thyroid carcinoma is constituted of aberrant, large and firm structure with considerable distortion of thyroid and non-thyroid surrounding tissue.

3.10.2.4 Treatment

As prognosis is poor, treatment should be initiated with proper consideration of feasibility. Aggressive surgical intervention may be done for thyroidectomy followed by chemotherapy and radiation.

3.10.3 Squamous Cell Carcinoma

Squamous cell carcinoma (SCC) is the cancer of a kind of epithelial cells and it is comprised of undifferentiated or poorly differentiated squamous cells. Although it is mostly detected in integumentary system including beak, skin, phalanges, wings however, its occurrence is also detected in the muco-epithelial cell lining of gastrointestinal system (oesophagus, oral cavity, crop), respiratory system (nasal or infraorbital sinus). Metastasis occurs rarely but not impossible. However local invasion occurs very frequently.

There is no known definite cause for SCC, however, constant irritation or chronic inflammation may be a triggering factor for it. Similarly chronic exposure to sunlight or UV radiation may be another responsible factor.

3.10.3.1 Clinical Findings

The SCC develops most commonly at skin, beak, wings, and phalanges, in the oral cavity, oesophagus, crop and sinuses (Fig. 3.16). In the skin or outer surface there may be formation of tumours and bleeding ulcers which are not amenable to treatment leading to necrosis. The beaks may be deformed with overgrowth. The

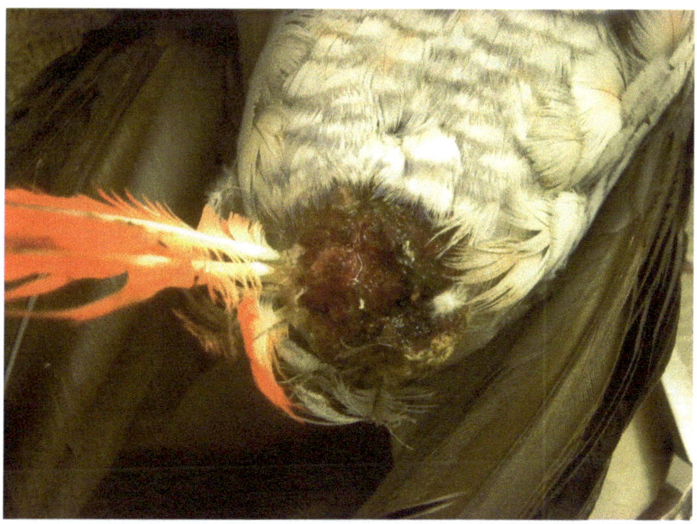

Fig. 3.16 Squamous cell carcinoma in an African grey parrot (*Courtesy* Kenneth R. Welle, Clinical Assistant Professor, University of Illinois, United States)

tumours in oral cavity, oesophagus and sinuses cause severe haemorrhagic and necrotic nodules. The birds may suffer from dysphagia, anorexia, regurgitation, dyspnoea, depression, keratoconjunctivitis with exophthalmos and severe occulo-nasal discharges. Secondary bacterial or fungal infection is possible in these lesions.

3.10.3.2 Diagnosis

Confirm diagnosis can be done by collection of fine needle aspirate from the suspected tissue sample and biopsy of the lesion. Microscopically, the SCC appears as well differentiated or poorly differentiated cells that form nests and cords with keratin centres. Cells are detected to be diffusely infiltrative. SCC may affect the uropygial gland where it needs to be differentiated from adenoma. In adenoma where solitary firm nodule or mass is developed and histologically well differentiated cells are observed. However, the carcinomatous growth tends to be constituted of poorly or undifferentiated cells and is usually, inflamed, haemorrhagic and infected with secondary bacterial invasion.

3.10.3.3 Treatment

Surgical excision of the mass/growth is the best option for treatment. It may be accompanied by radiation and chemotherapy. Radiation therapy with strontium-90 probe has given some promising result in SCC involving uropygial gland. Similarly cobalt-60 has been used with intra-lesional chemotherapy using carboplatin and cisplatin.

3.10.4 Xanthoma

Xanthoma is the non-cancerous tumours which appear in the form of diffuse thickening or little masses and dimple over the integuments. Psittacine birds including cockatiels and budgerigars are commonly affected. High fat deposition is common cause of xanthomas in birds.

3.10.4.1 Clinical Findings

The lesion can appear in any part of the body. However, wing tips, breast and legs, cloaca, vent and ventral abdominal regions are most likely noticed with xanthomas. The areas are extremely irritating and often cause pruritus in the birds. This leads to erosion and bleeding in the self-traumatised areas. Secondary bacterial infection may complicate the condition further. In case of budgerigars the xanthomas tend to expand and extend to the other areas like neck, breast, side of the wings and joints and restrict the activity of the bird. Infiltrated areas are friable and may be easily damaged. It may become ulcerated and easily infected further. Moreover, due to high vascularization of the areas and destruction of capillaries with the enlarged mass often led to sudden haemorrhagic episodes and sudden death.

3.10.4.2 Treatment

No particular therapy is available for the condition. Nutritional supplementation particularly with vitamin A may relieve the condition to some extent. In several cases surgical intervention to excise the portion is only effective remedy. However, if the xanthomas are big enough, surgical amputation is not recommended as it is difficult to close the lesion and control bleeding. Seed based diet with high cholesterol content is considered to play an important role for xanthoma. Therefore, affected birds should be kept on a balanced diet with more supplementation of fruits and vegetables. This may reverse the hyperlipidaemia and help to reduce its accumulation in the cutaneous or subcutaneous layer.

3.10.5 Fibrosarcomas

Fibrosarcoma is the neoplastic cells originated from mesenchymal cells or fibroblasts with the ability to produce collagen and usually involve the soft tissues of wing, leg, phalanges, head, beak, cere and trunk. This is a common neoplastic disorder detected among a variety of caged birds like macaws, budgerigars, cockatiels and parrots.

3.10.5.1 Clinical Findings

Clinically fibrosarcomas appear as solitary or multiple nodules with roots and they tend to be ulcerative, haemorrhagic and prone to secondary infection. Visceral form of fibrosarcoma involves multiple organs like liver, spleen, pancreas, abdominal cavity, proventriculus, small intestine, testes and ovary. They are locally invasive and may metastasize to other organs and even to muscles and bones. The visceral form develops symptoms depending upon the organs and degree of involvement.

3.10.5.2 Treatment

Surgery with radiation therapy is recommended. Superficial form may be surgically removed; however, prognosis is guarded for visceral form.

3.10.6 Lipoma and Liposarcoma

Lipoma is the most frequently observed benign tumour detected among the pet birds. It is the neoplastic growth developed from the adipocytes or lipocytes. Liposarcomas are malignant form and arise from immature adipocytes and lipoblast. Lipoma is soft and smooth round shaped where as liposarcomas are firm, highly vascular and poorly encapsulated.

3.10.6.1 Risk Factors

It is very common among the budgerigars followed by cockatoos and amazon parrots. Possibly in budgerigars there is some degree of genetic predisposition. Besides, high fat/energy diet and obesity are the other factors that may lead to such lipoma. Liposarcomas are found in sternum and uropygial gland area. The liposarcomas are locally invasive and can metastasize to other areas including skeletal muscle, liver or other organs of abdominal cavity. Liposarcomas were described in cockatiels, budgerigars, conure, African grey parrot, and quaker parakeet.

3.10.6.2 Clinical Findings

The birds are usually presented with solitary or multiple nodules in the skin or subcutaneous tissue particularly in the sternal or abdominal region. Besides, lipoma may be present in wings, legs, back, neck and uropygial gland region. The mass usually varies from 0.3 to 4 cm in diameter. Arising from thoracic or mesenteric fat the lipoma may be developed in the internal organ like ovary and liver. The mass may be itchy, traumatized and ulcerated and secondary bacterial or fungal infection is not impossible.

3.10.6.3 Treatment

Dietary management is necessary. The birds should be kept on low energy and low fat diet. Affected birds should be also encouraged for exercise. This can be substantially helpful for utilization of fat and reduction of the size of lipoma. Surgical excise of lipoma may be executed if it causes any clinical problem for the birds. Large lipoma affecting air sac may cause respiratory difficulty in birds. Besides, lipoma affecting wings and limbs may interfere with their movement. Generally lipoma and liposarcoma are highly vascular, therefore proper care must be taken before surgical intervention.

3.10.7 Neoplasia of Liver

Neoplasia of liver in birds may be of primary origin or it may be secondary to metastasis from other organs. Among the primary hepatic neoplasia, hepatocellular carcinoma and bile duct carcinoma are very frequent. Cholangiocarcinoma are the most common hepatic neoplasia reported among the captive and free ranging birds. This originates from the bile duct epithelial cells causing obstruction of the bile duct. Further involvement of hepatic parenchyma is also very common.

Bile duct neoplasia may be detected in the intra-hepatic, extrahepatic or peri-hilar region of liver. Although not clearly known, chronic inflammatory process with stasis of the bile duct leading to hyperplasia or metaplasia of bile duct may lead to cholagiocellular carcinoma.

Clinical symptoms include abdominal distention, change in liver function test, pain, generalized itching, mal-digestion, weight loss and emaciation. Neurological symptoms like somnolence, ataxia and seizures may be exhibited by the affected birds due to hepatic encephalopathy. Haematogenous spread of cholangiocellular carcinoma to distant organs like lung, kidney and pleura has been reported. It is potentially fatal disease with very little therapeutic option except rapid identification and surgical excision of the primary neoplasia.

Hepatocellular carcinoma or malignant hepatoma is also detected among the birds. Viral infection or chronic exposure to mycotoxin may be triggering factor for initiation of HCC. Clinical symptoms include, abdominal enlargement with fluid accumulation, painful palpable enlarged hepatic lobes, regurgitation and offensive foul smelling diarrhoea. The affected birds are often presented with debilitated and emaciated stage. Metastasis is rarely detected to occur in lung. Treatment is not effective however; hepatectomy to remove the affected part may be tried.

3.10.8 Pancreatic Neoplasia

Neoplasia of pancreas—pancreatic adenoma and adenocarcinoma were reported in Amazon parrots, macaws and large psittacine birds. The human pancreatic cancer was linked to various risk factors like over consumption of alcohol, meat, smoking however, such association was not established in birds. The pancreatic cancer mostly involves the neoplasia of cells of the gland that produces digestive enzymes. The pancreatic duct epithelium as well as acinar cells is involved. Clinical symptoms include, unexplained weight loss, improper digestion and abdominal enlargement or effusion.

3.10.9 Renal Neoplasm

Renal neoplasm is occasionally detected in the budgerigars. It is generally associated with articular gout, abdominal distension and inability to perch or ambulate. The exact etiology is not known; however, it is generally associated with and originated from embryonic nest cells in kidney. Renal carcinoma may metastasize to adjacent bone, muscle and liver. Due to compression over the sacral plexus clinical posture related symptoms are exhibited. Due to aggressive metastasis, it is quite difficult to treat renal neoplasm. Surgical excision of the affected part/kidney and planting radioisotope may be tried.

3.10.10 Lymphosarcoma

Lymphosarcoma (malignant lymphoma) is the cancer of lymphoid organs and is a common neoplasm of psittacines and passerines. The cancer tissue may affect multiple organs and most common site of metastasis is liver followed by kidney and spleen. Although many of leukaemia were linked to viral infection in domestic animals like feline pan-leukaemia virus, such association was not found in pet birds. Other areas like GI tract, skin, bone, oviduct, lungs, sinuses, thymus, testes, brain, mesentery, trachea, and pancreas may be involved with metastatic growth of lyphosarcoma.

3.10.10.1 Clinical Findings and Pathology

The clinical symptoms vary with the organ/tissue involved. The involved visceral organs like liver, pancreas, spleen and kidney are abruptly enlarged. Thickening and opacity are noticed following serosal infiltration of neoplastic lymphoid cells which ultimately causes enlargement and paleness of the internal organs. In case of cutaneous involvement, lesions are noticed in head and neck region. Other symptoms include anorexia, weight loss, depression, coelomic distention, paresis, lameness, blindness, regurgitation, or dyspnoea. In many cases of exophthalmos in psittacines particularly among young African grey parrots have been identified as a sequel to retrobulbar lymphoma. The neoplastic lymphocytes can be microscopically identified with little or moderate amounts of amphophilic to eosinophilic cytoplasm and a central nucleus containing reticulated to coarse chromatin.

3.10.10.2 Diagnosis

Clinical symptoms may not provide sufficient clues for diagnosis of lymphosarcoma. However, the cases may be suspected on the basis of persistent anaemia (PCV $< 35\%$), marked leucocytosis and lymphocytosis in complete blood count accompanied by symptoms like fatigue, depression, weight loss, respiratory difficulty, abnormal and enlarged mass in the coelomic cavity (particularly in visceral organs like liver, pancreas, kidney), abdominal pain, peripheral lymphadenopathy. Imaging technique like radiograph and USG may be helpful to detect such masses. Fine needle aspiration and biopsy of such masses will be helpful for confirmatory diagnosis.

3.10.10.3 Treatment

Surgical removal of the affected part of organ or mass followed by chemotherapy using the drugs like chlorambucil, doxorubicin, vincristine sulphate and cyclophosphamide constitute the major part of therapy with variable results. To control severe anaphylactic or allergic reaction anti-allergic drugs like diphenhydramine hydrochloride and corticosteroid like dexamethasone are advocated.

3.10.11 Ovarian Neoplasia

Various forms of ovarian neoplasms were reported in pet and free ranging birds of which gonadal neoplasm of stromal cell origin is the most frequent. These neoplasms appear as solitary or multiple nodular, pale masses with haemorrhage and necrosis. The second most frequently reported ovarian neoplasm is the ovarian carcinoma or adenocarcinoma which usually appears as large firm, pedunculated or lobulated masses. Ovarian carcinoma may metastasize to remote organs like mesenteric lymph nodes, liver, pancreas, intestine etc. Clinical symptoms may vary depending upon the involvement of organs and spread. The birds may reflect symptoms like abdominal distension, ascites, enlargement of abdominal masses, paralysis or paresis, weight loss, depression etc. Left leg paresis or paralysis of wings is seen often with occasional development of paraneoplastic symptoms like exostosis of bone and cartilage.

3.10.12 Chemotherapy and Radiotherapy in Treatment of Neoplastic Diseases in Caged Birds

Due to paucity of study very limited information is available regarding use of chemotherapy or radiation treatment for treatment of neoplastic diseases in birds. In general most of the studies indicated that chemotherapy is well tolerated by the birds, however, neoplasia is not responsive to chemotherapeutic treatment at optimum level. Similarly, radiation therapy works slowly over the tumour in avian system than that in mammalian counterparts. Before initiation of chemotherapy or radiotherapy, a thorough understanding of biology of tumours and effect of the respective treatment is necessary.

Vincristine, also known as leurocristine, is widely used for treatment of transmissible venereal granuloma in dogs. It binds with tubulin protein and causes disruption of mitotic spindle. Thus the separation of the dividing cell from its chromosome is inhibited during metaphase arresting the cell cycle. Thus the cells undergo apoptosis. It has been reported as a chemotherapeutic agent for treatment of malignant lympho-reticular neoplasia in African grey parrot, lymphoma in a Moluccan cockatoo and duck with lymphoma and leukaemia. However, as vincristine affects a wide range of dividing or proliferating cells, it may cause myelosuppression leading to pancytopenia, immunosuppression and neurotoxicity apart for gastrointestinal irritation.

Cisplatin, carboplatin and oxaliplatin are the platinum containing anticancer drugs which interfere with DNA replication by cross linking with them and the cell division or mitosis is interrupted. This ultimately leads to cell division arrest and cells undergo apoptosis. Cisplatin was reported to be used in fibrosarcoma, squamous cell carcinoma, pancreatic duct and bile duct carcinoma in pet birds. A dose of 1 mg/kg infused over 1 h was recommended by previous study conducted in cockatoos. Several adverse reactions like bone marrow suppression, regurgitation symptoms and weight loss were reported following cisplatin therapy.

$_L$asparaginase was reported for treating the case of lymphoma in a Moluccan cockatoo. It is an enzyme of bacterial origin. This enzyme causes hydrolysis of L-asparagine to $_L$aspartic acid and ammonia particularly in leukemic cells. Deprivation of asparagine leads to inhibition of protein synthesis and arrest of cell cycle leading to cell death. Adverse reaction like loss of weight, lethargy, in-appetence and regurgitation symptoms was reported in birds. Doxorubicin was reported to treat the cases of osteosarcoma in parrot and lymphoma in a Moluccan cockatoo. It acts as an intercalating agent and inhibits the topoisomerase II in diving cell population like cancer cells. Similarly cyclophosphamide and chlorambucil were also reported for treatment of lymphoma in birds.

Radiation therapy undergoes through a revolutionary changes in recent years. Previously radiation was given through orthovoltage X-ray machines or very large activities of 64-cobalt and 137-cesium. However, due to extreme adverse reactions these are no longer used. In veterinary practices linear accelerators are used as ionizing radiation to treat neoplasia either by powerful X-ray machines for deep seated tumours or electron beams to treat integumental neoplasia.

3.11 Hepatic, Pancreatic and Enteric Diseases

3.11.1 Hepatopathy

Hepatitis, the inflammation of liver is very common in pet birds.

3.11.1.1 Etiology

Although infection is primary cause of hepatitis, trauma either by direct injury from motor vehicle or by deposition of uric acid crystals may also cause severe hepatopathy. Several hepatotoxic drugs, aflatoxin, gossypol, ergot, pesticides, ingestion of heavy metals are often responsible for causing hepatopathy. Infectious hepatitis may occur by three routes—direct penetration, hematogenous route and ascending through biliary system. As the liver receives blood from two routes—arterial blood from hepatic artery and blood from GI tract via portal vein, infection often starts through hematogenous route. Inflammation may involve only liver parenchyma—hepatitis or only biliary duct—cholangitis or involves the both—cholangiohepatitis depending upon the cause of hepatic insult or organism.

Several haemoprotozoa (*Haemoproteus*, *Plasmodium*, and *Leucocytozoon*) are known to cause severe hepatitis, however they are not too frequent among the pet birds. Similarly other protozoan infections like *Trichomonas, Histomonas* several trematodes and nematodes can cause hepatopathy. Several bacterial pathogens which are known to cause enteritis and septicemia like *Salmonella* spp., *E. coli, Pseudomonas* spp., *Yersinia* spp. and *Campylobacter* spp. may also cause hepatomegaly, hepatic necrosis with infiltration of PMN cells especially heterophils. Bacterial pathogens could be detected from kupffer cells and hepatic macrophages. Endotoxins liberated from virulent bacteria in GI tract may affect the liver by entering through portal circulation.

Among the viral pathogens which are responsible for causing hepatic insufficiency are Paramyxovirus, Circovirus, Reovirus, Adenovirus, Polyomavirus and Herpes virus. Depending on the type of strains involved the severity of the lesions varies. In general, they are responsible for multifocal necrosis with several spots of hemorrhagic changes. *Rickettsia* spp. like *Aegyptianella pullorum* and fungal pathogen *Aspergillus* sp. may cause hepatitis.

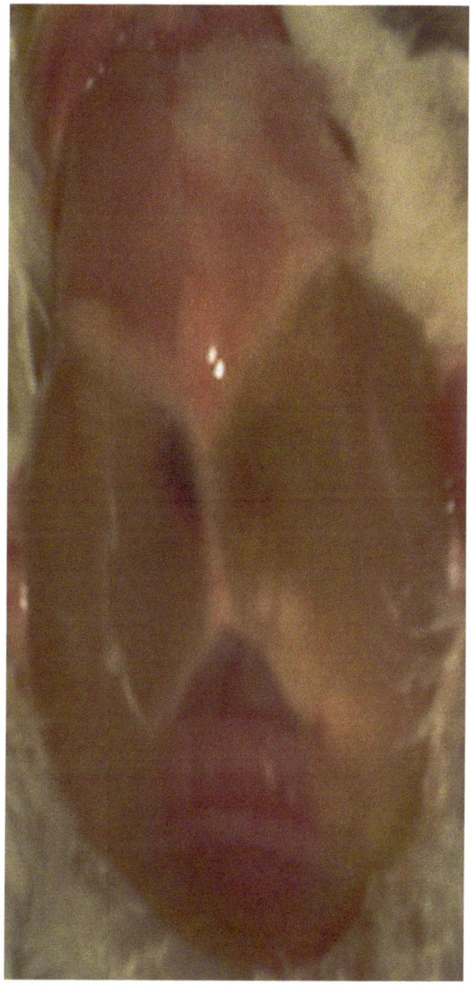

Fig. 3.17 Ascites in a bird due to hepatopathy (*Courtesy* Amrit Dhara)

3.11.1.2 Clinical Findings and Pathology

During hepatic insufficiency the liver fails to synthesize optimum amount of albumin which maintains the colloidal osmotic pressure. Therefore, there will be effusion of fluid due to overpowered hydrostatic pressure with accumulation of fluid, hydro-peritoneum, ascites, hydro-pericardium and anasarca (Fig. 3.17). Portal hypertension may also significantly contribute to such condition.

Due to failure to synthesize coagulation factors, petechial hemorrhages and coagulopathies are noticed among the affected birds. It is not uncommon to notice sudden death of the affected birds due to extensive internal hemorrhages. Increased serum level of ammonia may lead to hepatic encephalopathy with signs like seizures and behavioral abnormalities. With swelling of the hepatic cells and cholangitis, the normal flow of bile may be hampered leading to cholestasis.

Affected birds develop biliverdinaemia which may cause biliverdin stained wastage a green discoloration. The birds may suffer from indigestion, diarrhea and weight loss. Hepatic enlargement can be felt by palpation over the area and it may incite significant painful reaction.

The hepatic failure to clear out the toxic metabolite and accumulation of bile salt may lead to extreme pruritus and skin reactions. The affected bird may develop dyspnea due to pressure exerted by enlarged liver capsule or by accumulation of fluid in air-sac.

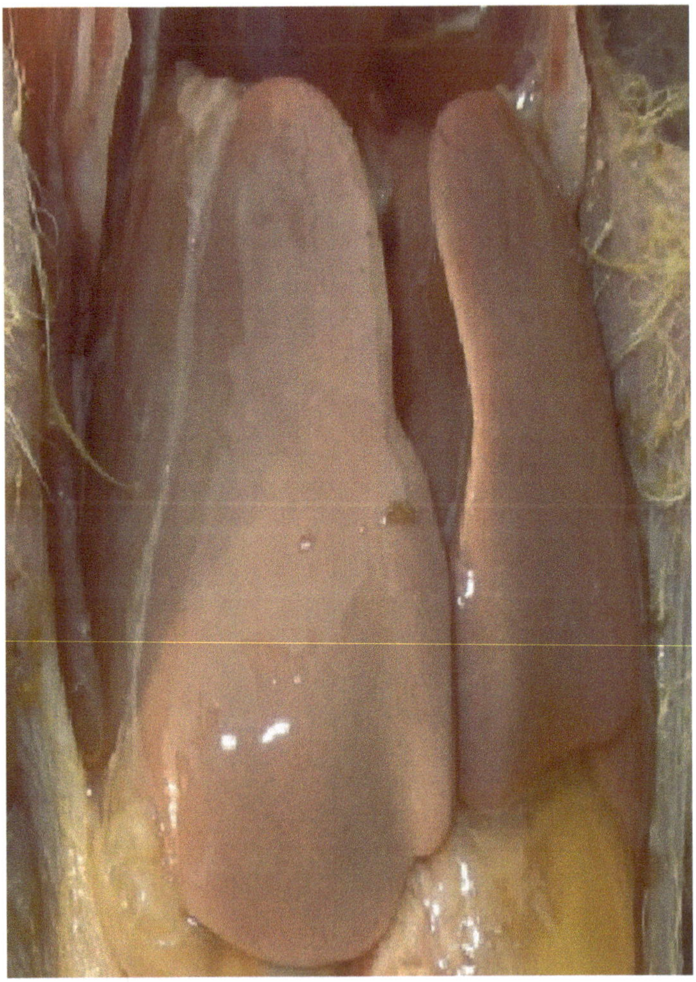

Fig. 3.18 Toxic hepatopathy in a bird

3.11.1.3 Diagnosis

The diagnosis is mainly done by serum analysis for any indication towards elevation of liver specific enzymes like aspartate amino transferase (AST), gamma glutamyl transferase (GGT) and glutamate dehydrogenase. Blood bilirubin level must be taken into consideration. Radiography and ultrasonography may indicate the presence of changes like accumulation of fluid, hepatomegaly and cirrhosis of liver. Toxic hepatopathy may reveal a gross abnormality with discoloration (Fig. 3.18).

3.11.1.4 Treatment and Management

Special and modified diet is required for management of hepatic failure in birds. High fiber diet may help to eliminate the toxins as well as to reduce the production of ammonia. Similarly to support the protein loss, quality protein may be give, however, a due consideration must be given not to facilitate the ammonia production. Readily available glucose is required in hepatopathy as it helps in regeneration of damaged liver as well as to expel out the toxin by supplying glucuronic acid. Diuretics are helpful to reduce the burden of accumulated fluid or edema aided by abdominocentesis. Adequate vitamin A, E and D may be given for detoxification and reduce the oxidative damage to liver. Silymarin is an effective drug to help the liver recover. When cholestasis is suspected ursodeoxycholic acid may be given. Respective antimicrobial or anti-parasitic therapy may be given to control the infection.

3.11.2 Pancreatitis

Inflammation of pancreas, pancreatitis may be acute or chronic. In birds it may be primary or secondary due to some bacterial or viral infection, heavy metal toxicity (zinc and selenium) and neoplasia. A major predisposing factor contributing to pancreatitis is high fat diet. Sometimes pancreatitis may be of idiopathic origin.

3.11.2.1 Pathophysiology and Clinical Findings

Pancreatic injury often releases activated digestive enzymes in the pancreatic parenchyma which further causes auto-digestion of the tissue and inflammatory changes in pancreatic parenchyma. Inflammatory mediators with infiltration of polymorph nuclear (PMN) cells cause further damage to pancreatic tissue. The enzymes like phospholipase A, elastase, trypsin and chymotrypsin play an important role for digestion of pancreatic tissue. Mild inflammation often causes edema of pancreas. However, acute hemorrhagic changes are not uncommon due to sudden inflammation.

Pancreatitis is always life-threatening if appropriate management and therapeutic intervention is not taken. The symptoms are vague and not indicative for confirmatory diagnosis. The affected birds may reveal the signs of indigestion, abdominal pain, frequent regurgitation, polydipsia, polyphagia, weight loss and abdominal discomfort.

3.11.2.2 Diagnosis

Elevation of serum amylase and lipase is indicative of pancreatitis. However, further investigation is needed for confirmatory diagnosis.

3.11.2.3 Treatment

Treatment of pancreatitis is very difficult. It requires an appropriate dietary management like low fat or fat free diet. Antibiotics and analgesics may be given along with fluid therapy for managing the cases as and when required. The sources of toxicity leading to such condition must be identified and removed.

3.11.3 Bacterial Enteritis

Like other mammals and domestic animals pet birds are frequently suffering from bacterial enteric infections where the organisms like *E. coli, Clostridium* spp., *Chlamydophila psittaci* and *Salmonella* spp. are frequent perpetrators. Among them *E. coli* infection is the most common as the organism is present in GI tract of the healthy birds as a commensal and may cause disease either as primary pathogen during conducive environment or as a secondary invader following viral, protozoal of fungal infection. Enterotoxigenic *E. coli* and the endotoxin liberated from these organisms may cause severe life-threatening infection. Infection by Gram positive anaerobic spore forming *Clostridium perfringens* type A and C can cause severe infection in young birds following abrupt changes in diet. They are responsible for foul smelling diarrhea with occasional hemorrhagic spots in fecal samples, necrotic enteritis, malignant edema and gangrenous dermatitis. Diagnosis can be done by gram stating the fecal samples revealing boat shaped Gram positive spore forming bacillus. The small intestine reveals presence of petechial hemorrhagic spots. Treatment with apple cider vinegar (10 ml/l of drinking water), penicillin and metronidazole (40–80 mg/l water) can be effective. Both the *E. coli* and *Clostridial* infections are common in psittacine birds particularly among cockatoos. *Salmonella* is another food and water borne infectious bacterial pathogen which is responsible for necrotic and ulcerative enteritis, diarrhea, respiratory infection, airsacculitis, and sinusitis, septic arthritis associated with hepatomegaly, splenomegaly, orchitis and oophoritis. This is a common pathogen detected in all the items present in close contact of the birds and adults may serve as asymptomatic carrier. Treatment is usually done by oral or systemic antibiotic especially aminoglycosides like amikacin (20 mg/kg, IM) and enrofloxacin (30 mg/kg, IM) depending upon the severity of infection. Plenty of water and electrolyte must be given to affected birds to prevent their collapse due to excessive dehydration.

3.11.4 Protozoal Enteritis

Cryptosporidia is small form of coccidian parasite which affects a large number of hosts including birds. The parasites live in the epithelial cell lining of the GI tract and are responsible for destruction of the villi of the enterocytes. It causes chronic diarrhea, in-appetence, weight loss and periodic hemorrhagic droppings. In many of the birds it may cause asymptomatic colonization. Infection is acquired through faeco-oral route. Infective oocysts are discharged via feces and susceptible birds may be infected by consuming such cysts. The feces may be stained by Giemsa stain for demonstration of the oocysts. Treatment is done by paromomycin sulphate.

Like other animals and poultry birds, pet birds are also susceptible to the coccidia which mainly infect the intestinal mucosa. Two of the genera are important—*Eimeria* and *Isospora*. The infective oocysts are shed with the feces and infection is generally acquired by ingestion of sporulated oocysts. Sporozoites released thereby, penetrate the intestinal epithelial cells to cause the disease. With the multiplication and destruction of the epithelial lining the affected birds manifest symptoms for their inability to digest and absorb nutrients. Young birds are mostly affected with symptoms like weight loss and diarrhea. The infection may be detected and confirmed based on the clinical symptoms and demonstration of oocysts in fecal smear. Affected birds may be treated successfully with anti-coccidial drugs like sulphadimethoxine and amprolium.

Hexamita columbae and *H. meleagridis* are known to cause anorexia, weight loss, severe diarrhea in pigeons, game and love birds, cockatiels, lorikeets and parrots. Fresh fecal samples should be examined for presence of protozoa. Nitroimidazole is generally effective however, periodic examination is necessary to check the efficacy of treatment. Affected birds should be given oral glucose solution to check hypoglycemia in affected birds.

3.11.5 Microsporidiosis (Encephalitozoonosis)

This is an immunosuppressive disease characterized by anorexia, lethargy, weakness, diarrhea, stunting, ruffled feathers, weight loss and neonatal mortality due to infection by acid fast Gram positive obligate intracellular protozoa. This is commonly detected among parrots, Amazons, lovebirds, budgerigars and finches. Many of the affected birds may exhibit ophthalmic symptoms like blepharospasms, corneal edema, chronic conjunctivitis and sinusitis. Treatment involved oral administration of albendazole @ 50 mg/kg once daily for 5 days.

3.12 Deficiency, Nutritional and Metabolic Diseases

Avian nutrition has gone through a revolutionary change when it comes to commercial poultry farming. However, such development is not true in every aspect when it concerns to the companion or cage birds. Despite the availability of formulated diets and practice of hand feeding, the birds used to suffer from deficiency, nutritional and metabolic diseases. One of the main causes of such deficiency or metabolic problem is that owners prefer to give the birds only what they like. This leads to feeding of the bird only seeds or pellets. Many times the birds are enthusiastically fed with nuts and other fat rich diets. Such selective consumption may lead to nutritional deficit and metabolic upset.

3.12.1 Obesity

Like human beings and pet dogs, modern life style and food-habits make the caged and pet birds prone to obesity. Obesity is a real problem which invites other diseases and act as risk factor for the life-threating emergency like sudden cardiac failure or syncope with respiratory failure or circulatory collapse. A bird is considered obese when its body weight is approximately 20% higher than the normal or usual one or when it attains a body condition or keel score of 4 out of 5.

3.12.1.1 Risk Factors

There are certain risk factors which makes the caged birds obese.

1. Galahs, macaws, Amazon parrots, and quaker parrots are usually more likely to be obese.
2. Over feeding of the birds or exuberance of food.
3. Sedentary life style and lack of exercise mostly because of small cage size with little opportunity to move
4. High fat diet like seeds, nuts, table food etc.

3.12.1.2 Consequences

The obese birds are prone to some disease conditions like lameness and arthritis due to constant overwork of the limbs to bear over-weight. The high fat diet and little body movement give little scope to metabolize the lipid leading to hyperlipidaemia and deposition of cholesterol or lipid in the blood vessels and liver. Thus, such birds usually suffer from fatty liver syndrome and atherosclerosis. The coronary thrombosis and sudden cardiac failure is not uncommon. Birds may also succumb to severe blood loss due to sudden rupture of the major arteries with atherosclerosis and aneurism.

3.12.1.3 Management

There is not medication which can directly help to control obesity. The birds should be given pelleted diet and fibre rich food with little or no fat. Ample opportunity to exercise should be given to the affected bird providing a larger cage with stir to climb or walk and flight cage. Similarly, rope or spiral rope may help them to exercise. Placing multiple food bowls help them to move and feed with enough scope of exercise.

3.12.2 Metabolic Bone Disease (MBD)

This is a complex of symptoms associated with loss of structural and functional integrity of bone, cartilages and skeleton deformity. This is also associated with other forms of bone diseases like rickets (in young), osteomalacia (in adults), secondary nutritional hyperparathyroidism, fibrous osteodystrophy, osteoporosis and cage layer paralysis or fatigue syndromes. The disease complex is generally associated with vitamin D, calcium and phosphorous deficiency in diet.

3.12.2.1 Etiology

The can occur in both the very young and old birds and mostly linked to calcium, phosphorous and vitamin D deficiency. Birds which are kept only on seed diet are more prone to the condition as calcium to phosphorus ratio in most seeds is poor due to high phosphorous and low calcium content. Seeds with high oil content are responsible for triggering such condition. As the caged birds have low exposure of sunlight, they are more susceptible to vitamin D3 deficiency. In African grey parrots such condition is characterized by fibrous osteodystrophy and is commonly associated with hypocalcaemia, hypovitaminosis D3 with deformities and curvature of long bones and vertebras. The birds kept on all meat diet (raptors) are known to develop the condition within 2–4 weeks.

3.12.2.2 Clinical Findings

In young birds the MBD is reflected as stunted growth, bowing of bones, spontaneous fracture, vertebral deformities, inability to perch and improper plumage formation.

Adult birds develop symptoms like osteomalacia with frequent fractures with the slightest trauma, abnormal mounting, egg binding, or cloacal prolapse. Egg production drops significantly and the birds may lay soft shelled eggs. With the fall of hatchability and egg production, embryonic death may also be noticed very frequently. Other symptoms include depression, lethargy, polydipsia, polyuria, regurgitation, ruffled feather with feather pricking and diarrhoea. Diarrhoea occurs usually due to polydipsia. Due to persisting hypocalcaemia the birds may exhibit frequent nervous disorders like seizures, paresis, paralysis and hypercalcaemic tetany.

3.12.2.3 Pathophysiology

MBD is usually associated with dietary supply of proper calcium, phosphorous and availability of vitamin D for utilization of them. Exposure of sunlight is essential for conversion of vitamin D3 in active form—1, 25 dihydrocholecalciferol. This active form is instrumental for absorption of calcium from diet through GI tract in the blood and their deposition in bone. Sunlight is absorbed and utilized in the bare surfaces of feet and shanks for activation of vitamin D. Lack of vitamin D may lead to improper mineralization of bone and skeletal deformity. On the other hand, lack of proper calcium phosphorous ratio (2:1) may lead to failure of calcium absorption.

High phosphorous level impairs the calcium absorption with formation of insoluble calcium phosphate complex.

3.12.2.4 Diagnosis

The case of MBD may be diagnosed on the basis of clinical finding and diet history. Serum calcium level (total and ionized) and 25-hydroxycholecalciferol are usually low. Radiographic investigation may reveal improper bone mineralization, decreased radio-density of bones and evidence of fractures.

3.12.2.5 Treatment

The birds should be placed with proper diet with rich supply of calcium and phosphorous in proper ratio, exposure to sunlight/ultraviolet light for sufficient period. In case of fractures it should be repaired with splinting or bandaging along with pain management with NSAID drugs. The birds should be given ample scope for exercise.

3.12.3 Hypovitaminosis A

Vitamin is a major neutraceutical and ingredient to maintain the structural and functional integrity of the epithelial tissue and it is crucial to provide a healthy immune system.

3.12.3.1 Etiology

In formulated diet for poultry or caged birds maize or other green plant is the only portion to supplement the requirement for vitamin A or its precursor carotenoids. Therefore, when the caged birds are provided only seed diets or diets with a mixture of seeds and pellets, they may suffer from vitamin A deficiency. Even some owner over-enthusiastically feed the birds with nuts and table food which suppresses the appetite of their pets to take maize or green foods at optimum level.

3.12.3.2 Clinical Findings and Pathology

Deficiency of vitamin A leads to metaplasia of epithelial linings in the respiratory, urogenital, gastrointestinal and integumentary system. The mucus secretary cells of oropharynx, choana, nasal sinuses, trachea and conjunctiva are keratinized and stop to function. This causes inflammatory exudate and cellular debris leading to watery nasal discharge and accumulation of caseous exudate in nasal sinuses, eye and trachea. White or yellowish pustules may form in the trachea or esophagus. As vitamin A deficiency leads to loss of mucosal protection and barrier and suppression of immunoglobulin production, secondary bacterial or viral infection may be observed in respiratory, gastrointestinal or urogenital system. Loss of integrity to the renal tubular system may lead to loss of urinary clearance and urate crystal deposition. This may cause creamy white deposition on kidney with loss of lobulation. Due to the loss of production of acid muco polysaccharides there may be

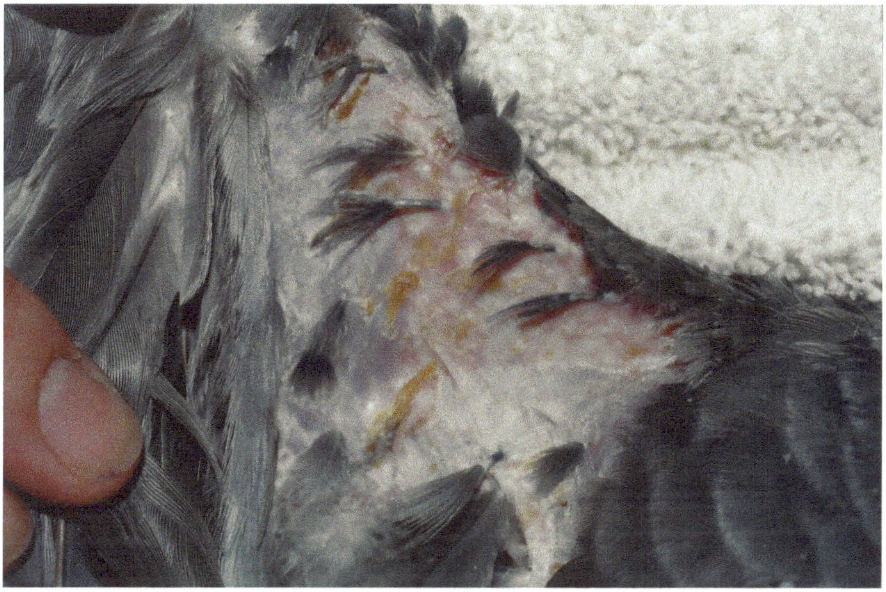

Fig. 3.19 Hypovitaminosis in African Grey parrot (*Courtesy* Petra Maria Burgmann, Canada)

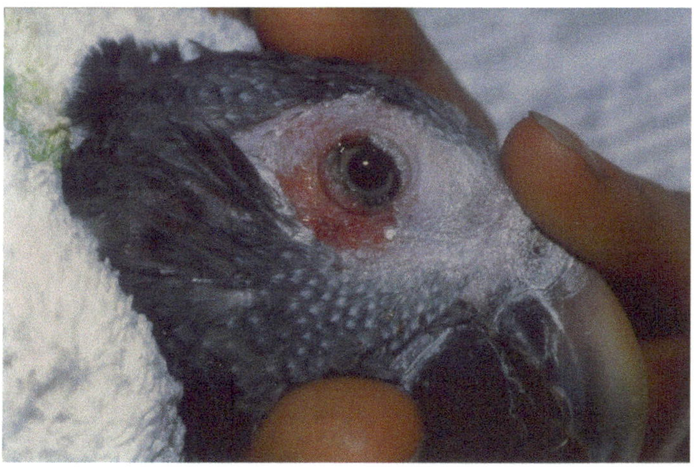

Fig. 3.20 Hypovitaminosis in African Grey parrot (*Courtesy* Petra Maria Burgmann, Canada)

total derangement of bone and cartilage formation particularly in the young birds. Vision is impaired due to keratinization of the conjunctiva (xeropthalmia). In general, the birds exhibit conjunctivitis, peri-orbital swelling, occulo-nasal discharge, sinusitis, polydipsia, polyuria, ruffled feather with poor feather quality, pododermatitis and loss of appetite (Figs. 3.19 and 3.20).

3.12.3.3 Treatment

Change in diet is an essential requirement to supply vitamin A. Good quality pelleted diet with maize or greens may be a good source. Dry formulation of vitamin A in acetate or palmitate along with an antioxidant (ethoxyquin) may be mixed with diet. Carotenoid rich spirolina is another option to use in deficient diet. In deficient population parenteral vitamin A can be given @ 100,000 U/kg, IM to treat the cases of hypovitaminosis A.

3.12.4 Excess Iron Deposition in Liver

This disease is also referred as iron storage disease or hemochromatosis which is characterized by excessive deposition of iron in the liver. This has been recorded in few psittacine species, mynahs, toucans and in some zoo birds.

3.12.4.1 Etiology

This condition is associated with excessive intake of dietary iron. Some genetic predisposition also plays a role for such development. Intake of citrus food or the foods rich in vitamin C facilitate the absorption of dietary iron.

3.12.4.2 Clinical Findings and Pathophysiology

The disease develops when the hepatic storage of iron is increased manifold and the lysosomes of the hepatocytes are unable to store iron anymore and release the iron reserve in the cytosol. This causes excessive oxidative damage of cell membrane and cellular proteins and nucleic acids.

In most of the cases affected birds develop and exhibit the symptoms simulating hepatic damage. The birds are anorectic and gradually lose their body weight. The abdomen is distended due to hepatic enlargement and developing ascites possibly due to decreased albumin synthesis and consequent drop in the colloidal osmotic pressure. Biliverdinaemia and biliverdinuria are seen in most of the birds. Post-mortem examination shows abnormality in liver, spleen and heart.

3.12.4.3 Diagnosis

The condition can be diagnosed on the basis of history, clinical symptom, and post-mortem examination. Liver is seen enlarged, golden brown in colour with scattered dark foci. Iron storage can be detected in hepatocytes and kupffer cells on histopathological examination of the collected tissue samples through biopsy or necropsy. This is associated with hepatic inflammation with infiltration of heterophils and lymphocytes.

3.12.4.4 Treatment

As the condition is associated with more iron in the body, efforts must be taken to reduce iron intake, reduce the absorption of dietary iron and facilitate the iron elimination from body. Affected birds must be given a food formulation with low iron level (within 50–100 ppm). Ascorbic acid present in the citrus food usually transform iron from its ferric (Fe^{3+}) to readily absorbable ferrous (Fe^{2+}) form, therefore, such citrus food must be avoided to minimize iron absorption. Owners should be advised so that they avoid giving the affected birds citrus fruits. Tanin, diets rich in fiber and phytate may be supplemented to the birds. To eliminate iron storage in the body periodic phlebotomy may be practiced (10% of body weight, once in a week). Iron chelating agent deferoxamine @ 100 mg/kg may be given PO/IM/SC once a day.

3.12.5 Fatty Liver Syndrome

This syndrome is more common in caged laying birds. The birds which are kept on high energy diet rich in carbohydrate and fat like all seed diets or nuts and the birds which lead a sedentary life style with limited scope of exercise are more prone to such condition.

Liver in birds are metabolically very active during egg laying and is the main site for lipogenesis. However, high energy diet, limited or no exercise, high environmental temperature lead to deposition of excessive fat in liver and abdominal cavity (Figs. 3.21 and 3.22).

Affected birds may die of hepatic enlargement and rupture following haemorrhage. Microscopically, enlarged hepatocytes are seen containing large fat globules. The condition can be managed with low energy diet, exercise and supplementation of lipotropic agents like choline, inositol and manganese.

3.12.6 Avian Amyloidosis

Amyloidosis is a disease mainly of the adult birds and caused by heavy extracellular deposition of normally autologous soluble amyloid proteins or their fragments in different important visceral organs and in the joints in characteristic fibrillary form. Once deposited, they become insoluble and resistant to proteolytic enzyme under normal physiological condition. This makes such amyloidosis an irreversible phenomenon.

Depending of the composition or characteristic of the fibrillary protein the common amyloid may be of four types—AL, AA, ATTR, and Aβ2 M. In animal, caged, wild and domesticated birds majority of the amyloid cases were found as AA type.

3.12.6.1 Etiology

The exact cause of amyloidosis is still obscure. Pathogenesis of AA-amyloidosis is complex and of multifactorial origin. In general it is thought that the syndrome is associated with abnormal protein metabolism, chronic debilitating diseases like avian tuberculosis, inflammations or tumours which may trigger a strong elevation in serum of serum amyloid A (apoSAA), an important acute phase reactant and the

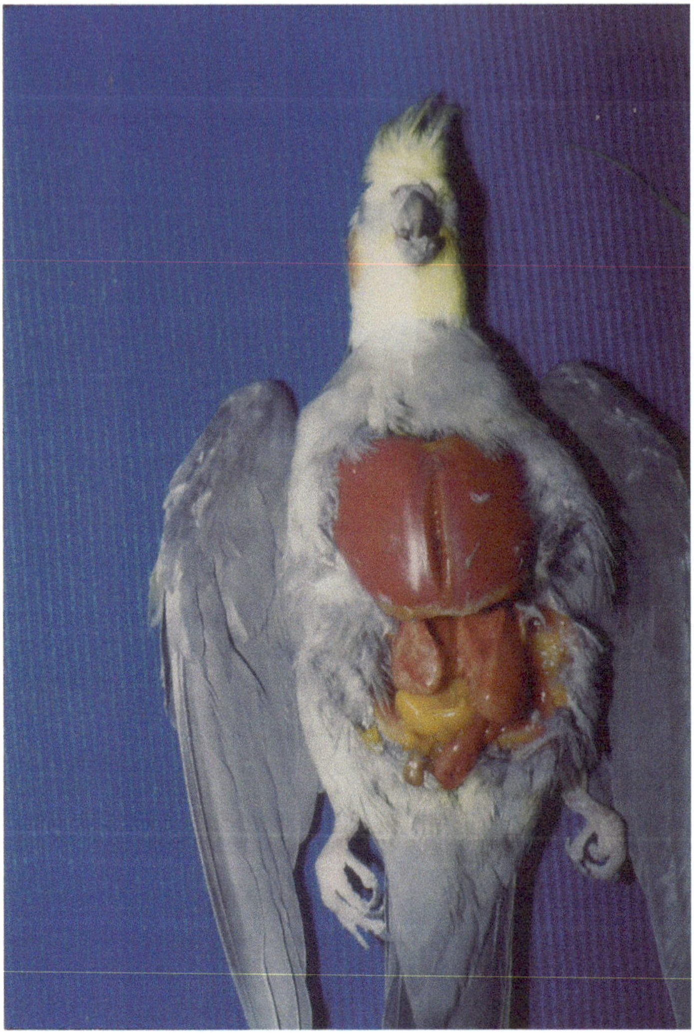

Fig. 3.21 Fatty liver in a cockatiel (*Courtesy* Petra Maria Burgmann, Canada)

AA precursor protein. This precursor protein is necessary for amyloid AA type development and certain amino acid substitutions may facilitate amyloidogenicity for unstable intermediate protein conformations required for the whole process. Other favouring factors include—amyloid-enhancing factor, an altered basement membrane protein and the presence of certain inorganic ions such as calcium and sulphate. Recent studies have suggested that vaccination may also be a triggering factor for induction of amyloidosis in commercial birds. In general it is thought that amyloid may be of primary origin where no triggering or inciting cause can be identified or it may be secondary to prolong infection with a persistent antigenic

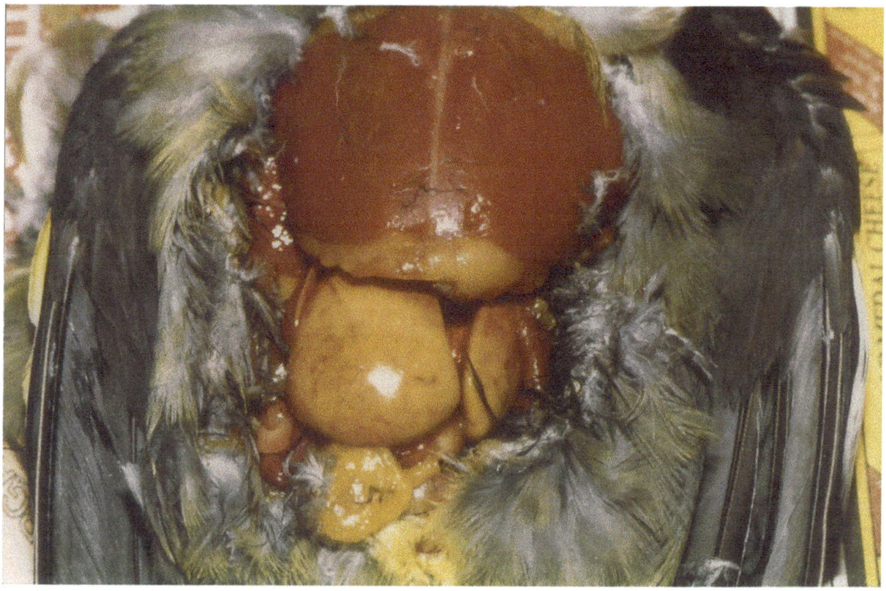

Fig. 3.22 Fatty liver in an African grey parrot (*Courtesy* Petra Maria Burgmann, Canada)

stimulation. In case of ducks persistent stress is generally thought as an inciting factor for amyloidosis. Amyloid arthropathy was reported with *Enterococcus faecalis* and *Mycoplasma synoviae* infection. Some degree of genetic predisposition too was reported to be associated with amyloidosis.

3.12.6.2 Clinical Findings and Pathology

The clinical presentation of the disease is non-specific and vague. However, most of the internal organs are affected. Among different visceral organs, liver, spleen and kidneys undergo characteristic changes like extensive enlargement, loss of lobules with rounded margins and pale discoloration (Fig. 3.23). Organs like proventriculus, large intestine, heart, gonads and endocrine organs are less affected. In case of ducks, swelling of feet and legs is very common with oedema. In few instances the affected ducks develop enlargement of ventral abdomen with ascites. This is probably due to circulatory collapse associated with hepatic amyloidosis. In such cases liver is usually enlarged and friable. It becomes brown to tan coloured. In chickens hepatic amyloidosis causes decreased egg production and sudden mortality due to ruptures of liver and haemorrhage. In amyloid arthropathy, the joints are swollen and painful and the affected birds show lameness and they are disinclined to move. Synovial fluid is thick and contains several yellow to green coloured deposits.

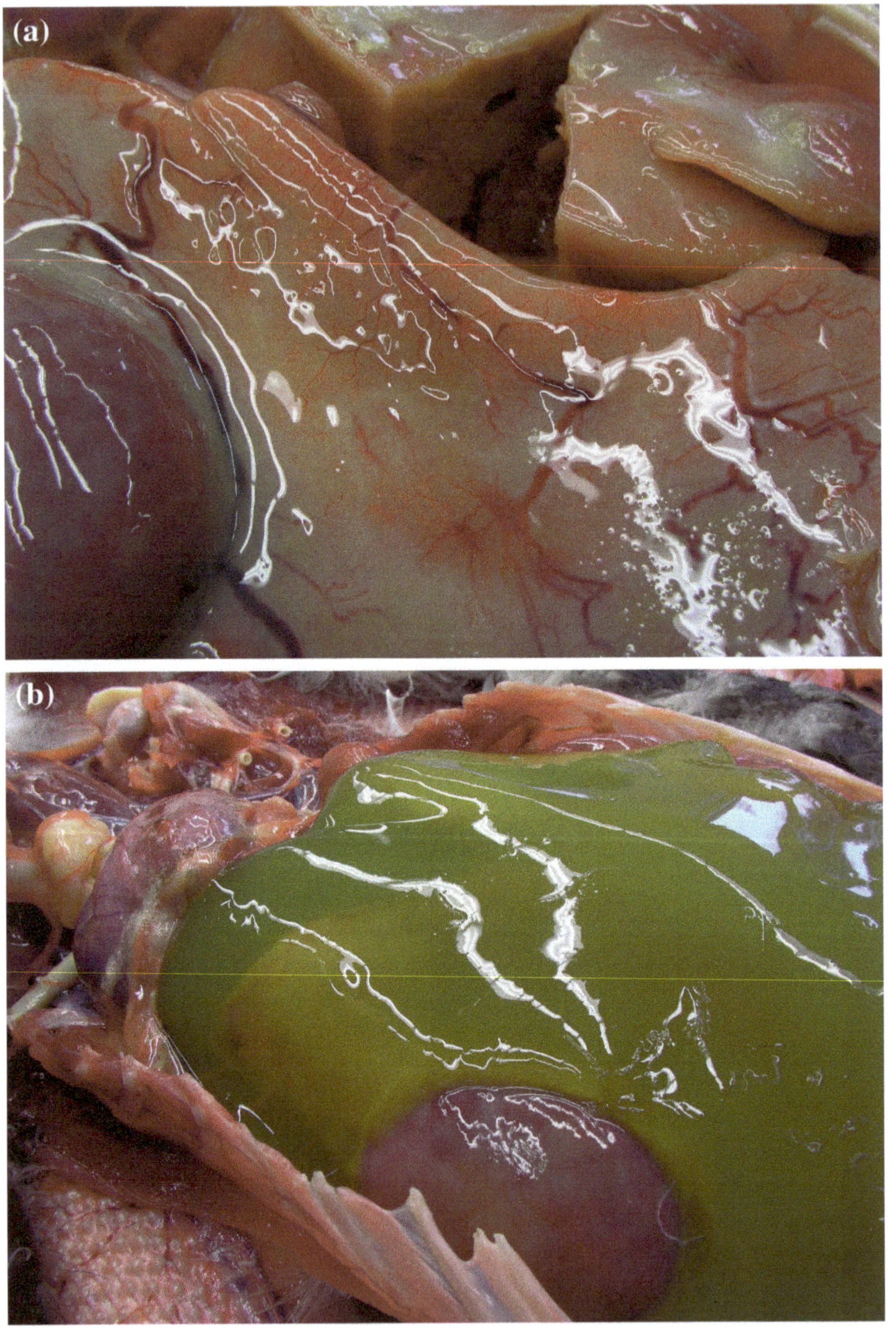

Fig. 3.23 Amyloidosis in bird affecting different organs (*Courtesy* Prof. Richard Hoop, University of Zurich, Switzerland)

3.12.6.3 Diagnosis

Due to vague and nonspecific clinical symptoms, it is rather difficult to diagnose the disease ante-mortem. Diagnosis is based on post-mortem finding. Tissue biopsy sample through cloaca or subcutaneous tissue aspirate may be tried to collect for histopathological examination. The tissues are usually examined via congo red staining or immune-histochemical staining using specific anti-sera. In hepatic tissue, histologically, amyloid is seen as pale-pink material and generally deposited in the space between the sinusoids and the hepatocytes. Hepatic chords may be atrophied despite hepatomegaly due to amyloidosis. A recent study in avian arthropathy in chickens revealed significant elevation in serum amyloid A (SAA) and serum amino acids like serine, glycine, isoleucine and phenylalanine.

3.12.6.4 Treatment

Currently there is no available therapeutic protocol for avian amyloidosis. However, one experimental study revealed ameliorative effect of methylprednisolone in amyloid arthopathy. In human patients anti-mitotic and anti-inflammatory drugs, like colchicine and dimethyl sulphoxide (DMSO) were used with successful outcome. However, their effect is still to be studied in birds.

Bibliography

Abrams GA, Paul-Murphy J, Murphy CJ (2002) Conjunctivitis in birds. Vet Clin N Am Small Anim Pract 5:287–309

Andreasen JR Jr, Andreasen CB, Latimer KS et al (1995) Thoracoabdominal myelolipomas and carcinoma in a lovebird (*Agapomis* sp.). J Vet Diagn Investig 7:271–272

Appleby RC (1984) Diabetes mellitus in a budgerigar (*Melopsittacus undulatus*). Vet Rec 115:652–653

Baker JR (1996a) Causes of mortality and morbidity in exhibition budgerigars in the United Kingdom. Vet Rec 139:156–162

Baker JR (1996b) Survey of feather diseases of exhibition budgerigars in the United Kingdom. Vet Rec 139:590–594

Bates G, Tucker RL, Ford S et al (1999) Thyroid adenocarcinoma in a bald eagle (*Haliaeetus leukocephalus*). J Zoo Wildl Med 30:439–442

Baumeister RF, Scher SJ (1988) Self-defeating behavior patterns among normal individuals: review and analysis of common self-destructive tendencies. Psychol Bull 104:3–22

Bavelaar F, Beynen A (2003) Severity of atherosclerosis in parrots in relation to the intake of α-linolenic acid. Avian Dis 47:566–577

Bavelaar F, Beynen A (2004) Atherosclerosis in parrots. A review. Vet Q 26:50–60

Beaufrère H (2013) Avian atherosclerosis: parrots and beyond. J Exot Pet Med 22:336–347

Beaufrère H, Ammersbach M, Reavill DR et al (2013) Prevalence of and risk factors associated with atherosclerosis in psittacine birds. J Am Vet Med Assoc 242:1696–1704

Beaufrère H, Nevarez J, Tully TN Jr (2010) Cloacolith in a blue-fronted amazon parrot (*Amazona aestiva*). J Avian Med Surg 24:142–145

Bennett R, Mader D (1996) Cloacal prolapse. In: Mader DR (ed) Reptile medicine and surgery. WB Saunders Company, Philadelphia, pp 355–358

Bermudez AJ, Hopkins BA (1995) Hemoglobinuric nephrosis in a rhea (*Rhea americana*). Avian Dis 39:661–665

Bezuidenhout A, Penrith M, Burger W (1993) Prolapse of the phallus and cloaca in the ostrich (*Struthio camelus*). J S Afr Vet Assoc 64:156–158
Blackmore D (1966) The clinical approach to tumours in cage birds. J Small Anim Pract 7:217–223
Blaxland J, Borland ED, Siller W et al (1980) An investigation of urolithiasis in two flocks of laying fowls. Avian Pathol 9:5–19
Blus LJ, Locke LN, Stafford CJ (1977) Ruptured-yolk peritonitis and organochlorine residues in a royal tern. Avian Dis 21:445–447
Bohorquez F, Stout C (1972) Aortic atherosclerosis in exotic avians. Exp Mol Pathol 17:261–273
Bowles H, Lichtenberger M, Lennox A (2007) Emergency and critical care of pet birds. Vet Clin N Am Small Anim Pract 10:345–394
Bowles HL (2002) Reproductive diseases of pet bird species. Vet Clin N Am Small Anim Pract 5:489–506
Burgos-Rodríguez AG (2010) Avian renal system: clinical implications. Vet Clin N Am Small Anim Pract 13:393–411
Burgos-Rodríguez AG, Garner M, Ritzman TK et al (2007) Cutaneous lymphosarcoma in a double yellow-headed Amazon parrot (*Amazona ochrocephala oratrix*). J Avian Med Surg 21:283–289
Butcher GD, Beck CC (1996) Avian Goiter (Thyroid Hyperplasia or Dysplasia). University of Florida Cooperative Extension Service, Institute of Food and Agriculture Sciences, EDIS. http://edis.ifas.ufl.edu/vm029. Accessed 7 Nov 2016
Candeletta SC, Homer BL, Garner MM et al (1993) Diabetes mellitus associated with chronic lymphocytic pancreatitis in an African grey parrot (*Psittacus erithacus erithacus*). J Assoc Avian Vet 7:39–43
Canny CJ, Ward DA, Patton S et al (1999) Microsporidian keratoconjunctivitis in a double yellow-headed Amazon parrot (*Amazona ochrocephala oratrix*). J Avian Med Surg 13:279–286
Caruso K, Cowell R, Meinkoth J et al (2002) Ab-dominal effusion in a bird [egg yolk peritonitis]. Vet Clin Pathol 31:127–128
Casares M, Enders F, Montoya J (2000) Comparative electrocardiography in four species of macaws (genera *Anodorhynchus* and *Ara*). J Vet Med Ser A 47:277–281
Chong K, Apostolov K (1982) The pathogenesis of nephritis in chickens induced by infectious bronchitis virus. J Comp Pathol 92:199–211
Circella E, Legretto M, Pugliese N et al (2014) Psittacine beak and feather disease-like illness in Gouldian finches (*Chloebia gouldiae*). Avian Dis 58:482–487
Clayton LA, Ritzman TK (2006) Egg binding in a cockatiel (*Nymphicus hollandicus*). Vet Clin N Am Small Anim Pract 9:511–518
Coleman CW (1991) Bile duct carcinoma and cloacal prolapse in an orange-winged Amazon parrot (*Amazona amazonica amazonica*). J Assoc Avian Vet 5:87–89
Cong W, Meng QF, Shan XF et al (2014) Seroprevalence and risk factors associated with hepatitis E virus infection in three species of pet birds in northwest China. Sci World J 2014:296285
Cosgrove A (1962) An apparently new disease of chickens: avian nephrosis. Avian Dis 6:385–389
Couvillion CE, Maslin WA, Montgomery RM (1990) Multiple feather follicle cysts in a wild turkey. J Wildl Dis 26:122–124
Cowan DF (1968) Avian amyloidosis I. General incidence in zoo birds. Pathol Vet Online 5:51–58
Cowen B, Wideman R, Rothenbacher H et al (1987) An outbreak of avian urolithiasis on a large commercial egg farm. Avian Dis 31:392–397
Crosta L, Gerlach H, Bürkle M et al (2003) Physiology, diagnosis, and diseases of the avian reproductive tract. Vet Clin N Am Small Anim Pract 6:57–83
de Matos R, Morrisey JK (2005) Emergency and critical care of small psittacines and passerines. Semin Avian Exotic Pet Med 14:90–105
de Wit M, Schoemaker NJ (2005) Clinical approach to avian cardiac disease. Semin Avian Exotic Pet Med 14:6–13

de Wit M, Schoemaker NJ, Kik MJ et al (2003) Hypercalcemia in two Amazon parrots with malignant lymphoma. Avian Dis 47:223–228

Dennis PM, Bennett RA (2000) Ureterotomy for removal of two ureteroliths in a parrot. J Am Vet Med Assoc 217:865–868

Desmarchelier A, Langlois I, Desmarchelier M (2015) Transient diabetes mellitus in a domestic ferret (*Mustela putorius furo*). Can Vet J 56:737–740

Desmarchelier M, Langlois I (2008) Diabetes mellitus in a nanday conure (Nandayus nenday). J Avian Med Surg 22:246–254

Donatti RV, Resende M, Ferreira FC et al (2014) Fatal proventricular dilatation disease in captive native psittacines in Brazil. Avian Dis 58:187–193

Doneley B (2016) Avian medicine and surgery in practice: companion and aviary birds. CRC Press, Boca Raton, Florida

Doukaki C, Kik MJ, Ducatelle R et al (2014) Thyroid adenocarcinoma in two parrots. Vet Rec 2: e000101

Effron M, Griner L, Benirschke K (1977) Nature and rate of neoplasia found in captive wild mammals, birds, and reptiles at necropsy. J Natl Cancer Inst (Bethesda) 59:185–198

El-Bahrawy A, Zaid A, Sunden Y et al (2015) Pathogenesis of pancreatitis in chickens after experimental Infection with 9a5b Newcastle disease virus mutant isolate. J Comp Pathol 153:315–323

Encinas-Nagel N, Enderlein D, Piepenbring A et al (2014) Avian bornavirus in free-ranging psittacine birds, Brazil. Emerg Infect Dis 20:2103–2106

Finlayson R (1965) Spontaneous arterial disease in exotic animals. J Zool (Lond) 147:239–343

Foerster SH, Gilson SD, Avery Bennett R (2000) Surgical correction of palatine bone luxation in a blue-and-gold macaw (*Ara ararauna*). J Avian Med Surg 14:118–121

Foster R, Smith M (2016) Birds: enlargement of thyroid gland. http://www.peteducation.com/article.cfm?c=15+1829&aid=2755. Accessed 5 Sept 2016

Fredrickson T (1987) Ovarian tumors of the hen. Environ Health Perspect 73:35

Fricke C, Schmidt V, Cramer K et al (2009) Characterization of atherosclerosis by histochemical and immunohistochemical methods in African grey parrots (*Psittacus erithacus*) and Amazon parrots (*Amazona* spp.). Avian Dis 53:466–472

Gallerstein G (1994) The complete bird owner's handbook. Howell Book House, New York

Gancz AY, Wellehan JF, Boutette J et al (2007) Diabetes mellitus concurrent with hepatic haemosiderosis in two macaws (*Ara severa, Ara militaris*). Avian Pathol 36:331–336

Gonçalves GAM, Salgado BS (2012) Post-mortem lesions of urolithiasis in a lesser seed finch (*Sporophila angolensis*). Acta Veterinaria Brasilica 6:52–55

Goodwin MA, Krabill VA (1989) Diarrhea associated with small-intestinal cryptosporidiosis in a budgerigar and in a cockatiel. Avian Dis 33:829–833

Gross W, Siegel P (1959) Coliform peritonitis of chickens. Avian Dis 3:370–373

Guerin J-L, Gelfi J, Dubois L et al (2000) A novel polyomavirus (goose hemorrhagic polyomavirus) is the agent of hemorrhagic nephritis enteritis of geese. J Virol 74:4523–4529

Guzman DS (2016) Avian soft tissue surgery. Vet Clin North Am Exot Anim Pract 19:133–157

Hadley TL (2010) Management of common psittacine reproductive disorders in clinical practice. Vet Clin N Am Exot Anim Pract 13:429–438

Hafner S, Harmon BG, Stewart RG et al (1993) Avian keratoacanthoma (dermal squamous cell carcinoma) in broiler chicken carcasses. Vet Pathol 30:265–270

Hakimuddin F, Abidi F, Jafer O et al (2016) Incidence and detection of beak and feather disease virus in psittacine birds in the UAE. Biomol Detect Quantif 6:27–32

Harcourt-Brown NH (1996) Torsion and displacement of the oviduct as a cause of egg-binding in four psittacine birds. J Avian Med Surg 262–267

Hartup BK, Dhondt AA, Sydenstricker KV et al (2001) Host range and dynamics of mycoplasmal conjunctivitis among birds in North America. J Wildl Dis 37:72–81

Hausmann JC, Mans C, Gosling A et al (2016) Bilateral Uveitis and Hyphema in a Catalina Macaw (*Ara ararauna* × *Ara macao*) with multicentric lymphoma. J Avian Med Surg 30:172–178
Heffels-Redmann U, Enderlein D, Herzog S et al (2011) Occurrence of avian bornavirus infection in captive psittacines in various European countries and its association with proventricular dilatation disease. Avian Pathol 40:419–426
Hess L (2000) Obesity and its consequences in an Amazon parrot (*Amazona aestiva aestiva*). In: Proceedings of the 21st annual conference and expo of the association of avian veterinarians, Portland, Oregon, USA, pp 119–120
Hochleithner M (1993) No therapy for feather duster budgerigars. J Assoc Avian Vet 7:98–98
Hollander W, Riddle O (1946) Goiter in domestic pigeons. Poult Sci 25:20–27
Hoppes S, Gurfield N, Flammer K et al (2000) Mycotic keratitis in a blue-fronted Amazon parrot (*Amazona aestiva*). J Avian Med Surg 14:185–189
Hoppes SM, Tizard I, Shivaprasad HL (2013) Avian bornavirus and proventricular dilatation disease: diagnostics, pathology, prevalence, and control. Vet Clin North Am Exot Anim Pract 16:339–355
Howard B (1992) Health risks of housing small psittacines in galvanized wire mesh cages. J Am Vet Med Assoc 200:1667–1674
Hubben K, Shirer JF, Enold GL (1964) Transmissible pituitary chromophobe adenoma in the parakeet. Avian Dis 8:203–208
Hvenegaard AP, Safatle A, Guimarães MB et al (2009) Retrospective study of ocular disorders in Amazon parrots. Pesqui Vet Bras 29:979–984
Ibi K, Murakami T, Goda WM et al (2015) Prevalence of amyloid deposition in mature healthy chickens in the flock that previously had outbreaks of vaccine-associated amyloidosis. J Vet Med Sci 77:1241–1245
Imada T, Yamaguchi S, Kawamura H (1979) Pathogenicity for baby chicks of the G-4260 strain of the picornavirus "avian nephritis virus". Avian Dis 41:582–588
Imada T, Yamaguchi S, Mase M et al (2000) Avian nephritis virus (ANV) as a new member of the family Astroviridae and construction of infectious ANV cDNA. J Virol 74:8487–8493
Ivanics E, Rudas P, Salyi G et al (1999) Massive goitre (*Struma parenchymatosa*) in geese. Acta Vet Hung 47:217–231
Jackson MK, Phillips SN (1996) Necrotizing hepatitis in pet birds associated with *Pseudomonas fluorescens*. Avian Dis 40:473–476
Julian R (1982) Water deprivation as a cause of renal disease in chickens. Avian Pathol 11:615–617
Kahn CM, Kahn M (2005) The Merck veterinary manual. http://www.merckvetmanual.com/mvm/search.html?st=pet+birds&start=1&context=%2Fmvm
Kaikabo A, Mustapha A, Yaroro I et al (2007) Occurrence of egg impaction and peritonitis in a flock of commercial laying hens in Damaturu Nigeria. Nig Vet J 28:56–58
Kaleta EF (1990) Herpesviruses of birds—a review. Avian Pathol 19:193–211
Keller KA, Beaufrere H, Brandão J et al (2013) Long-term management of ovarian neoplasia in two cockatiels (*Nymphicus hollandicus*). J Avian Med Surg 27:44–52
Keymer I (1980) Disorders of the avian female reproductive system. Avian Pathol 9:405–419
Krautwald-Junghanns M-E, Braun S, Pees M et al (2004) Research on the anatomy and pathology of the psittacine heart. J Avian Med Surg 18:2–11
Landman WJ (1999) Amyloid arthropathy in chickens. Vet Q 21:78–82
Landman WJ, Gruys E (1998) Amyloid arthropathy in an Indian peafowl. Vet Rec 142:90–91
Landman WJ, Gruys E, Gielkens AL (1998a) Avian amyloidosis. Avian Pathol 27:437–449
Landman WJ, vd Bogaard AE, Doornenbal P et al (1998b) The role of various agents in chicken amyloid arthropathy. Amyloid 5:266–278
Lewis JC, Kottke BA (1977) Endothelial damage and thrombocyte adhesion in pigeon atherosclerosis. Science 196:1007–1009
Lightfoot TL (2010) Geriatric psittacine medicine. Vet Clin N Am Exot Anim Pract 13:27–49

Lightfoot TL, Yeager JM (2008) Pet bird toxicity and related environmental concerns. Vet Clin N Am Exot Anim Pract 11:229–259

Loukopoulos P, Bautista AC, Puschner B et al (2015) An outbreak of thyroid hyperplasia (goiter) with high mortality in budgerigars (*Melopsittacus undulatus*). J Vet Diagn Investig 27:18–24

Lucké B, Schlumberger H (1949) Neoplasia in cold-blooded vertebrates. Physiol Rev 29:91–126

Lumeij J (2000) Pathophysiology, diagnosis and treatment of renal disorders in birds of prey. Rap Biomed 3:170–178

Luppi MM, de Melo AL, Motta RO et al (2007) Granulomatous nephritis in psittacines associated with parasitism by the trematode *Paratanaisia* spp. Vet Parasitol 146:363–366

Mans C, Brown CJ (2007) Radiographic evidence of atherosclerosis of the descending aorta in a grey-cheeked parakeet (*Brotogeris pyrrhopterus*). J Avian Med Surg 21:56–62

Mathey W (1967) Respiratory acariasis due to Sternostoma tracheacolum in the budgerigar. J Am Vet Med Assoc 150:777–780

Mayer J, Donnelly TM (2012) Clinical veterinary advisor, birds and exotic pets, 1: Clinical veterinary advisor. Elsevier Health Sciences

Millichamp NJ (2002) Ophthalmic disease in exotic species. Vet Clin N Am Exot Anim Pract 5:223–241

Murakami T, Inoshima Y, Sakamoto E et al (2013a) AA amyloidosis in vaccinated growing chickens. J Comp Pathol 149:291–297

Murakami T, Muhammad N, Inoshima Y et al (2013b) Experimental induction and oral transmission of avian AA amyloidosis in vaccinated white hens. Amyloid 20:80–85

Mutinelli F, Corro M, Catania S et al (2008) Multiple feather follicle cysts in a Moroseta hen (*Gallus gallus*). Avian Dis 52:345–347

Nakamura K, Ohtsu N, Nakamura T et al (2008) Pathologic and immunohistochemical studies of Newcastle disease (ND) in broiler chickens vaccinated with ND: severe nonpurulent encephalitis and necrotizing pancreatitis. Vet Pathol 45:928–933

Nap A, Lumeij J, Stokhof A (1992) Electrocardiogram of the African grey (*Psittacus erithacus*) and Amazon (*Amazona* spp.) parrot. Avian Pathol 21:45–53

Nedorost N, Maderner CA, Kolodziejek J et al (2012) Identification of mixed infections with different genotypes of avian bornaviruses in psittacine birds with proventricular dilatation disease. Avian Dis 56:414–417

Noonan BP, de Matos R, Butler BP et al (2014) Nasal adenocarcinoma and secondary chronic sinusitis in a hyacinth macaw (*Anodorhynchus hyacinthinus*). J Avian Med Surg 28:143–150

Nyholm NEI (1981) Evidence of involvement of aluminum in causation of defective formation of eggshells and of impaired breeding in wild passerine birds. Environ Res 26:363–371

Ogawa H, Sanada Y, Sanada N et al (2011) Proventricular dilatation disease associated with avian bornavirus infection in a Citron-crested Cockatoo that was born and hand-reared in Japan. J Vet Med Sci 73:837–840

Oglesbee BL, Hamlin RL, Klingaman H et al (2001) Electrocardiographic reference values for macaws (*Ara species*) and cockatoos (*Cacatua species*). J Avian Med Surg 15:17–22

Olsen GH (2003) Oral biology and beak disorders of birds. Vet Clin N Am Exot Anim Pract 6:505–521

Panigrahy B, Mathewson JJ, Hall CF et al (1981) Unusual disease conditions in pet and aviary birds. J Am Vet Med Assoc 178:394–395

Parisi L, Clubb S (2010) What is your diagnosis? Proventricular dilatation disease. J Avian Med Surg 24:152–154

Paterson S (2008) Skin diseases of exotic pets. Wiley, Oxford

Pees M, Straub J, Krautwald-Junghanns M-E (2004) Echocardiographic examinations of 60 African grey parrots and 30 other psittacine birds. Vet Rec 155

Perry RA, Gill J, Cross GM (1991) Disorders of the avian integument. Vet Clin N Am Small Anim Pract 21:1307–1327

Phalen DN, Logan KS, Snowden KF (2006) *Encephalitozoon hellem* infection as the cause of a unilateral chronic keratoconjunctivitis in an umbrella cockatoo (*Cacatua alba*). Vet Ophthalmol 9:59–63

Pilny AA (2008) The avian pancreas in health and disease. Vet Clin North Am Exot Anim Pract 11:25–34

Pilny AA, Luong R (2005) Diabetes mellitus in a chestnut-fronted macaw (*Ara severa*). J Avian Med Surg 19:297–302

Pinkster C (2016) T-cell Lymphosarcoma in a six-months old Indian Ringneck parrot. In: Annual conference of association of avian veterinarians Australian Committe Ltd. http://www.aavac.com.au/files/08-2016.pdf. Accessed 07 Nov 2016

Pollock C (2006) Diagnosis and treatment of avian renal disease. Vet Clin N Am Exot Anim Pract 9:107–128

Rae M (1995) Endocrine disease in pet birds. Semin Avian Exotic Pet Med 4:32–38

Raidal SR, Sarker S, Peters A (2015) Review of psittacine beak and feather disease and its effect on Australian endangered species. Aust Vet J 93:466–470

Ramis A, Gibert X, Majo N et al (1999) Metastatic oral squamous cell carcinoma in a Montagu's harrier (*Circus pigargus*). J Vet Diagn Investig 11:191–194

Ratcliffe HL (1933) Incidence and nature of tumors in captive wild mammals and birds. Am J Can 17:116–135

Reavill DR (2004) Tumors of pet birds. Vet Clin N Am Exot Anim Pract 7:537–560

Reavill DR, Dorrestein GM (2010) Pathology of aging psittacines. Vet Clin N Am Exot Anim Pract 13:135–150

Reece R (1992) Observations on naturally occurring neoplasms in birds in the state of Victoria, Australia. Avian Pathol 21:3–32

Rembert MS, Smith JA, Strickland KN et al (2008) Intermittent bradyarrhythmia in a Hispaniolan Amazon parrot (*Amazona ventralis*). J Avian Med Surg 22:31–40

Rich GA (2003) Syndromes and conditions of parrotlets, pionus parrots, poicephalus, and mynah birds. Semin Avian Exotic Pet Med 12:144–148

Rivera S, McClearen JR, Reavill DR (2009) Treatment of nonepitheliotropic cutaneous B-cell lymphoma in an umbrella cockatoo (*Cacatua alba*). J Avian Med Surg 23:294–302

Romagnano A (1996) Avian obstetrics. Semin Avian Exotic Pet Med 5:180–188

Rosen LB (2012) Avian reproductive disorders. J Exot Pet Med 21:124–131

Rosskopf W Jr, Woerpel R (1982) Egg yolk peritonitis in a California desert tortoise. Calif Vet 3:13–15

Rotstein DS, Flowers JR, Wolfe BA et al (2005) Renal trematodiasis in captive double-toothed barbets (*Lybius bidentatus*). J Zoo Wildl Med 36:124–126

Rubbenstroth D, Rinder M, Stein M et al (2013) Avian bornaviruses are widely distributed in canary birds (*Serinus canaria f. domestica*). Vet Microbiol 165:287–295

Rubinstein J, Lightfoot T (2014) Feather loss and feather destructive behavior in pet birds. Vet Clin North Am Exot Anim Pract 17:77–101

Rupley AE (1997) Manual of avian practice. Philadelphia WB Saunders

Russell W (1970) Iodide goiter in penguins. J Zoo Animal Med 24–27

Sarker S, Moylan KG, Ghorashi SA et al (2015) Evidence of a deep viral host switch event with beak and feather disease virus infection in rainbow bee-eaters (*Merops ornatus*). Sci Rep 5:14511

Sasipreeyajan J, Newman JA (1988) Goiter in a cockatiel (*Nymphicus hollandicus*). Avian Dis 32:169–172

Schlumberger HG (1954) Neoplasia in the parakeet I. Spontaneous chromophobe pituitary tumors. Cancer Res 14:237–245

Schmidt RE, Reavill DR (2002) Thyroid hyperplasia in birds. J Avian Med Surg 16:111–114

Schmidt RE, Reavill DR (2008) The avian thyroid gland. Vet Clin North Am Exot Anim Pract 11:15–23

Schmidt RE, Reavill DR, Phalen DN (2008) Pathology of pet and aviary birds. Wiley

Sedacca CD, Campbell TW, Bright JM et al (2009a) Chronic cor pulmonale secondary to pulmonary atherosclerosis in an African Grey parrot. J Am Vet Med Assoc 234:1055–1059
Sedacca CD, Campbell TW, Bright JM et al (2009b) Chronic cor pulmonale secondary to pulmonary atherosclerosis in an African Grey parrot. J Am Vet Med Assoc 234:1055–1059
Seibert LM (2006) Feather-picking disorder in pet birds. In: Manual of parrot behavior, pp 255–265
Sevimli A, Yagci A, Bulbul A (2013) Matrix metalloproteinase, vitamin A and methylprednisolone effects on experimentally induced amyloid arthropathy. Biotech Histochem 88:336–344
Sevimli A, Yalcin M, Cangul IT et al (2012) Investigation of serum amino acid and serum amyloid A concentrations in chickens with amyloid arthropathy. Vet Q 32:17–23
Shivaprasad HL, Phalen DN (2012) A novel herpesvirus associated with respiratory disease in Bourke's parrots (*Neopsephotus bourkii*). Avian Pathol 41:531–539
Shuster KA, Brock KL, Dysko RC et al (2012) Polytetrafluoroethylene toxicosis in recently hatched chickens (*Gallus domesticus*). Comp Med 62:49–52
Silvanose C, Bailey T, Naldo J et al (2001) Bacterial flora of the conjunctiva and nasal cavity in normal and diseased captive bustards. Avian Dis 45:447–451
Simone-Freilicher E (2007) Use of isoxsuprine for treatment of clinical signs associated with presumptive atherosclerosis in a yellow-naped Amazon parrot (*Amazona ochrocephala auropalliata*). J Avian Med Surg 21:215–219
Souza MJ, Newman SJ, Greenacre CB et al (2008) Diffuse intestinal T-cell lymphosarcoma in a yellow-naped Amazon parrot (*Amazona ochrocephala auropalliata*). J Vet Diagn Investig 20:656–660
Spadafori G, Speer B (1999) Birds for dummies. IDG Books Worldwide, Foster City, CA
Srinivasan P, Balasubramaniam GA, Murthy TRGK et al (2013) Bacteriological and pathological studies of egg peritonitis in commercial layer chicken in Namakkal area. Asian Pac J Trop Biomed 3:988–994
Steentjes A, Veldman KT, Mevius DJ et al (2002) Molecular epidemiology of unilateral amyloid arthropathy in broiler breeders associated with *Enterococcus faecalis*. Avian Pathol 31:31–39
Strunk A, Wilson GH (2003) Avian cardiology. Vet Clin N Am Exot Anim Pract 6:1–28
Tanabe Y (1965) Relation of thyroxine secretion rate to age and growth rate in the cockerel. Poult Sci 44:591–596
Tidemann SC, McOrist S, Woinarski JC et al (1992) Parasitism of wild Gouldian finches (*Erythrura gouldiae*) by the air-sac mite *Sternostoma tracheacolum*. J Wildl Dis 28:80–84
Tully TN, Dorrestein GM, Jones AK (2009) Handbook of Avian Medicine. W.B. Saunders, Edinburgh
Turrel JM, McMillan MC, Paul-Murphy J (1987) Diagnosis and treatment of tumors of companion birds II. AAV Today 1:159–165
Vice C (1992) Myocarditis as a component of psittacine proventricular dilatation syndrome in a Patagonian conure. Avian Dis 36:1117–1119
Vink-Nooteboom M, Schoemaker N, Kik M et al (1998) Clinical diagnosis of aneurysm of the right coronary artery in a white cockatoo (*Cacatua alba*). J Small Anim Pract 39:533–537
Wade LL, Newman SJ (2004) Hemoglobinuric nephrosis and hepatosplenic erythrophagocytosis in a dusky-headed conure (*Aratinga weddelli*) after ingestion of garlic (*Allium sativum*). J Avian Med Surg 18:155–161
Wadsworth P, Jones D (1979) Some abnormalities of the thyroid gland in non-domesticated birds. Avian Pathol 8:279–284
Walsh AL, Shivaprasad H (2013) Unusual lesions of atherosclerosis in psittacines. J Exot Pet Med 22:366–374
Web A (2014) Birds: Loss of voice/changes in voice. https://www.beautyofbirds.com/lossofvoice.html. Accessed 7 Nov 2016

Welchman Dde B, Ainsworth HL, Jensen TK et al (2013) Demonstration of *Ornithobacterium rhinotracheale* in pheasants (*Phasianus colchicus*) with pneumonia and airsacculitis. Avian Pathol 42:171–178

Wensman JJ, Jaderlund KH, Holst BS et al (2014) Borna disease virus infection in cats. Vet J 201:142–149

Wideman JRR, Rocish WB, Satnick JL et al (1989) Methionine hydroxy analog (free acid) reduces avian kidney damage and urolithiasis induced by excess dietary calcium. J Nutr 119

Wideman R, Closser J, Roush W et al (1985) Urolithiasis in pullets and laying hens: role of dietary calcium and phosphorus. Poult Sci 64:2300–2307

Willis AM, Wilkie DA (1999) Avian ophthalmology, Part 2: Review of ophthalmic diseases. J Avian Med Surg 13:245–251

Wisser J, Ippen R (1991) Goiter incidence among zoo birds. Wiener Tierarztliche Monatsschrift 78:362–365

Worell AB (2012) Current trends in avian pediatrics. J Exot Pet Med 21:115–123

Zandvliet MM (2005) Electrocardiography in psittacine birds and ferrets. Semin Avian Exotic Pet Med 14:34–51

Zhang W, Shen Q, Mou J et al (2008) Hepatitis E virus infection among domestic animals in eastern China. Zoonoses Public Health 55:291–298

Toxicity 4

Indranil Samanta and Samiran Bandyopadhyay

4.1 Heavy Metal Toxicity

Heavy metals (lead, zinc, mercury, cobalt, nickel, cadmium, selenium) are potent toxic substances and are included in the world health organization's (WHO) list of chemicals with public health concern. Toxicity of heavy metals depends on their oxidative state (redox-active and redox-inactive) in the environment and their reactions with other compounds. Redox-active metals (copper, iron) can generate reactive hydroxyl radicals which are toxic for cellular mitochondria, microsomes and peroxisomes. Redox-inactive metals (lead, nickel, cadmium) destroy cellular antioxidants (e.g. glutathione). Exposure to heavy metals for prolonged period can cause decline in carotenoid levels, survival period, reproductive performances, expression of singing behaviour and feather brightness; oxidative stress and genetic alterations in passerine birds such as in pied flycatchers (*Ficedula hypoleuca*) and great tits (*Parus major*). Offspring sex-ratio is not altered in great tits (*Parus major*) exposed to heavy metals.

Feather of birds (great tits, blackbirds, robins, blackcaps, raptors, seabirds) act as an indicator ('biomonitor') for accumulation of heavy metal in birds as well as in the local environment. Origin of heavy metal in feathers is either exogenous (from contaminated dust, sand, vegetables and water) or endogenous (from contaminated blood during formation of keratin). Lead and cadmium are mostly exogenous and mercury is considered as endogenous metal. The birds prefer to deposit or excrete the heavy metals in their feathers, uropygial glands, salt glands, eggs and egg shells.

4.1.1 Lead Toxicosis (Plumbism)

Lead toxicosis is reported from parrots, wild kea (*Nestor notabilis*), raptors and wild waterfowls. Ingestion is the major route of lead transfer into the birds from the environment. The plants in lead rich soils can accumulate the lead in their fruits and

I. Samanta and S. Bandyopadhyay, *Pet Bird Diseases and Care*,
DOI 10.1007/978-981-10-3674-3_4

leaves. Transfer of lead to the birds occurs through the consumption of fruits, insects and soils (ground foraging birds). In households, paint, lead batteries, wine bottle tops, curtain weights, windows, electrical clips, car exhaust fumes act as major source of lead toxicosis.

Ingested lead usually remains in the GI tract for a long time and following degradation in the proventriculus or gizzard, this is being slowly released in blood. Lead is deposited in soft tissues and bone over the time. Lead has severe impact over multiple organs. It causes necrosis of the epithelial lining of GI tract and damage to the endothelial cells. It also causes increased fragility of the erythrocytes and bone marrow suppression leading to reduced generation of RBCs and other blood cells. Degeneration of brain capillaries may cause severe cerebral oedema.

Lead concentration of 0.2 ppm and above in blood is considered as toxic. Acute lead poisoning causes seizures, aggressiveness, difficulties in walking, flying and landing, regurgitation, diarrhoea, polyuria, blood in faeces, green/yellow urates, and anaemia in captive parrots. Chronic lead poisoning is common in wild birds. Chronic toxicosis causes emaciation, paralysis of wings and legs, esophageal impaction and death. In some species of wild birds, such as, yellow wagtail (*Motacilla flava*), Italian sparrow (*Passer italiae*), American robin (*Turdus migratorius*), blackbirds (*Turdus merula*), gray catbird (*Dumatella carolinensis*), song sparrow (*Melospiza melodia*), northern cardinal (*Cardinalis cardinalis*), northern mockingbird (*Mimus polyglottos*), carolina wren (*Thryothorus ludovicianus*), clay-colored thrush (*Turdus grayi*), great-tailed grackle (*Quiscalus mexicanus*) and house sparrow (*Passer domesticus*), higher content of lead in blood than normal level (>0.2 ppm) is detected although the birds appeared healthy. The wild birds are efficient in excretion of lead from the circulation into their bone and feathers.

Whole blood samples (with heparin, not EDTA) from the suspected birds can be collected as clinical specimen. History of diet and clinical symptoms are important in diagnosis of lead toxicosis. Detection of lead content present in the collected blood samples can confirm the toxicosis (Fig. 4.1). Identification (by radiography)

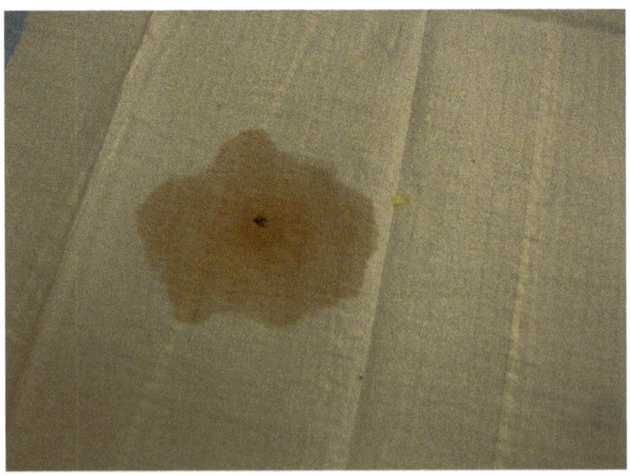

Fig. 4.1 Detection of lead toxicosis in blood samples collected from great bustard (*Otis tarda*) (*Courtesy* Ashley M. Zehnder)

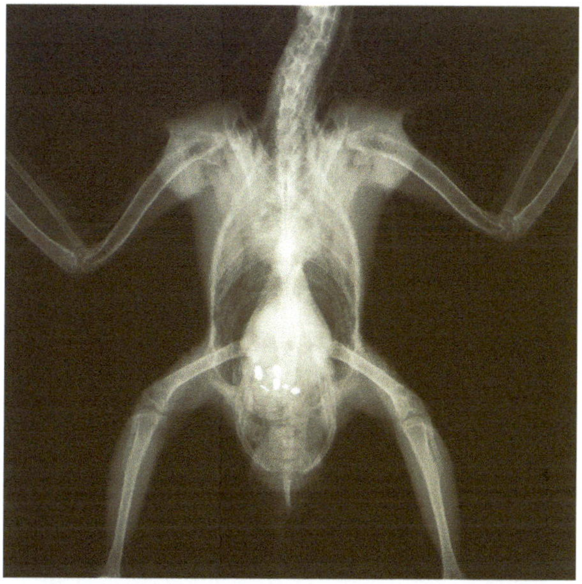

Fig. 4.2 Radiograph showing the presence of heavy metal (*Courtesy* Petra Maria Burgmann, Canada)

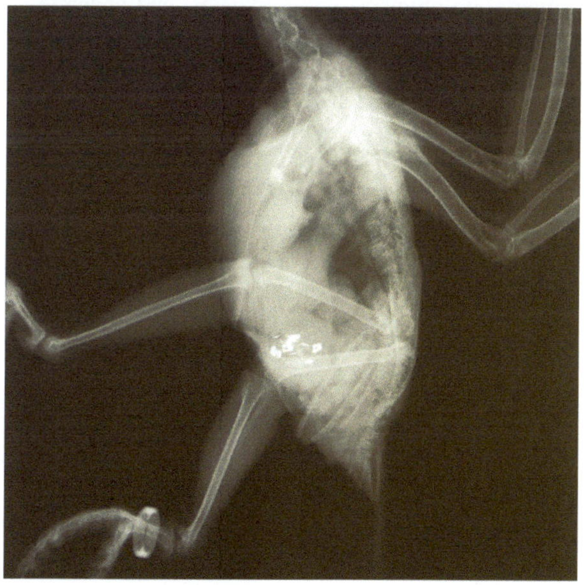

Fig. 4.2 (continued)

of radiodense particulate materials in the enteric tract of birds can further confirm the lead toxicosis (Fig. 4.2).

In confirmed cases of lead toxicosis, chelation therapy can be followed (calcium-EDTA, versenate, penicillamine) until the signs are resolved. Experimental study showed that dimercaptosuccinic acid (40–80 mg/kg body weight, oral,

12 h interval) and Ca-EDTA (40 mg/kg body weight, intramuscular, 12 h interval) are effective against lead toxicosis in cockatiels. Dimercaptosuccinic acid (DMSA) is detected to be more effective due to oral administration although margin of safety is low. Supportive treatment with anti-convulsion drugs (e.g. diazepam) and fluid therapy can be administered. After initial stabilization of birds, lead particles can be removed by endoscopic or surgical means.

4.1.2 Zinc Toxicosis (New Wire Disease)

Zinc toxicosis in birds is documented due to accidental ingestion of newly galvanized wires, cage coatings, hardwares, metallic toys, leg rings, pennies and fertilizers. Naturally occurring zinc toxicosis is reported in macaws, lovebirds, amazons, nicobar pigeon, gray-headed chachalaca and ducks. Zinc concentration (>2 ppm) in blood of psittacine birds (except cockatoos and Eclectus parrots) suggests about toxicosis. In cockatoos and Eclectus parrots, zinc is normally present as 2.5–3 ppm in blood.

Non-specific clinical signs such as feather picking or chewing, regurgitation, crop stasis, depression, lethargy, anorexia, weight loss, dark coloured faeces, polyuria, polydypsia, ataxia, paresis, and anaemia are developed due to zinc toxicosis (Fig. 4.3). Loss of pancreal acinar architecture, presence of hyaline bodies and tissue debris in pancreas, atrophy and necrosis of pancreatic cells, degeneration of proximal convoluted tubules and descending loops of Henle in kidneys, hepatic biliary retention, necrotizing ventriculitis, and necrotizing enteritis are characteristic necropsy findings in birds with zinc toxicosis.

Blood can be collected with plastic syringes without any contact with rubber (stoppers or syringe plungers) for detection of zinc level. Determination of zinc concentration in blood, pancreas and liver of suspected birds, histopathological examination of pancreas can diagnose zinc toxicosis.

Fig. 4.3 Feather picking behaviour in an African grey parrot (*Courtesy* Prof. Elena Circella, Avian Diseases Unit, Department of Veterinary Medicine, University of Bari, Italy)

In confirmed cases of zinc toxicosis, chelation therapy can be followed (calcium-EDTA, versenate, penicillamine) until the signs are resolved. Supportive treatment with anti-convulsion drugs (e.g. diazepam) and fluid therapy can be administered. Removal of zinc source from the surroundings can reduce the exposure and consequently the zinc level in the body.

4.1.3 Arsenic Toxicosis

Arsenic is a metalloid having the properties of both metals and non-metals. In the environment, coal burning, pesticides, wood preserving chemicals, mining act as major source of arsenic pollution. Different studies throughout the world detected the presence of arsenic in wild passerine birds (Table 4.1). Arsenic toxicosis is rarely reported from household pet birds except from the arsenic infested localities.

In wild passerine birds, arsenic toxicosis causes decreased clutch and brood size, reduced hatching, smaller eggs, growth abnormalities in legs and wings, aggressive behaviour, change in songs, reduced carotenoid colouration of plumage. In pet birds, hepatic lipidosis is detected in amazon parrots and cockatiels suffering with arsenic toxicosis. Nestling zebra finches (*Taeniopygia guttata*) experimentally fed with arsenical compound showed decreased tarsi and wing cord length.

Table 4.1 Arsenic detected in birds in different countries

Birds	Country
Pied flycatcher (*Ficedula hypoleuca*)	Sweden, Russia, Norway, Finland, Estonia, Latvia, Germany, UK, Netherlands, Spain
Great tits (*Parus major*)	Finland, Portugal, Sweden
Blue tit (*Cyanistes caeruleus*)	Finland
Rook (*Corvus frugilegus*)	Poland
House wren (*Troglodytes aedon*), American robin (*Turdus migratorius*), red-winged blackbird (*Agelaius phoeniceus*), marsh wren (*Cistothorus palustris*), tree swallow (*Tachycineta bicolor*), Florida scrub-jay (*Aphelocoma coerulescens*), yellow-breasted chat (*Icteria virens*), willow flycatcher (*Empidonax traillii*), yellow warbler (*Setophaga petechia*), common starling (*Sturnus vulgaris*), mountain bluebird (*Sialia currucoides*), black-capped chickadee (*Poecile atricapillus*)	United States of America
Greenfinch (*Chloris chloris*)	China
Mountain chickadee (*Poecile gambeli*), American dipper (*Cinclus mexicanus*)	Canada
Red-breasted nuthatch (*Sitta canadensis*)	Finland

4.1.4 Mercury Toxicosis

In wild passerines [tree swallow (*Tachycineta bicolor*); swamp sparrow (*Melospiza georgiana*)], raptors and piscivorous birds [fish-eating birds; for e.g. common loon (*Gavia immer*)] accumulation of mercury causes reduced egg production, poor hatching performances, immunosuppression, and abnormal singing. Nestlings or young birds are less affected due to sequestration of mercury in their feathers which prevents accumulation of mercury in blood and tissues in toxic level. Mercury has affinity for keratin and is stored in the developing feathers. However, when the feathers are fully developed, mercury level in blood and tissues may rise to produce toxicosis.

4.1.5 Copper and Iron Toxicity

In general copper toxicity is rarely detected among the birds. Painting with copper sulfate over the wire, pennies and copper ammunition are the major source of copper toxicity. Another important source of copper toxicity is acid metalliferous water bodies for the free ranging birds. Copper is an essential ingredient of Cu–Zn dependent superoxide dismutase which controls the intracellular antioxidant defence mechanism by controlling the reactive oxygen species including super oxides. However, excess copper leads to loss of zinc homeostasis and impairment of antioxidant defence mechanism. Clinically copper toxicity is associated with anaemia, lethargy, prostration, depression, weakness, anaemia, convulsions and coma. Pathological observation is prominent in GI tract including ulcerative and necrotic haemorrhages and congestion of proventricular and ventricular layer. Ulcerative haemorrhages are common in duodenum. Similarly Iron toxicity was also infrequently reported among the parrots with exposure to cast-iron feeding bowls with chipped enamel. Affected birds exhibits symptoms like chronic emaciation, lack of appetence, weight loss, lethargy and weakness.

4.2 Organophosphate Toxicity

Organophosphate insecticides (malathion) are used as spray in pet animals which may produce toxicity in pet birds if used in a room with poor ventilation. Insect repellent strips are also considered as potent source of organophosphate toxicity in pet birds. In agriculture (including fruit crops), the insecticides are also used to protect the crops. Insects killed by the organophosphate are often eaten by wild birds. Sometimes insecticides and fertilizers leach into groundwater and birds used to drink the contaminated water.

Fig. 4.4 Twisting of leg in a parrot (*Courtesy* Barun Dev Das, Animal Resources Development department, Government of West Bengal, India)

Organophosphates are potent inhibitor of cholinesterase enzymes present in the nervous system. Birds are more sensitive to the organophosphates than human and animals. Clinical signs of organophosphate toxicity include diarrhoea, vomition, weakness, twisting of legs and paralysis (Fig. 4.4). Delayed onset of clinical symptoms is also observed.

Diagnosis is based on history of organophosphate exposure and clinical symptoms. Plasma cholinesterase assay can confirm the toxicosis in birds. Atropine (0.1–0.6 ml/kg body weight, intramuscular or subcutaneous injection) and pralidoxime chloride (10–100 mg/kg body weight, intravenously, 24 h interval) are recommended for organophosphate toxicity in birds. Sedatives (doxepin @ 0.2 mg/kg body weight, intramuscular, 4 h interval) can be used as supportive treatment.

4.3 Organochlorine (Organohalogen Pollutant) Toxicity

Similar to organophosphates, organochlorine or organohalogen pollutants such as dichloro diphenyl trichloroethane (DDT), benzene hexachloride, aldrin, chlordane, toxaphene, methoxychlor, isodrin, endrin, and heptachlor are used in agriculture to prevent infestation of insects and fungi. Use of DDT is banned in United States, Canada, Mexico and several other countries since 1970s. Insects (earthworms), green mussels, snails, fishes, prawns, crabs, frogs killed by the organochlorine compounds are often eaten by wild birds. Inhalation of toxic fume generated during spraying of the insecticides in field is another possible way of transmission in birds. Wild and migratory passerine birds such as Eurasian tree sparrow (*Passer montanus*), common magpie (*Pica pica*), northern pintails (*Anas acuta*), swallow (*Petrochelidon* spp.), dusky flycatcher (*Empidonax oberholseri*), Bell's vireo (*Vireo bellii*), blue-gray gnatcatcher (*Polioptila caerulea*), orange-crowned warbler (*Vermivora celata*), spotted owlet (*Athene brama*), black-winged stilt (*Himantopus himantopus*), white-cheeked terns (*Sterna repressa*), little egret (*Egretta garzetta*), pond heron (*Ardeola grayii*), white-breasted kingfisher (*Halcyon smyrnensis*), little-ringed plover (*Charadrius dubius*) are reported to accumulate organochlorine compounds. Few of these birds are used as 'biomonitor' for organochlorine or organohalogen pollutant accumulation in local environment. In black-capped chickadees (*Poecile atricapillus*), organochlorine toxicity is detected to be associated with beak deformities. In song sparrows (*Melospiza melodia*), accumulation of polychlorinated biphenyls (PCBs) disrupts testosterone metabolism and causes abnormal song behaviour.

However, caged birds usually get the toxicity when they feed the treated seeds/feed ingredients. Toxicity due to consumption of contaminated food (with Aldrin, dieldrin, heptachlor, chlordane and endrin) were reported in birds. Similarly, topical spray may also cause poisoning in birds (toxaphene). Higher consumption often leads to sudden death of affected birds due to acute poisoning.

Organochlorine compounds are CNS stimulants and cause neurological manifestations like convulsion, seizures, tonic or clonic spasms, tremors, incoordination, behavioural abnormality and sudden falling by sides. Acute excitability is also accompanied by high rise of temperature and watery or mucosal discharge from nasal passage. Exposure to sub-lethal dose or less toxic compounds like DDT may cause reproductive impairment like reduced eggshell thickness, increased eggshell cracking, increased embryonic mortality, decreased survival of the new-born. Other signs of subacute and chronic intoxication include molting, dehydration and cyanosis of the comb, weight loss, and cessation of egg production.

The post-mortem examination usually reflects some characteristic changes like congestive changes in liver, kidney, lung and brain. Several haemorrhagic spots may be detected in pericardium. The birds died of chronic toxicity are usually emaciated with muscle wasting and prominence of keel bone. Other PM findings include—hepatomegaly, reduced spleen size and atrophy of other organs. Petechial haemorrhages are noticed in trachea, pleura and neck.

Diagnosis is confirmed on the basis of clinical signs of central nervous and reproductive systems and history indicating exposure to organochlorines and pathological lesions. In case of acute exposure, detection of organochlorine compounds in ingesta or viscera are helpful. Other organs like liver, brain and blood should be analysed for residue of such compounds in birds.

No specific treatment is well documented for chlorinated hydrocarbon poisoning in birds. When the affected bird is suspected to consume the poison gastric lavage and saline purgative are useful. Activated charcoal is given to prevent gastrointestinal absorption. Skin exposure by dipping or spraying may be taken care by thorough washing with water and detergent. To extenuate the tremors or convulsion, sedative drugs like phenobarbital/diazepam is advocated. Atropine sulphate may be used to control parasympathomimetic signs. Oral or parenteral rehydration is necessary to rejuvenate the affected bird along with calcium borogluconate and glucose to protect from against liver damage and hyperkalemia.

4.4 Polytetrafluroethyelene (Teflon) Toxicity

Polytetrafluroethyelene gas is generated at home during overheating (>250 °C) of teflon coated non-stick pans possibly in microwave oven (pyrolysis). The fume is inhaled by the pet birds kept in kitchens. This kind of accidental toxicity is reported from parakeets, cocatiels and budgerigars. Sudden death without any clinical symptom is the most common finding. If the birds survive for certain period wheezing, dyspnoea, ataxia and convulsions are observed. Lung is the primary target organ of polytetrafluroethyelene gas which causes necrotizing and hemorrhagic pneumonitis. Necropsy findings include hyperaemia of lung parenchyma with atelectasis. Accumulation of protein, debris, and blood cells is observed in lung alveoli and bronchi. Application of nebulized 1% calcium chloride, oxygen gas, steroids (nebulized prednisolone) and diuretics are possible ways of treatment in birds.

Bibliography

Albert C, Williams TD, Morrissey CA et al (2008) Tissue uptake, mortality, and sublethal effects of monomethylarsonic acid (MMA (V)) in nestling zebra finches (*Taeniopygia guttata*). J Toxicol Environ Hlth Part A 71:353–360

Blandford TB, Seamon PJ, Hughes R et al (1975) A case of polytetrafluoroethylene poisoning in cockatiels accompanied by polymer fume fever in the owner. Vet Rec 96:175–178

Costa RA, Petronilho JMS, Soares AMVM et al (2011) The use of passerine feathers to evaluate heavy metal pollution in central Portugal. Bull Environ Cont Toxicol 86:352–356

Denver MC, Tell LA, Galey FD et al (2000) Comparison of two heavy metal chelators for treatment of lead toxicosis in cockatiels. Am J Vet Res 61:935–940

Gillet AMTY, Seewagen CL (2014) Mercury exposure of a wetland songbird, Agelaius phoeniceus, in the New York metropolitan area and its effect on nestling growth rate. Environ Monitor Assess 186:4029–4036

Handel CM, Van Hemert C (2015) Environmental contaminants and chromosomal damage associated with beak deformities in a resident North American passerine. Environ Toxicol Chem 34:314–327

Leonzio C, Bianchi N, Gustin M et al (2009) Mercury, lead and copper in feathers and excreta of small passerine species in relation to foraging guilds and age of feathers. Bull Environ Cont Toxicol 83:693–697

McLelland JM, Reid C, McInnes K et al (2010) Evidence of lead exposure in a free-ranging population of kea (*Nestor notabilis*). J Wildl Dis 46:532–540

Mora MA (2008) Organochlorine pollutants and stable isotopes in resident and migrant passerine birds from northwest Michoacan, Mexico. Arch Environ Cont Toxicol 55:488–495

Puschner B, Judy LS, Galey FD (1999) Normal and toxic zinc concentrations in serum/plasma and liver of psittacines with respect to genus differences. J Vet Diag Invest 11:522–527

Rainio MJ, Kanerva M, Salminen JP et al (2013) Oxidative status in nestlings of three small passerine species exposed to metal pollution. Sci Total Environ 454:466–473

Sánchez-Virosta P, Espín S, García-Fernández AJ et al (2015) A review on exposure and effects of arsenic in passerine birds. Sci Total Environ 512:506–525

Senthilkumar K, Kannan K, Subramanian A et al (2001) Accumulation of organochlorine pesticides and polychlorinated biphenyls in sediments, aquatic organisms, birds, bird eggs and bat collected from south India. Environ Sci Pollut Res 8:35–47

Stoltz JH, Galey F, Johnson B (1992) Sudden death in ten psittacine birds associated with the operation of a self-cleaning oven. Vet Hum Toxicol 34:420–421

Wells RE, Slocombe RF (1982) Acute toxicosis of budgerigars (*Melopsittacus undulatus*) caused by pyrolysis products from heated polytetrafluoroethylene: microscopic study. Am J Vet Res 43:1243–1248

Zanen AL, Rietveld AP (1993) Inhalation trauma due to overheating in a microwave oven. Thorax 48:300–302

Diagnostic Techniques

5

Indranil Samanta

Most of the pet bird owners prefer to visit their veterinarians during crisis period when possibility of recovery becomes low. Instead, visit to veterinarians in a regular interval, recently termed as 'wellness examination' will provide a knowledge package consisting of infection if present, balanced diet, proper exercise, behaviour and husbandry practices to be followed. Inexperienced bird owners and owners of aged birds with chronic ailment should have regular wellness examination of their pets. To provide this knowledge package the veterinarian should also follow appropriate diagnostic approaches with proper medicinal, nutritional, and husbandry recommendations.

5.1 Anamnesis

Taking history from the owner regarding health, diet and husbandry practices of their pet birds is the most primitive but still considered as most valuable diagnostic technique. This is a two-way technique which depends on co-operation of the owners, questioning ability and patience of the veterinarians and moreover, proper record keeping. The veterinarian can know whether the ailment is new or progressed from earlier stage. Co-operation from owners with truthful replies is required for proper diagnosis.

The veterinarians should primarily focus on age, species, sex, physical activities, urination and droppings (colour, consistency and amount) of the birds. Exposure to environment such as soil, garden, toxic fume or smoke, bird shows/fairs, shelters, aviaries, other infected birds and wild birds should be enquired from the owners. Evaluation of earlier recommendations regarding diet, bedding materials is necessary. What exactly the bird is taking as feed, not what is provided to them is

I. Samanta and S. Bandyopadhyay, *Pet Bird Diseases and Care*,
DOI 10.1007/978-981-10-3674-3_5

important to know. Excess or deficiency of nutrients is associated with different diseases. Excess fat in diet (present in seeds) causes obesity and hepatic lipidosis. Deficiency of vitamin and calcium causes osteoporosis and impairment of vision. The veterinarian can judge the possibility and severity of a specific ailment depending on the species, age and exposure of the birds. Sub-clinical carriers of zoonotic infections should also be identified specially when the birds are reared by aged or immunosuppressed owners. To perform all these functions properly the veterinarians should have an updated knowledge other than experience.

5.2 Physical Examination

Physical examination of the bird should be performed in a clean perch or small enclosure. Gentle restraining of the birds is necessary not only to avoid injury of the birds but also to gain the trust of the owners. Bad handling always creates doubt about the ability of a veterinarian. Mild sedation with midazolam (0.2–0.5 mg/kg body weight, intramuscular) or butorphanol (1–2 mg/kg body weight, intramuscular) will help to restrain hyperactive birds.

Alertness, weight, posture (straight and upright in healthy birds), feather condition (ruffled or normal), walking (leg twisting), breathing sound, colour, consistency and odour of droppings and urine, vision, behaviour (nervousness, relationship with owner, biting tendency) should be judged primarily (Fig. 5.1).

Fig. 5.1 Ruffled feather in a sick parrot (*Courtesy* Lefeber vet)

5.3 Collection of Clinical Samples

5.3.1 Blood

The volume of collected blood sample should be 0.5–1.0% of the patient's body weight. EDTA is the most suitable anticoagulant except in crowned cranes, hornbills, eagle owl, laughing kookaburra, curassows and corvidae. In these species of birds, EDTA causes haemolysis and heparin is used as anticoagulant. Lithium heparin is suitable for blood biochemical parameters (except glucose and calcium), not for cellular study as it causes clumping of all avian blood cells. Blood is collected with a 23–25 s.w.g. needle bent at a single angle. The puncture site should be blocked with finger pressure immediately. Avian veins are fragile and possibility of haematoma formation is more. Blood can be collected from right jugular vein, brachial vein, medial metatarsal (caudal tibial) vein, external thoracic vein, and by cutting claws or direct heart puncture (Table 5.1).

Collection of avian serum in sufficient volume to conduct all the required tests is difficult due to presence of large fibrin clots which often decreases the volume. Heparinized plasma or whole blood is most common sample available from avian practitioners.

Table 5.1 Collection of blood samples from different avian veins

Blood collection sites	Species	Remarks
Right jugular vein	Flamingos, budgerigars, raptors, penguins, ostrich	Unlike mammals, avian jugular vein is not located in any furrow. It is present subcutaneously at right neck. Due to fat deposition in neck, jugular vein is unsuitable for blood collection from pigeons
Brachial vein	Most of the birds (for e.g. snow goose, gull, bald ibis, pigeon, bald eagle, Eurasian kestrel, hawk, thrush, crane, African grey parrot, rhea etc.)	Possibility of haematoma formation is more in small birds and racing pigeons
Medial metatarsal vein	Pigeons, raptors, ducks	Possibility of haematoma formation is low
Claw cutting	Small finches	After cleaning the claws with antiseptic it is cut and the blood is collected in capillary tubes. The bleeding can be stopped with silver nitrate or ferrous subsulphate

5.3.2 Droppings/Cloacal Swabs

Fresh droppings or faeces should be collected in sterile, screw capped short jars or plastic vials. Rubber caps should be avoided because the gas generated in the collected faeces may blow the cap. The faeces and urine should not be mixed together.

Cloacal swabs are collected by gently mopping the cloaca with sterile cotton swabs. The swabs should be moistened with transport medium before collection. The collected swabs are put into Stuart transport medium, 50% buffered glycerol, Sabouraud broth or BHI broth for transport (Figs. 5.2 and 5.3).

5.3.3 Crop/Proventriculus Washing

Normal saline (warm, 20 ml/kg body weight) is instilled into the crop or proventriculus with a blunt ended catheter or gavage tube. A part of the fluid is aspirated out with gentle negative pressure. For proventriculus washing, endotracheal tube can be used with general anaesthesia.

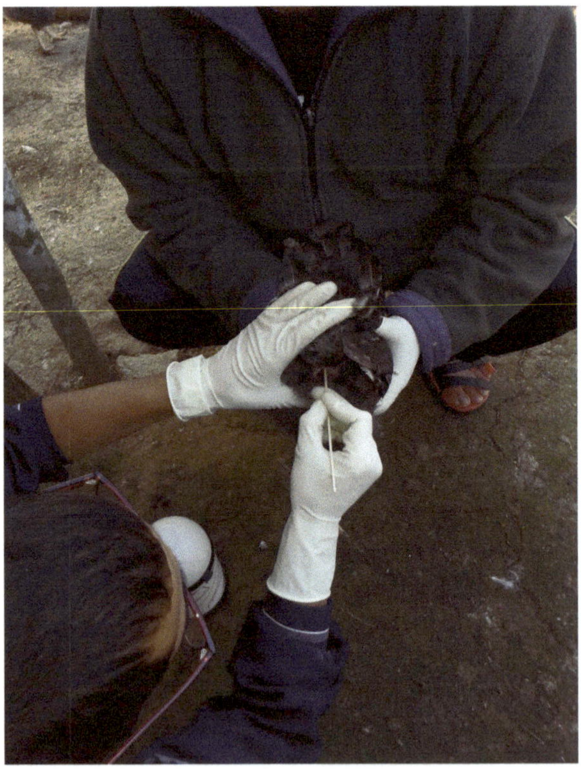

Fig. 5.2 Collection of cloacal swabs from birds (*Courtesy* Pratik Ghosh, Department of Veterinary Microbiology, WBUAFS, Kolkata, India)

Fig. 5.3 Preservation of collected cloacal swabs from birds (*Courtesy* Pratik Ghosh, Department of Veterinary Microbiology, WBUAFS, Kolkata, India)

5.3.4 Air Sac Washing

A sterile catheter attached with a syringe is inserted into the last intercostal space of a bird and sterile normal saline (3 ml for large birds) is injected and aspirated out immediately.

5.3.5 Tracheal Washing

Nylon tube (1 mm in diameter) or a canine catheter is inserted through the glottis and sterile normal saline (0.5–2 ml/kg body weight) is flushed and the washing is collected immediately. It is safe to collect tracheal washing from highly sedated birds.

5.3.6 Autopsy Samples from Vital Organs

The incised portion of the vital organ (liver) is wiped gently with a paper towel to remove excess blood. Indurated tissue is scraped with a sterile scalpel and the tissue is kept over the slide for further examination.

5.3.7 Bone Marrow

Bone marrow is collected from proximal tibiotarsal bone or keel of the sternum. Paediatric bone biopsy needle (23 s.w.g.) can be used for bone marrow collection. Bone marrow samples are useful for confirmation of leukaemia and non-regenerative anaemia.

5.4 Processing of Clinical Samples

Blood, exudates or aspirates from the lesions are required to mix with EDTA (or heparin) to prevent clotting. Physical parameters of the fluid samples such as colour (haemolysis), specific gravity, turbidity (presence of fat particles) are judged. Cellular part of the fluid is concentrated by low speed centrifugation (1500 rpm for 10 min).

5.5 Diagnostic Techniques Used in Laboratory

(a) Direct Examination

A smear can be prepared with the fluid sample for detection of microbes and parasites, morphology and differential counts of cellular components (heterophils, macrophages, leukocytes) and cells with malignancy.

Blood smear is primarily stained with Leishman's, Wright's or Giemsa stain. More time is required to stain avian blood cells than the mammalian. The pH of buffered water used for washing the slides should be more acidic (pH 5.0). Total and differential counting of avian blood samples requires expertise, as it is difficult to identify avian leukocytes, scattered throughout the microscopic field. Laboratory with expert technicians is required for estimation of blood biochemical parameters, enzyme and electrolyte concentration. Standard values of avian biochemical parameters, enzyme and electrolyte concentration are described in Table 5.2. These parameters although roughly indicate about occurrence of an infection or infestation or intoxication. It cannot confirm the diagnosis since wide variation exists in standard values between species, physiological conditions (moulting, egg laying, migration, age, sex etc.), and husbandry practices (nutrition, weather, wild or captive etc.). In some rare species no standard value is available to compare.

The smears prepared from tissue samples can be stained with Romanowsky, Wright's, Giemsa stain or fungal stains (Grocott) (Fig. 5.4). Sudan III or IV along with new methylene blue stain is used if the fat content is more in the collected tissues. Use of bacterial stains such as Gram's stain and acid fast stain is preferred in suspected clinical cases (Fig. 5.5).

Table 5.2 Standard values of blood cells, biochemical parameters, enzymes, electrolytes in birds

Parameter	Standard value	Remarks
Packed cell volume (PCV)	40–55%	Young birds (not fully fledged) have lower PCV
Haemoglobin	12.2–20 g/dl	Decreased haemoglobin and PCV (haemolytic anaemia): blood parasite infection (*Haemoproteus*, *Plasmodium*, *Leucocytozoon*); gastrointestinal parasitism (*Capillaria*, ascarids, coccidiosis, giardiasis); mite infestation; bacterial and yeast infection (salmonellosis, colibacillosis, yersiniosis, *Macrorhabdus*, proventricular ulceration, campylobacteriosis); viral infection (Pacheco's virus, Herpes virus)
Mean corpuscular volume (MCV)	121–200 fl (psittacines: 99 fl; cassowary: 280 fl)	Increased value indicates regenerative or macrocytic anaemia
Mean corpuscular haemoglobin concentration (MCHC)	28–38 g/dl	Reduced MCHC index (Chronic non-regenerative anaemia): chronic infection (chlamydophilosis, toxoplasmosis, aspergillosis, salmonellosis, yersiniosis, colibacteriosis, campylobacteriosis); toxicosis [lead, copper, zinc, chloramphenicol, pesticides (DDT, carbamates), aflatoxins]; starvation and malnutrition
Polychromatic index	Raptors: above 3.5%	–
Total erythrocyte count (TEC)	$2.1–5.5 \times 10^{12}$/l (Mean: 3.9×10^{12}/l)	Young birds (not fully fledged) have lower TEC; Higher in migratory birds during flying
Leukocyte count	$1–32 \times 10^{9}$/l [heterophils 20–75%, lymphocytes 20–65%, monocytes 2–5%, basophils 2.5–6%, eosinophils 1–4%]	Leucopenias-toxicosis Leucocytosis-infection with bacteria, fungi, parasite, neoplasia, trauma Heteropenia, lymphopenia-Viral infection Lymphocytosis-neoplasia
Thrombocyte count (platelet)	$20–30 \times 10^{9}$/l	Thrombocytosis-bacterial infection Thrombocytopenia-severe septicaemia

(continued)

Table 5.2 (continued)

Parameter	Standard value	Remarks
Total serum protein	3–5 g/dl	Hypoproteinaemia—Hepatopathy, gastrointestinal parasitism, nephritis, trauma, lead toxicosis, Pacheco's disease, anaemia, malnutrition etc. Hyperproteinaemia—prior to egg laying stage in physiological condition, dehydration, acute infection and shock
Total plasma protein	0.15 g/dl above the value of serum protein	–
Serum albumin	1.0–2.2 g/dl	–
Albumin/globulin ratio (A/G)	1.4–4.9	Decreased A/G ratio—acute and chronic infections (e.g. *Chlamydophila*, aspergillosis, mycobacteriosis)
Serum glucose	200–500 mg/dl (Emu: 158 mg/dl)	Normal value varies with age, diet, breeding season of the birds. The serum glucose level is decreased during day time and increases during night (reverse for nocturnal birds) *Hyperglycaemia*—stress, lead toxicosis, pancreatitis, Diabetes mellitus in granivorous birds, budgerigars, cockatoos, Amazon parrots, macaws, cockatiels, toucans *Hypoglycaemia*—starvation, hypovitaminosis, septicaemia
Uric acid	2–15 mg/dl	Normal value is more in carnivorous birds than the granivorous birds. High level of uric acid—starvation, gout, trauma, toxicity (due to excess gentamicin, sulphonamides, azole group antifungals), hypervitaminosis D_3, bacterial or viral infection
Urea	2.4–4.2 mg/dl	High level of plasma urea-dehydration, cardiopathy, cloacal impaction, neoplasm, blockage of renal tubules with urate crystals during salt poisoning

(continued)

Table 5.2 (continued)

Parameter	Standard value	Remarks
Cholesterol	108–330 mg/dl	High level—fatty degeneration of liver, xanthomatosis (yellowish cholesterol deposition in any tissue)
Bile acids [Most of the species produce biliverdin, not bilirubin. True jaundice (icterus) is not detected in birds (except in macaws when bilirubin concentration is >2.36 mg/dl)]	18–144 μmol/l (psittacines)	High bile acid level in plasma—hepatopathy
Aspartate amino transferase (AST)	52–270 IU/l	High level indicates hepatopathy, Pacheco's disease, chlamydophilosis, toxicosis due to pesticides, adverse reaction of drugs (doxycycline injection, azole group of antifungals)
Alanine amino transferase (ALT)	6.5–263 IU/l	High level—hepatopathy
Lactate dehydrogenase (LDH)	46–442 IU/l	High level—hepatopathy
Alkaline phosphatase (ALP)	42–479 IU/l	High ALP level—osteomyelitis, bone neoplasms, fractures, aflatoxin poisoning, rickets, hyperparathyroidism, physiologically high during egg laying period
Creatine phosphokinase (CPK)	110–480 IU/l	High CPK level—convulsions, lead toxicity, chlamydophilosis, bacterial septicaemia, vitamin E deficiency
Calcium	8–12 mg/dl (Budgerigars: 6.4–11.2 mg/dl; chicken: 13.2–23.7 mg/dl)	Hypercalcaemia: ovulation (physiological), dehydration, bone tumour Hypocalcaemia: muscular spasm, seizure, steroid therapy
Sodium	127–170 mEq/l	Hypernatraemia: salt poisoning in wild birds, feeding of excess salt with feed (peanuts, potato crisps) in pet birds

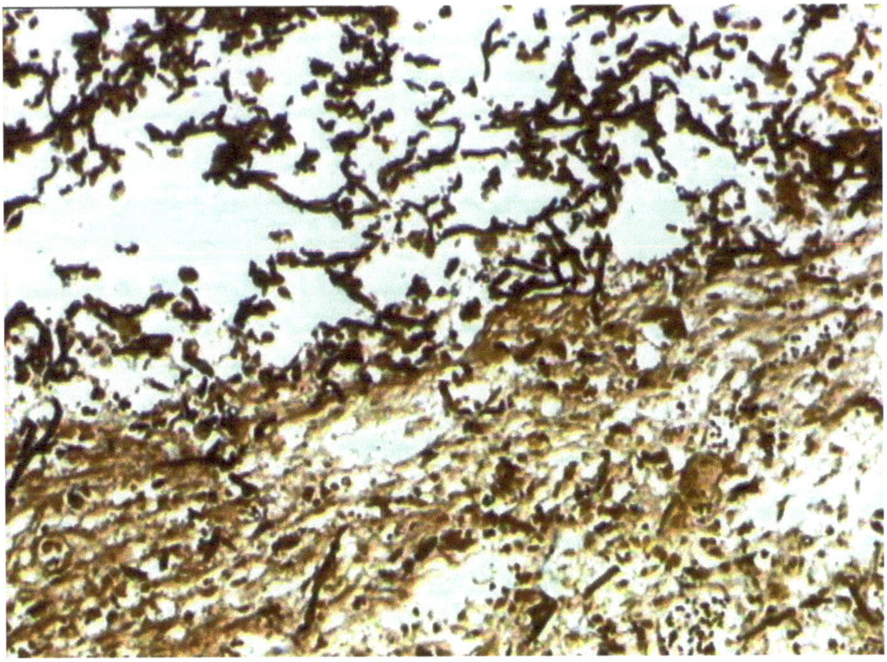

Fig. 5.4 *Penicillium* hyphae present in lung tissue of African grey parrot (Grocott stain) (*Courtesy* Dr. Giovanni Lanteri, University of Messina, Italy)

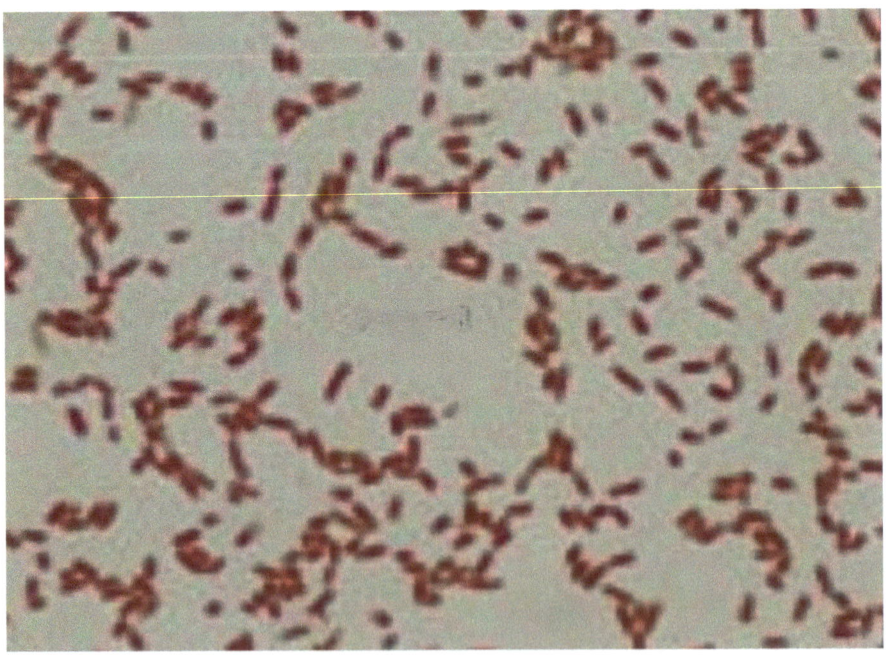

Fig. 5.5 *Klebsiella* spp. in gram stained smear isolated from birds (100×, *Courtesy* Achintya Mahanti, Department of Veterinary Microbiology, WBUAFS, Kolkata, India)

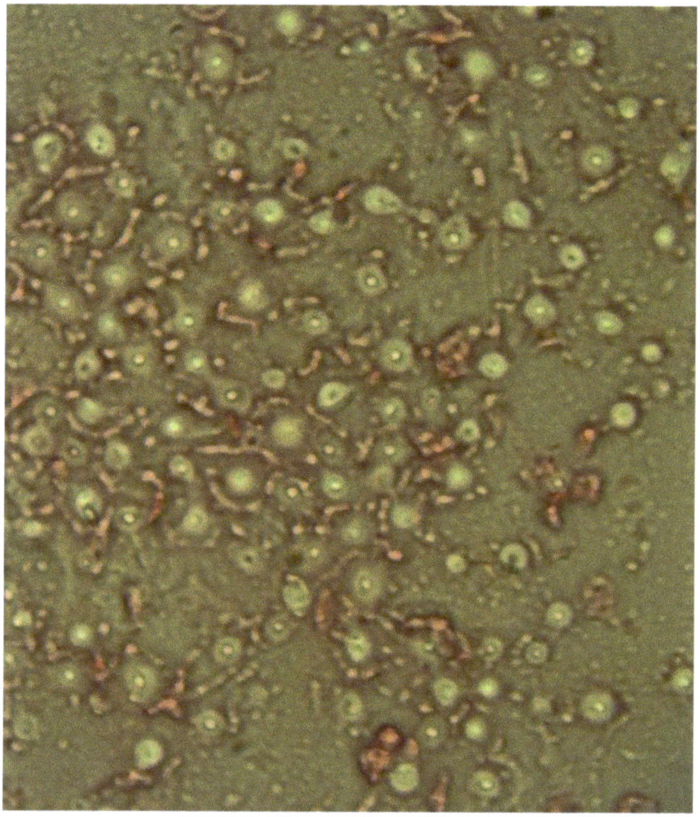

Fig. 5.6 Encapsulated yeast in fresh unstained droppings (*Courtesy* Prof. Adjunta Karin Werther, Universidade Estadual Paulista, Brazil)

Fresh faeces/droppings can be visualized unstained for detection of parasite eggs, yeast and few bacteria (Fig. 5.6).

(b) Different serological, immunological and molecular biology based tests can be performed for detection of avian infections, infestations or intoxications (Table 5.3). Lack of species specific reagents (polyclonal and monoclonal antibodies), standardized and reproducible techniques, considerable variations in results between laboratories are the major constraints in diagnostics of avian medicine. All the laboratory test reports should be correlated with signs and symptoms of the birds by the veterinarian.

Table 5.3 Diagnostic approaches for avian infections, infestations and intoxications

Avian infection	Diagnostic approaches
Avian aspergillosis	Antigen detection ELISA, PCR, isolation
Avian influenza	Hemagglutination-inhibition (HI), agar gel immunodiffusion (AGID), virus neutralization, enzyme-linked immunosorbent assay (ELISA), PCR, isolation of virus
Beak and feather disease	Haemagglutination inhibition (HI), blocking ELISA, histopathology, PCR
Avian Bornavirus infection (Proventricular dilatation disease)	ELISA, indirect immunofluorescence assay, isolation of virus, histopathology, reverse transcriptase-PCR, radiography
Campylobacteriosis	Direct examination, isolation of bacteria, ELISA
Avian chlamydophilosis	Direct examination, micro immunofluorescence (MIF) test, ELISA, CFT, elementary body agglutination test, isolation, PCR
Avian cryptococcosis	Direct examination, Detection of Cryptococcal antigen by ELISA, isolation, PCR
Avian cryptosporidiosis	Direct examination, Capture enzyme-linked immunoassays, PCR
Avian giardiasis	Direct examination, ELISA (antigen capture), immunofluorescence, PCR
Lyme disease	Direct Examination, isolation, ELISA, PCR
Avian mycobacteriosis	Direct examination for acid-fast organisms, histopathology, PCR, isolation
Newcastle disease	Virus neutralization test, plaque neutralization, hemagglutination-inhibition, agar gel immunodiffusion, enzyme-linked immunosorbent assay, isolation of virus, real-time reverse-transcriptase polymerase chain reaction
Salmonellosis	Isolation of bacteria, rapid whole blood/serum agglutination test, ELISA, PCR
Avian yersiniosis	Direct examination, isolation of bacteria
Mycoplasmosis	Direct examination with contrast phase microscopy Contrast phase microscopy, dark phase illumination techniques, serum plate agglutination test, PCR, isolation of *Mycoplasma*
Pasteurellosis	Direct examination, isolation of bacteria
Escherichia coli infection	Isolation of bacteria, PCR
Avian candidiasis	Wet mount with 10% KOH, histopathology, isolation, PCR
Mycotic proventriculitis (*Macrorhabdus ornithogaster*)	Direct examination by Gram's staining, histopathology
Avian sarcocysticosis	Visualization of sarcocystis in the muscles after post-mortem, muscle biopsy, indirect immunofluorescent assay, PCR
Avian trichomoniasis	Wet mount
Coccidiosis	Examination of faeces for oocysts or macrogametes

(continued)

Table 5.3 (continued)

Avian infection	Diagnostic approaches
Ascaridiasis	Detection of thick shelled Ascarid eggs in faecal sample
Pacheco's disease	Isolation of virus in cell lines, demonstration of virus in clinical samples by electron microscopy, detection of viral DNA by PCR or real-time PCR
Psittacine adenovirus infection	Isolation of virus, electron microscopy, PCR
Psittacine poxvirus infection	Clinical symptoms, electron microscopy, detection of Bollinger body in tissue samples and Psittacine Poxvirus specific-PCR
Avian Polyoma virus infection	Electron microscopy, virus neutralization test, immunofluorescent antibody staining, in situ hybridization, PCR
Psittacine Papillomavirus infection	Presence of typical cauliflower like growth in cloaca or oral mucosa, histological examination of biopsy samples, in situ hybridization, PCR
Avian Reovirus infection	Isolation of virus, electron microscopy, immunofluorescence staining, PCR
Infectious bronchitis	Isolation of virus, electron microscopy, pan-coronavirus reverse transcriptase-PCR.
Toxoplasmosis	Direct examination, histology, Modified agglutination test, *T. gondii*-specific PCR
Usutu virus infection	Isolation of virus, immunohistochemical (IHC) staining, PCR
West Nile virus infection	Isolation of virus, immunohistochemical (IHC) staining, PCR
Lead toxicosis	Detection of lead content in blood, radiography of enteric tract
Zinc toxicosis	Determination of zinc concentration in blood, histopathological examination of pancreas
Organophosphate toxicity	History, plasma cholinesterase assay
Polytetrafluroethyelene (Teflon) toxicity	Histopathology

5.6 Diagnostic Techniques in Avian Clinics or Hospitals

5.6.1 Radiography

Digital radiography is a useful diagnostic technique used currently in avian medicine for rapid detection of underlying hidden etiology. It is easy to perform in birds due to smaller size and one exposure is sufficient to take the image of the whole body. Avian air sac system acts as a negative contrast to the organs which make the interpretation easier for the radiologists. Other than high power X-ray apparatus, film-screen combination and developing system, expert technician is required for interpretation of avian radiographic images. Short exposure times (0.015–0.05 s) are recommended for taking avian images because image quality is detoriated due to high respiration rate of the birds even under anaesthesia. For taking images of internal organs and skeleton of birds, rare earth screens and mammography screens are recommended, respectively. In general, fine film screen combinations are preferred for avian images.

5.6.2 Ultrasonography (USG)

In ultrasonography, the image is produced with transmission and reflection of sound waves. In birds, use of USG is limited because the ultrasound cannot invade a gas filled air sac. Other hindering factors include circulatory and respiratory distress of birds during examination, application of coupling gels to make contact between transducer and avian skin, and lack of experienced veterinarians. Fasting of the birds for 2–4 h is required before USG examination. Confirmation of hepatomegaly, cardiac disease, disorders of kidney and reproductive tract, and ascites is possible through USG examination.

5.6.3 Computed Tomography (CT)

This technique generates a cross-sectional image for accurate visualization through the use of X-rays. General anaesthesia is required and the whole procedure takes 10–15 min time to be completed. Examination of avian skull, sinuses and lower respiratory tract is possible through CT.

5.6.4 Magnetic Resonance Imaging (MRI)

This technique also generates cross sectional images through strong external magnetic force. MRI can detect the presence of caseous plug, granuloma, mucocele, polyp in brain, spinal cord, coelomic organs and upper respiratory tract. Detection of accurate location of these lesions helps in surgery. MRI takes longer time than CT for examination.

5.6.5 Myelography

In larger birds (1 kg or more), images are taken after injection of non-ionic iodinated contrast medium (0.8–1.2 ml/kg body weight) into subarachnoid space at thoraco-synsacral junction. The technique is indicated for detection of compressive and traumatic lesions in the spinal cord.

5.6.6 Echocardiography

Echocardiography provides useful information of cardiac function and structure of the heart. Earlier it was difficult in birds due to presence of air sacs which block the passage of ultrasound waves. Currently echocardiograph machines with advanced technology (7.5 MHz or higher frequency, doppler function, more than 100 frames/s) is used successfully in avian medicine. Fasting for 2–12 h (psittacine, pigeons) or longer period (raptors) is recommended before echocardiography. Food filled enteric tract may create an obstruction between the machine and heart. Ventromedian (psittacines, raptors) and parasternal (pigeons) approaches are followed for avian echocardiography.

5.6.7 Electrocardiography (ECG)

ECG is used to observe cardiac function during anaesthesia and for recording cardiac stages (systole, diastole). Use of ECG in avian medicine is not frequent due to difficulties in making connection of leads with skin of the birds, lack of reference values and alterations of ECG values under stress or anaesthesia.

5.6.8 Endoscopy

Avian endoscopy helps the clinicians to examine internal organs (lungs, air sacs, heart, intestinal tract, liver, kidneys, adrenal glands, spleen, pancreas, gonads, oviduct, and shell gland) through a small and single incision. Endoscopy of oral cavity also allows examination of esophagus, crop, proventriculus, glottis and trachea. In addition to physical examination, clinicians can collect tissue biopsies from vital organs, coelomic musculature and abnormal soft tissue structures.

Treatment, Management and Care 6

Indranil Samanta

6.1 Body Weight Measurement

Measurement of bird's body weight is crucial for detection of medicine dosage. The body weight of small birds (passerines) is measured by electronic balance. Medium sized birds (duck, geese, turkey) can be weighed in a weighing balance by putting them in a sac or box with their head (nostrils) outside. For larger birds (falcon) the birds are placed on a perch attached to weighing scales. Standard body weight of pet birds is described in Table 6.1. The clinicians should consider at least 25% variation in both sides of the mean body weight specially during diseased condition.

6.2 Administration of Medicines

6.2.1 Oral

Individual oral administration can be done in small sized flocks maintained in households and small scale aviaries. Accurate dosing is possible by oral administration of drugs to the individual bird by dropper, cocktail stick, gavage tube, and rubber tube attached with hypodermic syringe. In parrots, mouth gag or stomach tubes are used for oral administration. For larger flocks, drugs are administered through soft food and drinking water. Dosage of medicines in drinking water is based on daily consumption of 150 ml water per kilogram of avian body weight (Table 6.2). Consumption of water although varies with species (mynah, budgerigars, raptors drink less amount of water), climate, body condition and diet. Due to intake of water in high or low quantity, therapeutic drug concentration is not maintained properly in birds. Medicines added in drinking water sometimes change the colour and taste of the water which is not preferred by the birds. Colour vision of pet birds is although a debatable issue, taste buds (bitter and salty) have detected

I. Samanta and S. Bandyopadhyay, *Pet Bird Diseases and Care*,
DOI 10.1007/978-981-10-3674-3_6

Table 6.1 Standard body weight of pet birds

Bird	Standard body weight (kg)
Blue and gold macaw (*Ara ararauna*)	0.8–2
Scarlet macaw (*Ara macao*)	0.81–1.1
Orange-winged Amazon (*Amazona amazonica*)	0.44–0.47
Blue-fronted Amazon (*Amazona aestiva*)	0.27–0.51
Yellow-fronted Amazon (*Amazona ochrocephala*)	0.26–0.47
African grey parrot (*Psittacus erithacus*)	0.3–0.45
Orange-bellied Senegal parrot (*Poicephalus senegalus*)	0.12–0.15
Lesser sulphur-crested cockatoo (*Cacatua sulphurea*)	0.22–0.31
Greater sulphur-crested cockatoo (*Cacatua galerita galerita*)	0.67–0.8
Umbrella cockatoo or great white cockatoo (*Cacatua alba*)	0.5–0.6
Dusky-headed conure (*Aratinga acuticaudata*)	0.15–0.18
Yellow rosella parakeet (*Platycercus flaveolus*)	0.1–0.12
Blue-headed parrot/Pionus parrot (*Pionus menstruus*)	0.23–0.27
Cockatiel (*Nymphicus hollandicus*)	0.07–0.1
Budgerigar (*Melopsittacus undulatus*)	0.035–0.085
Blue-winged grass parakeet (*Neophema chrysostoma*)	0.05
Fisher's lovebird (*Agapornis fi scheri*)	0.04–0.05
Song thrush (*Turdus philomelos*)	0.08
Blackbird (*Turdus merula*)	0.05
American robin (*Turdus migratorius*)	0.06–0.08
European robin (*Erithacus rubecula*)	0.02–0.03
Starling (*Sturnus vulgaris*)	0.064
Great Indian Hill mynah (*Gracula religiosa intermedia*)	0.18–0.25
House sparrow (*Passer domesticus*)	0.025–0.03
Zebra finch (*Poephila guttata*)	0.01–0.015

(continued)

Table 6.1 (continued)

Bird	Standard body weight (kg)
European goldfinch (*Carduelis carduelis*)	0.015–0.02
Great tit (*Parus major*)	0.017–0.02
Canary (*Serinus canaria*)	0.012–0.25
eagle owl (*Bubo bubo*)	1.6–2.5
Barn owl (*Tyto alba*)	0.25–0.6
Goshawk (*Accipiter gentilis gentilis*)	0.6–1.2
Common buzzard (*Buteo buteo*)	0.6–1.1
Red-tailed hawk (*Buteo jamaicensis*)	0.7–1.3
Saker falcon (*Falco cherrug*)	0.7–1.3
Domestic/mallard duck (*Anas platyrhynchos*)	0.9–3.5
Muscovy duck (*Cairina moschata*)	3.5–5.0
Herring gull (*Larus argentatus*)	0.69–1.5
Common/arctic tern (*Sterna hirundo/paradisaea*)	0.09–0.1
Heron (*Ardea cinerea*)	1.3
Racing pigeon (*Columba livia*)	0.23–0.54
Wood pigeon (*Columba palumbus*)	0.45–0.69
Collared dove (*Streptopelia decaocto*)	0.15–0.22

in birds specially in pigeons. The parrots can take tablet and capsules in powder form spread over sweet biscuits or bread with sugar or honey. Coating a seed with medicines for feeding ailing birds should not be practiced because most of them remove the seed coat before eating. The soft home-made food (boiled rice mixed with sugar, seed, and water) is a better alternative for administration of medicine.

Absorption of drugs from gastrointestinal tract is hampered in presence of parasites and hypovitaminosis-A. Absorption of certain antibiotics is reduced in presence of food items (penicillin, ampicillin, lincomycin), and calcium-magnesium (oxytetracycline).

Table 6.2 Dosage of common drugs in food and drinking water for small sized birds

Drugs	Dosage in food (mg/kg)	Dosage in drinking water (mg/l)
Ampicillin	2000–3000	1000–2000
Amoxicillin	300–500	200–400
Amphotericin B	100–200	100–200
Chlortetracycline	1500	1500
Doxycycline	1000	250
Enrofloxacin	200	200
Fenbendazole	25	25
Furazolidone	300	300
Metronidazole	100	100
Neomycin	200	200
Nystatin	200,000 IU	200,000 IU
Spectinomycin	400	400
Spiramycin	400	400
Sulfadimidine	–	150
Trimethoprim/Sulfonamide	200	200
Tylosin	400	400

6.2.2 Parenteral

In pet birds medicines can be administered through intramuscular, subcutaneous, intravenous, intratracheal, intraosseous and intracoelomic routes (Table 6.3).

Intramuscular
In small sized birds (passerines), intramuscular route is preferred into the pectoral muscle (cranial third). Pectoral muscle is not preferred by the owners of racing pigeons and falcons because they believe that flight of the birds is adversely affected. Special care should be taken in fledglings in which bones are not fully ossified. The needle during injection of pectoral muscle may accidentally go through the soft bone and penetrate underlying viscera. Other muscles such as iliotibialis lateralis, and biceps femoris of leg can also be used for intramuscular injection.

Subcutaneous
Avian skin is not sufficiently elastic and the medicine injected through subcutaneous route often leaks out through the point of needle puncture. Precrural fold (in front of the leg or thigh) is the preferred site for subcutaneous injection because leg movement of birds helps in distribution of drug after administration. In dehydrated small birds fluid is administered through the inguinal skin fold.

Intravenous
Intravenous route is used for drug administration during emergency. Brachial vein, superficial metatarsal vein on medial surface of leg are common veins used for injection. In small passerine birds, jugular vein is preferred.

Table 6.3 Parenteral and oral dosage of common drugs for pet birds

Drugs	Dosage (mg/kg body weight)	Remarks
Amoxicillin/Clavulanic acid	125 mg/kg, oral, 6 h interval	Recommended for psittacine birds
Amikacin	10–15 mg/kg, IM or IV, 8–12 h interval	Recommended for most species of birds
Cefotaxime	75–100 mg/kg, IM or IV, 4–8 h interval	Recommended for most species of birds
Ciprofloxacin	15–20 mg/kg, IM, 12 h interval	Recommended for most species of birds
Clindamycin	25 mg/kg, oral, 8 h interval	Recommended for most species of birds
Doxycycline	25–50 mg/kg, IM for 5–7 days	Recommended for psittacine birds
Fluconazole	5–15 mg/kg, oral, 12 h interval for 15–60 days	Recommended for most species of birds
Itraconazole	10 mg/kg, oral, 24 h interval for 15–90 days	Recommended for psittacine birds (except African grey parrot)
Piperacillin/Tazobactam	100 mg/kg, IM, 8–12 h interval	Recommended for psittacine birds
Sulfamethoxazole/Trimethoprim	20 mg/kg, oral, 8–12 h interval	Recommended for psittacine birds
Ticarcillin/Clavulanic acid	100 mg/kg, IM, 12 h interval	Recommended for most species of birds

Intratracheal

Respiratory infection sometimes requires intratracheal administration of drugs. A human intravenous catheter attached with syringe can be used for intratracheal administration. The catheter is inserted through the glottis present in oral floor. Tongue is restrained and the neck is held vertically during insertion. The whole procedure should be done under deep sedation.

Intraosseous

Intraosseous injection in birds can be performed with a cannula needle (20–22 s.w.g.) attached with a stylet. Distal extremity of the ulna or proximal tibiotarsal bone can be used for this purpose.

Intracoelomic

Treatment of air sac infection requires intracoelomic administration of drugs to get early response. The skin and underlying muscle at right of the midline is picked up and the needle (16 s.w.g.) is inserted carefully into the fold. The procedure should be done under deep sedation.

6.2.3 Topical

The ointments and creams are applied topically by a sterile cotton swab or ear buds. Tincture and dimethylsulfoxide (DMSO) act as useful carrier of topical medicines because they are rapidly absorbed by avian skin.

6.2.4 Ophthalmic

Use of eye drops requires repeated handling of the affected birds. Misting the eye with fine spray of sterile water mixed with medicine is another option. Injection of small amount (0.01–0.05 ml) of antibiotic or steroid can be done under the conjunctiva of upper eyelid.

6.3 Management Practices

(a) Birds are often stressed due to high metabolic rates (higher in passerine than psittacine birds), exposure to loud human voice or noise, barking of dogs and rapid movement of other creatures. Wild birds are more stressed because they are not accustomed to human contact. The stress can be reduced by placing the birds in perches or small cages above human eye level preferably in a separate room. A flock of birds can be kept in a larger cage with perches made of tree branches (Fig. 6.1). During visit to clinics or hospitals, the birds become more stressed due to unfamiliar environment.

(b) The bird's room should be well ventilated so that infection cannot spread through inhalation route.

(c) Frequent feeding of small pet birds with balanced diet is required to minimize weight loss. Due to high metabolic rate there is rapid heat loss and utilization of energy reserves (glycogen and fat) which causes weight loss in birds. The weight loss becomes more rapid in temperate countries with low ambient temperature. In raptors and other larger birds, feeding twice in a day is sufficient to maintain the body weight.

(d) The cages should be cleaned and disinfected in regular interval. The date of last cleaning may be written in a white board attached with the cage. The cages should be equipped with removable tray so that the birds have least disturbance during cleaning operation. Fogging with disinfectant fume effectively cleans the hidden corners of the room.

(e) Continuous bouncing against the bars of a cage is a typical vice of some birds. Placing a cloth within the cage can prevent the vice and protect the birds from injuries.

Fig. 6.1 Large cage with perches (*Courtesy* Sanjay Shit, ARD Department, Government of West Bengal, India)

Fig. 6.1 (continued)

(f) Some pet birds (parrots, budgerigars) prefer the company of birds of similar species. Loneliness can generate several vices such as feather picking (non-infectious etiology).

(g) Parrots, owls, raptors prefer to take bath. The birds can be sprayed with mist of water by a garden hose pipe or automatic sprinkler (used in larger aviary). Bathing cleans the feathers and helps to conserve body temperature. Young birds are not bathed regularly.

6.4 Care of Pet Birds

(a) Commercial dilute feed is offered in every 2 h interval during 1st and 2nd day of life after hatching, and in every 3 h interval on 3rd and 4th day. The crop should be empty at night. From 5th day of life, undiluted commercial feed is offered 4–5 times a day. Gradually the number of feeding is decreased depending on body weight (slow in larger birds).

(b) Probiotics (*Lactobacillus* preparations) and antifungals should be prescribed along with antibiotics in chicks. Fluid after warming is offered to the chicks mostly by subcutaneous route. The catheter is used through jugular vein in severely dehydrated chicks.

(c) Yolk sac should be retracted and absorbed within the body after hatching of the chicks from the eggs. Due to infection (*E. coli*) and hyperthermia sometimes the yolk sac is not fully retracted. Proper antibiotic treatment and surgery can check the ailment.

(d) Lack of proper nutrition and deficiency of vitamin D, calcium can generate stunting of growth, leg and toe deformities (splay leg). The chicks should be placed on paper towels and surgery is required in severe condition.

(e) Crop stasis due to chronic candidiasis is the most common paediatric problem. It is diagnosed by clinical symptoms such as static oversized crop, regurgitation and dehydration, and biopsy. Crop washing and systemic treatment with antifungal (Sect. 2.4.3) is recommended (Fig. 6.2).

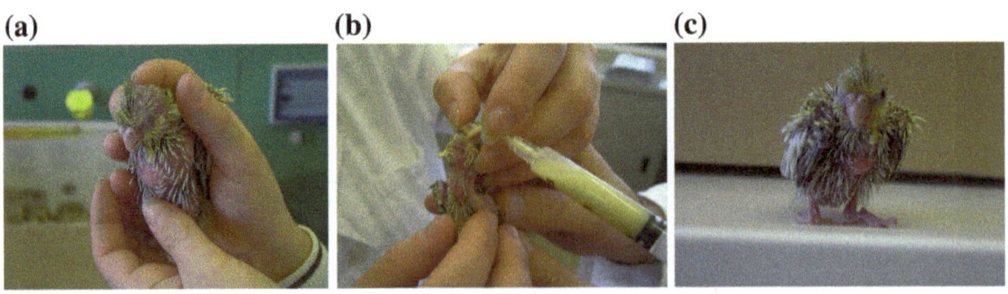

Fig. 6.2 **a** Occurrence of crop stasis in a baby cockatiel **b** removal of food and crop washing **c** healthy cockatiel chick after treatment (*Courtesy* Prof. Elena Circella, Avian Diseases Unit, Department of Veterinary Medicine, University of Bari, Italy)

(f) Regular administration of antibiotic in sub-therapeutic dosage in drinking water of adult birds or chicks should be avoided. It can destroy normal beneficial flora present in the avian enteric tract and moreover, it generates resistant bugs, unresponsive to further antibiotic treatment.

(g) Vitamin (A and B complex) should be prescribed as supportive treatment because most of the captive birds are chronically deficient in vitamin-A and B-vitamins are rapidly metabolized in avian body. Hypovitaminosis impedes absorption of drugs from gastrointestinal tract.

Appendix

List of Zoonotic Diseases from Pet Birds

Bacterial diseases

(a) Campylobacteriosis
(b) Chlamydophilosis
(c) *Escherichia coli* infection
(d) *Klebsiella* spp. infection
(e) Lyme disease
(f) Pasteurellosis
(g) Salmonellosis
(h) Tuberculosis
(i) Yersiniosis

Viral diseases

(a) Avian Influenza
(b) Newcastle disease
(c) West Nile fever

Fungal diseases

(a) Cryptococcosis

Parasitic diseases

(a) Cryptosporidiosis
(b) Giardiasis
(c) Toxoplasmosis

Guideline for Prevention of Pet Bird Zoonosis

Following guidelines can be offered to the pet bird owners to prevent transmission of zoonotic infection:

I. Samanta and S. Bandyopadhyay, *Pet Bird Diseases and Care*,
DOI 10.1007/978-981-10-3674-3

(a) Cages, water and feeding trays, perches should be cleaned regularly
(b) Washing hands with soap for a sufficient period of time is required before and after handling the birds, cleaning of cages, trays, drinking water pots etc.
(c) Returning home from any bird show or gatherings with other birds, the clothes and shoes of the owners should be cleaned properly
(d) Bathing pot should be available for 1–2 h per day to the birds. Excess water after bathing may act as source of pathogens
(e) The foods (commercial) should be stored in a clean pot. Home made food should be prepared instantly before giving to the birds
(f) Mites can inhabit perches and trays which should be checked weekly. The dirty spots present in the reverse side of tray and perches can be collected and observed for presence of mites. Ectoparasites act as indicator of poor hygiene
(g) The birds imported from other country should be accompanied with a veterinarian's fitness certificate and seller's certificate regarding their legal import
(h) Newly purchased birds should be quarantined for a specific period as suggested by the owner's veterinarian (30–45 days recommended by CDC)
(i) Regular health check up (wellness examination) of the birds should be carried out by a qualified veterinarian. General symptoms of illness such as lying at the bottom of the cage, ruffled feathers, nasal or ocular discharge, conjunctivitis, respiratory distress can be monitored by the owners.

Index

I. Samanta and S. Bandyopadhyay, *Pet Bird Diseases and Care*,
DOI 10.1007/978-981-10-3674-3

Kolibri 360 GmbH
Ferdinand-Jühlke-Straße 7
99095 Erfurt, Deutschland
produktsicherheit@kolibri360.de